Field-Effect and Bipolar Power Transistor Physics

Field-Effect and Bipolar Power Transistor Physics

Adolph Blicher

FORMERLY WITH
RCA Laboratories
Solid State Technology Center
Somerville, New Jersey

1981

ACADEMIC PRESS
A Subsidiary of Harcourt Brace Jovanovich, Publishers
New York London Toronto Sydney San Francisco

ACADEMIC PRESS, INC.
111 Fifth Avenue, New York, New York 10003

United Kingdom Edition published by
ACADEMIC PRESS, INC. (LONDON) LTD.
24/28 Oval Road, London NW1 7DX

Library of Congress Cataloging in Publication Data

Blicher, Adolph.
Field-effect and bipolar power transistor physics.

Includes bibliographical references and index.
1. Field-effect transistors. 2. Power transistors.
3. Bipolar transistors. I. Title.
TK7871.95.B56 621.3815'28'0153 80-68565
ISBN 0-12-105850-6 AACR2

PRINTED IN THE UNITED STATES OF AMERICA

81 82 83 84 9 8 7 6 5 4 3 2 1

To the memory of my parents

JOSEPH AND PAULINE BLICHER

CONTENTS

PART I GENERAL CONSIDERATIONS

Chapter 1 **Semiconductor Surface Theory Concepts**

Chapter 2 Semiconductor Properties at High Carrier Concentrations

Chapter 3 Avalanche Breakdown

Chapter 4 Avalanche Breakdown Improvement Methods

Chapter 5 Selected Fabrication Techniques

PART II BIPOLAR POWER TRANSISTORS

Chapter 6 Power Transistor Structures. Bipolar Transistor Models

Chapter 7 Current Gain at High Carrier Concentrations

Chapter 8 Current–Voltage Characteristics of Power Transistors

Chapter 9 Frequency Response. Switching Transient. Microwave Transistors

Chapter 10 Transistor Thermal Properties. Instabilities

PART III FIELD-EFFECT (UNIPOLAR) TRANSISTORS

Chapter 11 **Junction Field-Effect Transistors (JFETs)**

Chapter 12 **Insulated Gate Field-Effect Transistors**

Chapter 13 **MOS Power Transistor Structures and Design Considerations**

PREFACE

This volume is an introduction to the physics of operation of power transistors. It deals with bipolar devices as well as field-effect power transistors, recently introduced commercially. In short, the book provides an up-to-date account of the progress made in power transistor design. It is written for electrical engineers who design power transistor circuits, device physicists and designers, and university students. The reader should have some familiarity with small signal transistor physics; the presentation is at the senior undergraduate or first-year graduate level. Stress is given to phenomenological aspects, and the mathematical background needed does not go beyond elementary differential equations.

The volume consists of three parts. The first part is devoted to general considerations and includes a review of semiconductor surface theory as a background to understanding surface phenomena and, notably, to understanding surface field-effect device operation; a discussion of the effect of high carrier concentration on the semiconductor properties; and a survey of various junction avalanche breakdown improvement techniques, such as beveling, depletion region shaping, and resistive plate usage.

Selected fabrication methods described in Part I deal mainly with some of the more recent techniques, such as ion implantation and neutron transmutation doping. It is beyond the scope of this book to discuss fully all presently used device techniques. There are many texts available that include the description of these techniques.

In Part II we deal with bipolar transistors and start with a review of the basic structures of power transistors. This is followed by a discussion of several electrical models of bipolar transistors and the models' limitations.

Special attention is given in Part II to the behavior of current gain at high carrier concentrations. The effects of energy gap narrowing, Auger recombination, base widening, and emitter current crowding are fully studied. The reader is also introduced to several methods of gain improvements, such as interdigitation and heterojunction-type emitters. Saturation and quasi-saturation of bipolar transistors, the frequency response, and microwave transistors are also considered in Part II. The last chapter of this part is devoted to thermal phenomena and electrical–thermal instabilities such as the forward and reverse second breakdown and their improvement methods.

Part III of the book is devoted entirely to junction field-effect and surface field-effect transistors. The theory of operation of the junction field-effect power transistor and its various structures is reviewed in Chapter 11. Both pentodelike and triodelike device characteristics are considered. The description of the metal Schottky-barrier field effect transistor (MESFET) is also included in this chapter. The last two chapters are devoted to the insulated gate field-effect transistors (IGFET) and specifically to the subclass known as the metal–oxide–semiconductor field-effect transistor (MOSFET or MOST). Chapter 12 discusses the static and dynamic properties of MOS transistors having pentodelike and triodelike characteristics. Reviewed in Chapter 13 are two- and three-dimensional metal–oxide–semiconductor (MOS) power transistor structures, which include, for example, the high voltage DMOS transistor and the V-groove MOS transistor with vertical drains. A large portion of this chapter is directed toward MOS power transistor design considerations.

ACKNOWLEDGMENTS

I wish to express my thanks to Dr. H. Kressel of the David Sarnoff Research Center, Princeton, N.J., for suggesting this book and to Messrs. B. A. Jacoby, C. R. Turner, and B. V. Vonderschmitt of the RCA Solid State Division, Somerville, N.J., for partial support of this work. I also wish to express my appreciation for the many interesting discussions I had while writing this book with Messrs. H. Becke, R. Denning, J. Olmstead, and C. F. Wheatley, all of RCA Laboratories, Solid State Technology Center, Somerville, N.J.

I am greatly indebted to Mrs. B. J. Peterson and Ms. N. Parks of RCA Laboratories, Solid State Technology Center, Somerville, N.J., for their excellent typing and secretarial services. My special thanks are given to Mr. D. Russell of the Engineering Publications Department, RCA Solid State Division, Somerville, N.J., for the very effective assistance with graphics and to Miss B. McCoy, RCA Solid State Division, Somerville, N.J., for expert library services.

SELECTED LIST OF SYMBOLS

Symbol	*Description*	*First used in Section*
A	area, emitter area	—
A_C	collector area	7.2
A_E	metallurgical emitter area	10.5
A_n, A_p	constants	3.1
a_s	screening radius	2.5
B_n, B_p	constants	3.1
BV	breakdown voltage	3.1
BV_{CBO}	collector breakdown voltage with emitter open	8.2
BV_{CEO}	collector breakdown voltage with base open	8.2
$BV_{CEO(sus)}$	collector-to-emitter sustained breakdown voltage (base open)	4.7
BV_D	drain breakdown voltage	11.5
BV_{FP}	breakdown voltage with field plate	4.1
BV_{planar}	planar junction breakdown	4.1
BV_{plane}	plane junction breakdown	4.1
BV_{SiO_2}	SiO_2 dielectric breakdown	12.9
C	gate capacitance	1.7
C	constant	2.5
C_B	substrate impurity concentration	3.1
C_D	diffusion capacitance	6.2
C_G	gate total capacitance	11.2
C_{gd}	gate-to-drain capacitance	11.2
C_{gs}	gate-to-source capacitance	11.2
C_i	gate oxide capacitance per unit area	1.1

Symbol	*Description*	*First used in Section*
C_{in}	input capacitance	11.2
C_{SE}	surface concentration of arsenic-doped emitter	7.6
C_{si}	capacitance defined by Eq. (1.37)	1.7
C_{ss}	capacitance defined by Eq. (1.38)	1.7
C_{TC}	collector junction capacitance	9.1
C_{TC}	channel-to-bulk capacitance	12.7
C_{TD}	drain-to-bulk capacitance	12.7
C_{TE}	emitter junction capacitance	9.1
C_{TS}	source-to-bulk capacitance	12.7
C_{th}	thermal capacitance	10.1
C_0	surface concentration	3.1
d	metallurgical channel width	11.1
D_a	ambipolar diffusivity	2.1
D_B	diffusion constant in the base	10.5
D_c	high injection value of the electron diffusivity constant	7.3
D_n	electron diffusivity	7.3
D_p	hole diffusivity	2.1
D_{th}	thermal diffusivity	10.2
E	electric field	3.1
E_b	junction field at breakdown	4.9
E_c	energy at edge of conduction band	1.1
E_{crit}	critical field for avalanche breakdown	13.7
$E_{crit(v)}$	critical field for carrier velocity saturation	11.6
E_F	Fermi level	1.2
E_g	energy gap	2.5
E_{go}	energy gap extrapolated to $t = 0°K$	2.5
E_i	semiconductor intrinsic energy level	1.2
E_i	gate electric field	12.1
E_s	surface field	1.3
$E_{s/b}$	reverse second breakdown	10.5
E_t	energy level of a recombination center	2.4
E_T	average transverse electric field at the $Si–SiO_2$ interface	12.6
E_V	energy at edge of valence band	1.1
$E_0(r)$	unscreened field at the distance r from origin	2.5
f	frequency	1.2
f_T	gain–bandwidth product	6.2
g	incremental conductance	1.1
g'_{be}	defined by Eq. (6.71)	6.2
g_c	incremental collector conductance	6.2
g_d	incremental drain conductance	11.1
g_{d0}	incremental drain conductance in the JFET linear region	11.1
g_m	transconductance	6.2
g_{msat}	transconductance in the saturation region	11.1
G_0	conductance, Eq. (11.10)	11.1
$h = x_d$	JFET depletion region width	11.1
h_{CR}	crossover gain	7.3
h_D	JFET depletion region width at the drain side	11.1

Symbol	*Description*	*First used in Section*
h_{fe}	common emitter small signal current gain	6.2
h_{FE}	common emitter dc current gain	6.2
h_{FEO}	common emitter peak current gain in the active region	8.2
h_{RE}	reverse common emitter current gain	6.2
h_S	JFET depletion region width at the source side	11.1
h_0	depletion layer width with zero gate bias	11.1
I_B	base current	6.2
I_C	collector current	6.2
I_{CBO}	collector leakage current	6.2
I_{CR}	crossover current	7.3
I_D	drain current	11.1
I_{DO}	drain current for zero gate bias	11.1
I_E	emitter current	6.2
I_G	excess gate current	1.3
I_n^*	ionization integral for electrons	3.1
I_p^*	ionization integral for holes	3.1
I_r	recombination current	6.2
$I_{s/b}$	forward second breakdown	10.5
I_{sub}	excess substrate current	12.10
I_0	collector current at $t = t_0$	9.2
J_{gen}	space-charge generated current density	2.6
J_n	electron current density	6.2
$J_n(0)$	electron current density at the emitter–base edge	6.3
J_p	hole current density	6.2
$J_p(0)$	hole current density at the emitter–base edge	6.2
J_{rec}	space charge recombination current density	2.6
J_s	entropy flow	10.4
J_S	emitter saturation current density	7.4
J_{SCL}	space-charge limited current	12.9
J_0	critical current density for base widening of the n^+–p–n^+ transistor	7.2
J_0'	critical current for base widening in the n^+–p–v–n^+ structure (high field case)	7.2
J_0''	critical current density for base widening in the n^+–p–v–n^+ structure (low field case)	7.2
k	Boltzmann's constant	1.2
k	thermal conductivity	10.2
K	radiation damage coefficient	5.4
L	impurity diffusion length	7.1
L	channel length	12.3
L'	drift region length	12.8
L'	center-to-center spacing for spreading resistance calculation	13.4
L_E	one-half of the emitter stripe width	7.2
L_n	electron diffusion length	6.2
M	avalanche multiplication factor	3.1
n	electron density	2.1
n'	excess electron density	2.1
N	impurity concentration	3.3

Symbol	*Description*	*First used in Section*
N_A	*p*-type silicon acceptor concentration per cubic centimeter	1.1
N_b	background doping concentration	5.2
N_{B0}	base impurity concentration under the emitter	10.5
N_D	*n*-type silicon doping donor concentration per cubic centimeter	1.1
N_{dose}	total implanted dose per unit area	5.2
N_{E0}	emitter surface concentration	10.5
N_{FS}	fast surface states	1.8
n_i	intrinsic carrier concentration	1.2
n_{ie}	effective intrinsic concentration	2.5
$N_I(x)$	implanted concentration per unit area as a function of depth x	5.2
n_s	electron density at the surface	1.2
n_0	equilibrium bulk electron density	1.2
N_0	surface concentration	7.1
p'	excess hole density	2.1
p_s	hole density of the surface	1.2
P_T	dissipated power	10.1
p_0	equilibrium hole density in bulk silicon	1.2
$p'(0)$	injected hole density at $y = 0$	8.2
$P(0)$	total hole density at the edge of the *p*-*v* junction	8.2
q	electron charge	1.1
q_B	total base excess charge	6.2
Q_B	total majority carrier charge in the base	6.2
Q_{B0}	total majority carrier charge in the base in equilibrium	6.2
Q_{B0}/qAD_n	Gummel number	6.2
Q_d	charge in the surface depletion region	1.3
Q_{dmax}	maximum value of Q_d	1.3
Q'_{dmax}	charge in the surface depletion region under nonequilibrium condition	1.5
q_E	total emitter excess charge	6.2
Q_{eff}	lumped equivalent of the oxide distributed charges	1.2
Q_{E0}	equilibrium majority carrier charge in the emitter	7.4
Q_F	excess majority carrier charge in the base due to forward conduction	6.2
Q_G	gate charge per unit area	1.2
Q'_G	total gate charge	12.7
Q_H	electric charge defined by Fig. 4.30	4.6
Q_I	implanted dose	4.9
Q_L	electric charge defined by Fig. 4.30	4.6
Q_n	charge in the *n*-type channel	1.3
$q\phi_m$	metal work function	1.1
$q\phi_{ms}$	metal–silicon work function	1.1
$q\phi_{si}$	silicon work function	1.1
$q\phi_{ss}$	silicon–silicon work function	1.4
Q_R	electric charge defined by Fig. 4.30	4.6
Q_{R}	excess majority base charge due to the inverted transistor action	6.2
Q_s	total silicon surface charge density per unit area	1.3
Q_{ss}	fixed positive oxide charge per unit area	1.1

Symbol	*Description*	*First used in Section*
Q_{TC}	charge in the collector–base transition capacitance	6.2
Q_{TE}	charge in the emitter–base junction capacitance	6.2
q_{th}	heat generated per unit volume	10.2
r	distance from origin	2.4
R_B	base series resistance	10.5
R_C	collector resistance	9.2
R_d	drain series resistance	11.1
R_E	emitter series resistance	10.5
R_{gen}	current generator resistance	13.2
r_j	junction radius and depth	12.3
R_L	load resistance	9.2
R_{on}	MOS transistor on-resistance	13.4
R_p	range of the implant	5.2
R_s	source series resistance	11.1
R_{sth}	thermal resistance of heat sink to air	10.3
R_{sub}	substrate parasitic resistance	12.10
$R_{th}(t)$	thermal resistance	10.1
R_0	metallurgical channel resistance	11.1
S	entropy	10.4
S	thermal stability factor	10.5
S_G	entropy generation rate	10.4
T_A	heat sink (or ambient) temperature	10.3
t_d	delay time	13.2
t_f	fall time	13.2
t_r	rise time	13.2
t_s	storage time	9.2
T_S	surface temperature	10.3
$\mathrm{T}_1, \mathrm{T}_2, \mathrm{T}_3$	channel transit time	12.8
U	excess carrier recombination rate	2.4
v	carrier velocity	9.1
V_A	Early voltage	6.2
V_{Ap}	applied voltage	6.1
V_B	reverse Early Voltage	6.2
V_{BC}	collector–base voltage	6.2
V_{BE}	emitter–base voltage	7.6
V_{BG}	back-gate voltage	12.1
V_{CC}	collector supply voltage	9.1
V_D	drain-to-bulk voltage	11.1
V_{Dsat}	saturation drain voltage	11.1
V_{FB}	flat-band voltage	1.1
V_G	gate voltage	1.2
V_{GG}	gate supply voltage	13.2
V_i	built-in junction voltage	9.1
V_{ox}	voltage across the oxide	1.3
V_{pt}	punch-through voltage	10.5
V'_{pt}	punch-through voltage under current flow	10.5
V_R	reverse biased junction voltage	3.1

Symbol	*Description*	*First used in Section*
v_s	saturated drift velocity	2.4
V_{SBR}	voltage at second breakdown	13.7
V_T	equilibrium threshold voltage	1.3
V'_T	nonequilibrium threshold voltage	1.5
V_{1i}, V_{2i}	heterojunction built-in voltages	7.6
W	space-charge width	2.6
W	energy	13.2
W	measure of disorder (entropy)	10.4
W_B	base width	6.2
W'_B	base width for the Van der Ziel–Agourdis model	7.2
W_{Beff}	total base width under base widening conditions	7.3
W_{CIB}	collector-induced base	7.2
W_E	effective emitter width	6.2
W_H	depletion layer width for highly doped silicon	4.6
W_L	depletion layer width for lightly doped silicon	4.7
W_n	width of the low concentration collector $v(n^-)$-region	7.2
X	etch depth	4.8
x_d	junction depletion region width	3.1
x_{dmax}	channel-bulk maximum depletion region thickness	1.3
x_{dmaxp}	maximum depletion layer width of the plane junction	4.1
x_i	gate-oxide thickness	12.1
x_{inv}	distance below surface where the semiconductor becomes intrinsic	1.1
x_j	junction depth	3.4
x_o	oxide thickness	4.1
x_0	crossover point (Fig. 7.2)	7.1
Y_E	etch extent	4.8
y_0	characteristic length for current crowding	7.4
Z	emitter width in the z direction	7.3
Z	channel width in the z direction	11.1
α	transistor common base current gain	6.2
α_{eff}	effective ionization rate	3.1
α_F	forward alpha	6.2
α_n	electron ionization rate	3.1
α_p	hole ionization rate	3.1
α_R	inverted transistor alpha	6.2
α_s	small signal common base current gain	6.2
α_T	base transport factor	6.2
γ	ratio of the junction depth to the depletion region width	3.4
γ	emitter efficiency	6.2
γ'	defined by Eq. (4.8)	4.1
γ_E	emitter efficiency in the common emitter configuration	6.2
δ	current transport defect factor	7.7
ΔE	activation energy	10.5
ϵ_i	insulator permittivity	1.3
ϵ_s	silicon permittivity	1.3
ζ	edge-to-center current ratio	7.3
η	factor dependent on device structure	11.5

Symbol	*Description*	*First used in Section*
θ	effective bevel angle defined by Eq. (4.13)	4.6
θ	thermal time constant	10.3
$\theta(x)$	current spreading distance Fig. (13.30)	13.7
λ	space constant	4.6
μ	carrier mobility	2.1
μ^*	transistor amplification factor	11.5
μ_a	ambipolar mobility	2.1
μ_A	ambipolar mobility as defined by Eq. (2.3)	2.1
μ_{cc}	carrier–carrier scattering mobility	2.2
μ_{eff}	effective surface mobility	1.11
μ_n	electron mobility	1.11
μ_p	hole mobility	2.1
μ_0	undoped semiconductor mobility	2.3
$v = n^-$	low concentration *n*-type silicon	6.1
ρ	electric charge	7.2
ρ	resistivity	8.2
ρ_B	base average resistivity	7.3
ρ_C	v-region resistivity	7.2
$\rho(n)$	charge density per unit area at position n	11.1
ρ_s	material resistivity	10.4
σ	conductivity	11.1
σ_n	electron capture cross section	2.4
σ_p	hole capture cross section	2.4
τ	minority carrier lifetime	2.1
τ_A	Auger recombination lifetime	7.1
τ_{EC}	total time delay between the emitter and collector	9.1
τ_{eff}	effective base region lifetime	6.2
τ_F	base transit time	6.2
τ_F'	base transit time with base widening	9.2
τ_{HL}	high-level lifetime	5.4
τ_{LL}	low-level lifetime	5.4
τ_n	electron lifetime	2.4
τ_{n0}	low-level electron lifetime	2.4
τ_p	hole lifetime	2.4
τ_{p0}	low-level hole lifetime	2.4
τ_{SRH}	Shockley–Read–Hall recombination lifetime	7.1
τ_x	transit time through the space charge region	9.1
τ_0	carrier lifetime in the space charge region	2.4
τ_1, τ_2	time constants	9.2
ϕ	electron radiation influence	5.4
ϕ	amplitude of the incomplete elliptic integral of the first kind	7.3
ϕ_F	Fermi potential	1.3
ϕ_{FG}	Fermi potential of the silicon gate	1.4
ϕ_n	Quasi-Fermi potential of the *n*-channel	1.5
ϕ_p	Quasi-Fermi potential in the *p*-channel	1.5
ϕ_s	surface potential	1.2
$\omega_T = 2\pi f_T$	angular gain–bandwidth product	9.1

PART I

General Considerations

CHAPTER 1

SEMICONDUCTOR SURFACE THEORY CONCEPTS

The properties of the semiconductor surface are essential to the understanding of all types of semiconductor devices, i.e., those that operate due to the phenomena occurring primarily at the semiconductor surface, like metal–oxide semiconductors (MOS) transistors, and those whose performance is mainly due to the bulk properties such as planar and mesa-type bipolar transistors. The presence of the various surface charges and states has a profound effect on the electrical characteristics of both power MOS and power bipolar transistors. Our review is limited to the silicon crystal surface properties as being the best understood and most important from the practical point of view.

1.1 Flat-Band Voltage

A MOS diode, known also as a MOS capacitor (Fig. 1), consisting of a metallic electrode (gate) and SiO_2 layer on a silicon substrate, is a useful tool for surface properties studies.

Under ideal conditions, i.e., when there is no work function difference between the metal used and silicon and there are no electric charges in the

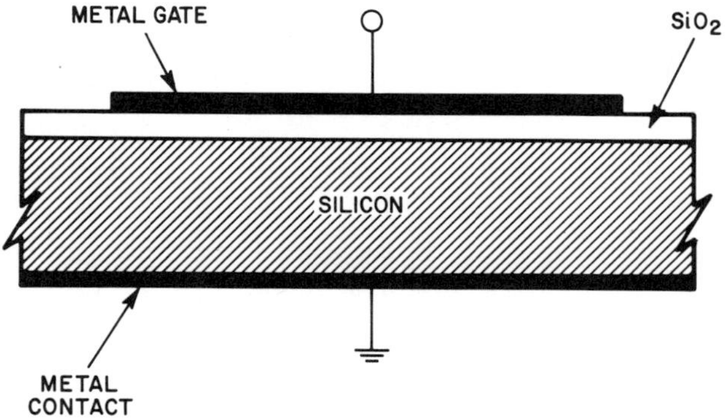

Fig. 1. A MOS diode (capacitor) with metallic gate.

silicon dioxide, the energy band diagram for *p*-type silicon can be represented by Fig. 2 where $q\phi_m$ and $q\phi_{si}$ are the metal and silicon work functions, respectively. The ideal case also assumes that the dielectric's resistance is infinite. Under these conditions, the device is in thermal equilibrium, the metal and silicon Fermi levels are aligned, and the energy bands are flat.

When the difference of the work functions is not equal to zero, the silicon energy bands become bent as in Fig. 3. In the case of Fig. 3, the metal work function $q\phi_m$ (where q is the electron charge) is smaller than that of silicon

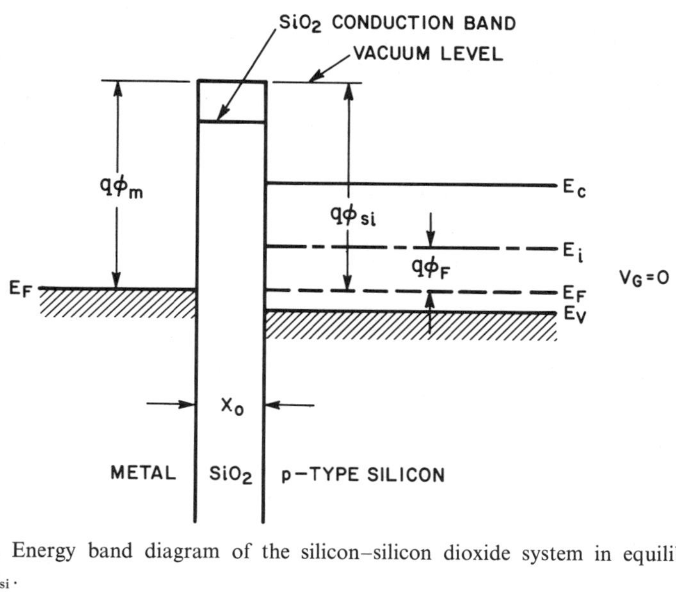

Fig. 2. Energy band diagram of the silicon–silicon dioxide system in equilibrium for $q\phi_m = q\phi_{si}$.

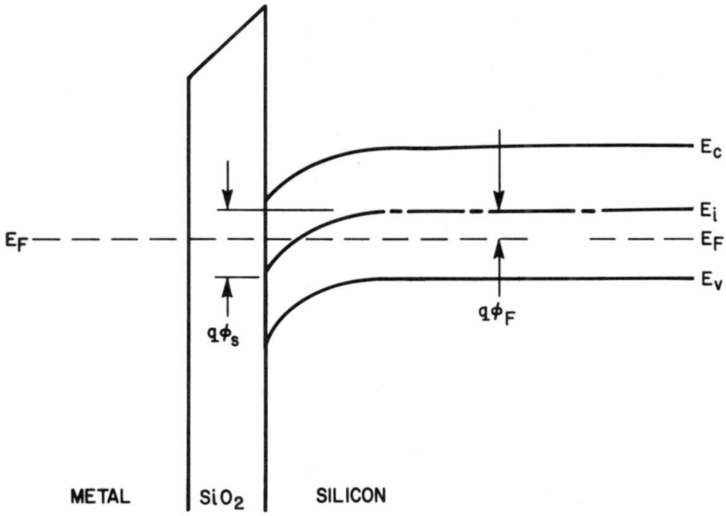

Fig. 3. Energy band bending for $|q\phi_m| < |q\phi_{si}|$.

$q\phi_{si}$; consequently, some electrons leave the metal and reach the p-type silicon crystal by a path external to the diode and more conductive than the silicon dioxide. The metal gate becomes, therefore, positively charged and repels holes causing the appearance of uncompensated negative acceptors forming a space-charge region at the semiconductor surface. Because the number of holes at the semiconductor surface is smaller than that expected from its doping level, the silicon energy bands bend so that the Fermi level E_F comes closer to the intrinsic energy level E_i. The metal and semiconductor Fermi levels are aligned, after the initial transient, because of the existence of the external conducting path.

Due to charge separation, a potential difference appears that is shared by the oxide and the space-charge region. This voltage is equal in magnitude but opposite in sign to the metal–oxide semiconductor work function ϕ_{ms}.

In order to revert the energy bands to the straight flat-band condition, it is necessary to apply to the metal gate a voltage known as the flat-band voltage given by

$$V_{FB1} = \phi_m - \phi_{si} = \phi_{ms}. \tag{1}$$

Under the flat-band conditions, the MOS device is not in thermal equilibrium.

Fig. 4 [1] gives the values of ϕ_{ms} (with SiO_2 present) as a function of silicon doping for aluminum and gold. For example, $\phi_{ms} = -0.97$ V for aluminum and p-type silicon doped to the acceptor density $N_A = 2 \times 10^{16}$

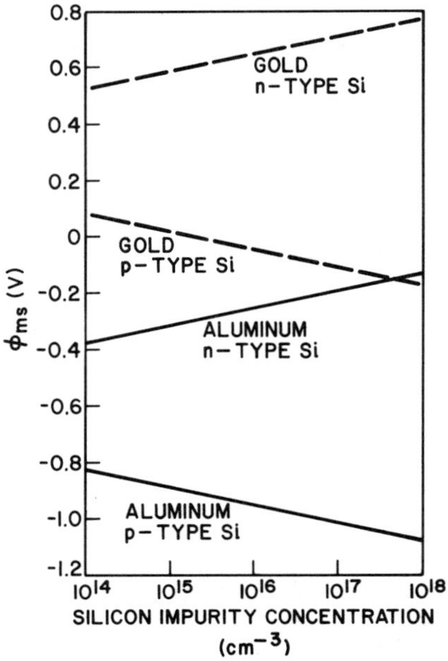

Fig. 4. ϕ_{ms} as a function of silicon impurity concentration for aluminum and gold. (From Deal *et al.* [1].)

cm^{-3}; $\phi_{ms} = -0.24$ V for aluminum and *n*-type silicon with donor density $N_D = 2 \times 10^{16}\ cm^{-3}$.

In all practical situations the silicon dioxide contains electrical charges to be discussed in some detail in Section 1.8. These charges are predominantly positive and usually consist of fixed charges Q_{ss} located in SiO_2 at about 20 nm from the silicon interface and other charges which are fixed and/or mobile, but distributed in the oxide volume. The positive charge Q_{ss} induces a voltage across the oxide which can be expressed approximately as $V_{FB2} = Q_{ss}/C_i$, where C_i is the oxide capacitance per unit area. In the well-processed oxide, Q_{ss}/q is less than $10^{10}\ cm^{-2}$. The distributed ionic charges also induce a voltage across the oxide which can be represented approximately by $V_{FB3} = -Q_{eff}/C_i$, where Q_{eff} is the lumped equivalent of the distributed charges. For good devices, Q_{eff}/q should be less than $10^{10}\ cm^{-2}$. The flat-band voltage, which accounts for the metal–silicon work function difference and charge induced voltages, is therefore represented by

$$V_{FB} = V_{FB1} + V_{FB2} + V_{FB3} = \phi_{ms} - Q_{ss}/C_i - Q_{eff}/C_i. \tag{2}$$

This expression is valid for the *p*- and *n*-substrates.

1.2 Surface Potential in Equilibrium

The surface potential ϕ_s (Fig. 3) is the measure of the energy band bending. It represents the silicon potential at the surface measured from the bulk intrinsic energy level E_i. ϕ_s is positive when the band is bent downward and vice versa. The concept of the surface potential is helpful in the determination of the surface mobile carrier concentration, which can be expressed as [2, 3]

$$n_s = n_0 \exp(q\phi_s/kT) \tag{3a}$$

and

$$p_s = p_0 \exp(-q\phi_s/kT), \tag{3b}$$

where n_s and p_s are the electron and hole densities at the surface, respectively; n_0 and p_0 are the equilibrium densities of electrons and holes in the bulk crystal, respectively; q is the electronic charge; k is Boltzmann's constant; and T is absolute temperature. With $E_i = 0$

$$n_0 = n_i \exp(E_F/kT), \tag{4a}$$

and

$$p_0 = n_i \exp(-E_F/kT), \tag{4b}$$

where n_i is the silicon crystal intrinsic concentration. Expressions (3a) and (3b) are valid for the p- and n-type silicon. When $q\phi_s = E_F$, these equations indicate that the surface becomes intrinsic, i.e., $n_s = p_s = n_i$.

The silicon surface is neutral when $\phi_s = 0$ for the flat-band condition. In order to induce charges at the silicon surface, it is necessary to apply a sufficiently high voltage to the gate. An application of a small external positive gate potential $V_G > V_{FB}$ (a p-type substrate is considered) results in the repulsion of holes from the surface and in the surface depletion region formation. This is called surface depletion for which the bands are bent downward and ϕ_s is positive (Figs. 3 and 5 [3]). Further increase in the positive gate voltage with $V_G \gg V_{FB}$ results in the attraction of electrons to the silicon surface so that it behaves like an n-type although the substrate is p-type. This is called surface inversion. In this case the energy diagram shows further downward band bending so that the $E_F - E_i$ at the surface changes its sign (Figs. 3 and 6a [3]) and becomes positive at the surface. Strong inversion occurs when the electron density at the surface exceeds the acceptor bulk density, i.e., when $n_s > N_A$. This takes place when $q\phi_s \simeq 2E_F$ with the reference level $E_i = 0$. The inverted surface region (the n-channel) creates a depletion region with the p-substrate in complete analogy with the n–p junction.

A sufficiently high negative potential applied to the gate will attract

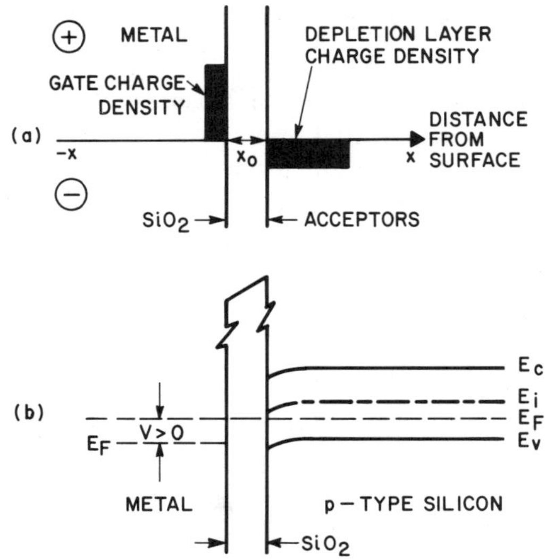

Fig. 5. (a) Charge distribution and (b) energy band diagram for a depleted surface. (From Grove *et al.* [3].)

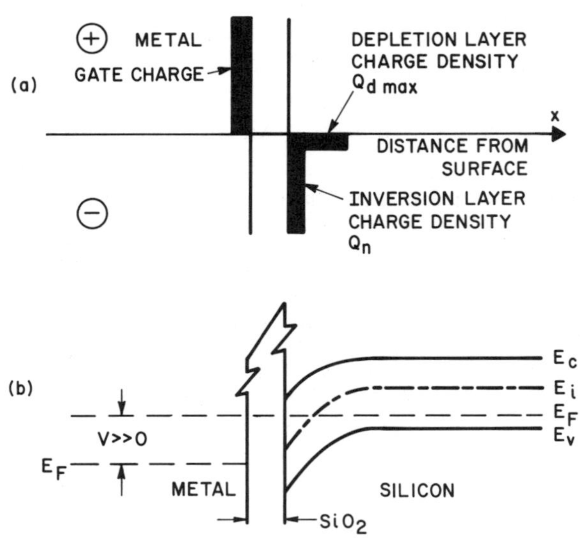

Fig. 6. Charge distribution and energy band diagram for surface inversion. (From Grove *et al.* [3].)

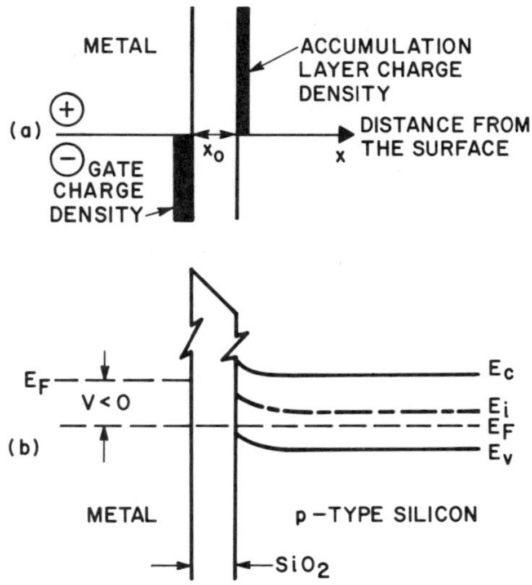

Fig. 7. (a) Charge distribution and (b) energy band diagram for surface accumulation. (From Grove *et al.* [3].)

mobile holes to the silicon surface; hence more holes will be available than in the bulk silicon. This is called surface accumulation (Fig. 7 [3]). A higher concentration of holes is equivalent to the existence of a more *p*-doped crystal. The energy bands are bent, therefore, upward, and the Fermi level E_F comes closer to the valence band as expected for more *p*-type semiconductors. The surface potential ϕ_s is negative for the accumulation case.

Table I summarizes the state of the surface for various gate voltages.

TABLE I

p-TYPE SILICON SURFACE AT VARIOUS GATE VOLTAGES

V_G	$V_G < V_{FB}$	$V_G = V_{FB}$	$V_G > V_{FB}$	$V_G > F_{FB}$	$V_G \gg F_B$
ϕ_s	Negative	0, flat-band condition	Positive, $q\phi_s < \lvert E_F \rvert$	Positive, $q\phi_s = \lvert E_F \rvert$	Positive, $q\phi_s > \lvert E_F \rvert$
Condition at the surface	Hole accumulation	Neutral (no net charge)	Depletion of holes	Intrinsic, $n_s = p_s = n_i$	Inversion, appearance of the *n*-channel
Bands	Bent upward	Flat	Bent downward	Bent downward	Bent downward

1.3 Surface Charges in Strong Inversion[1]

By Gauss's law, the total silicon surface charge density per unit area is

$$Q_s = -\epsilon_s E_s = -\epsilon_i E_{ox}, \tag{5}$$

where ϵ_s and ϵ_i are the silicon and SiO_2 permittivities, respectively; $E_s = -d\phi_s/dx$ is the surface field; E_{ox} is the oxide field; and Q_s is the sum of the charge in the inversion layer Q_n and the charge in the depletion region Q_d which forms between the induced *n*-layer and the *p*-type silicon substrate:

$$Q_s = Q_n + Q_d. \tag{6}$$

The surface space-charge density consists of negative ions (acceptors) for the *p*-type silicon substrate and of positive ions (donors) for the *n*-type silicon. For $\phi_s = \phi_F$ just before inversion, the surface is intrinsic, $Q_n = 0$, and

$$Q_s = Q_d. \tag{7}$$

As the positive surface potential is increased, the negative charge Q_n in the inversion layer continues increasing; but the charge Q_d in the depletion region reaches its maximum value Q_{dmax} when the surface becomes strongly inverted, i.e., when the surface potential $\phi_s \geqslant 2\phi_F$, with $\phi_F = -E_F/q$.

The value of Q_{dmax} can easily be obtained by observing that the junction between the inverted *n*-layer and the *p*-substrate can be considered as an n^+–p one-sided step junction reverse biased with $2|\phi_F|$ V. Its depletion region thickness is therefore

$$x_{dmax} = (4\epsilon_s|\phi_F|/qN_A)^{1/2}, \tag{8a}$$

where

$$|\phi_F| = (kT/q)\ln(N_A/n_i). \tag{8b}$$

For example, for $N_A = 2 \times 10^{16}$ cm^{-3}, $x_{dmax} \simeq 0.2$ μm. Analogous expressions are valid for the *n*-type substrate.

The total charge stored in the depletion region can be expressed as

$$Q_{dmax} = -qN_A x_{dmax} = -(4\epsilon_s q N_A \phi_F)^{1/2}. \tag{9}$$

For $N_A = 2 \times 10^{16}$ cm^{-2}, $Q_{dmax} \simeq 4 \times 10^{11}$ cm^{-2}. For strong inversion,

$$Q_s = Q_{max} + Q_n. \tag{10}$$

The presence of the inverted surface with its Q_n charge is utilized for current conduction in MOS transistors (see Chapter 12).

[1] See Grove *et al.* [3].

1.4 Threshold Voltage for Strong Inversion[2]

The threshold voltage $V_G = V_T$ is the gate voltage required to obtain strong inversion with $\phi_s = 2\phi_F$. The gate voltage can be expressed as the sum of the voltage appearing across the oxide, the work function potential, and the surface potential, equal to $2\phi_F$:

$$V_G = V_T = V_{ox} + \phi_{ms} + 2\phi_F, \tag{11}$$

where V_{ox} is the voltage across the oxide. The charge on the metal gate because of the charge neutrality requirement can be expressed as (Fig. 8)

$$Q_G = -Q_{ss} - Q_{dmax} - Q_n. \tag{12}$$

Just prior to the strong inversion Q_n is small and can be neglected; therefore

$$Q_G = -Q_{ss} - Q_{dmax}. \tag{13}$$

The charge on the gate is also given by Gauss's law as

$$Q_G = \epsilon_i V_{ox}/x_i, \tag{14a}$$

$$Q_G = C_i V_{ox}, \tag{14b}$$

where ϵ_i is the silicon dioxide permittivity, x_i the oxide thickness per unit area, and C_i the oxide capacitance per unit area. Combining (11)–(14), we obtain an expression for the threshold voltage

$$V_T = \phi_{ms} + 2\phi_F - Q_{ss}/C_i - Q_{dmax}/C_i \tag{15}$$

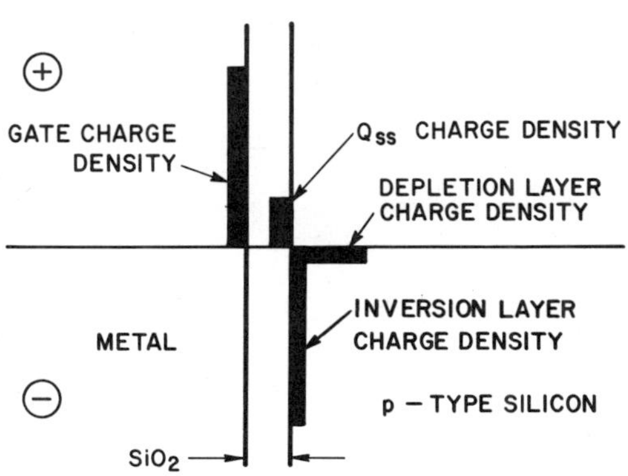

Fig. 8. Charge distribution. (From Grove *et al.* [3].)

[2] See [4] and [5].

valid for the n- and p-channel devices. If the distributed ionic charges in SiO_2 are taken into account, the threshold voltage becomes

$$V_T = \phi_{ms} + 2\phi_F - Q_{ss}/C_i - Q_{dmax}/C_i - Q_{eff}/C_i, \tag{16}$$

where Q_{eff} is the effective ionic charge density in the silicon dioxide per unit area. The threshold voltage expression (16) can also be written as

$$V_T = V_{FB} + 2\phi_F - Q_{dmax}/C_i \tag{17}$$

by utilizing the flat-band expression (2). For a device with an aluminum gate, $N_A = 2 \times 10^{16}$ cm^{-3}, $(Q_{ss} + Q_{eff})/q = 10^{10}$ cm^{-2}, $x_i = 100$ nm, and the threshold voltage V_T will be 1.6 V at 25°C.

Large, always positive Q_{ss} makes the fabrication of the enhancement-type n-channel devices more difficult than the fabrication of the p-channel devices because V_T will always be negative for low doped substrates and the surface may become inverted for $V_G = 0$ before any positive gate voltage is applied unless special processing techniques are utilized to minimize Q_{ss}. Ion implantation to increase the crystal concentration at the surface is a very useful method for the V_T value adjustment.

1.4.1 Silicon Gate

To obtain self-aligned gates with no drain or source overlap, it is advantageous to use highly doped polycrystalline silicon gate electrodes in place of metallic ones [6]. The work function ϕ_{ss} for the silicon gate is

$$\phi_{ss} = \phi_{FG} - \phi_F, \tag{18}$$

where ϕ_{FG} and ϕ_F are the Fermi potentials of the polysilicon gate and bulk

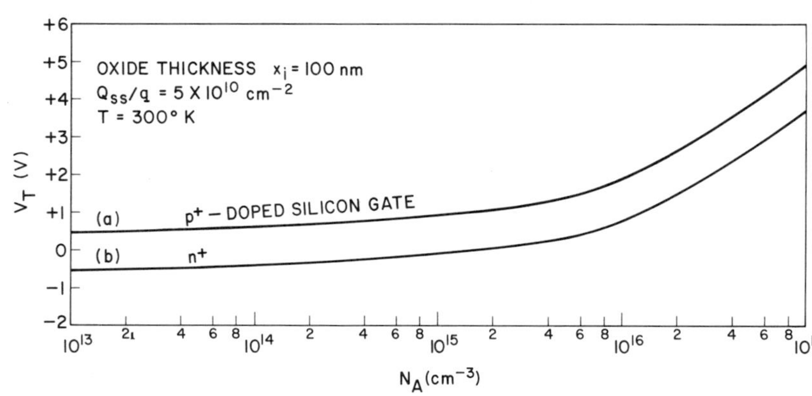

Fig. 9. n-Channel silicon gate threshold voltage as a function of the p-type silicon bulk concentration (a) p^+-doped silicon gate (b) n^+-doped silicon gate.

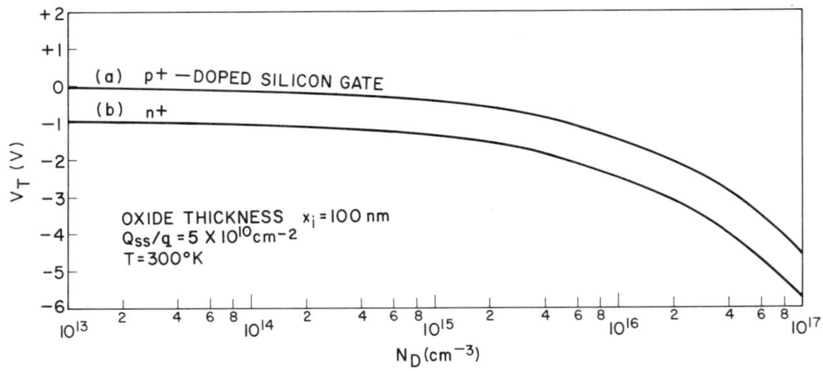

Fig. 10. *p*-channel silicon gate threshold voltage as a function of the *n*-type silicon bulk concentration (a) p^+-doped silicon gate and (b) n^+-doped silicon gate.

silicon, respectively [7]. For the heavily doped *p*-type polysilicon gate and highly doped n^+-substrate, $\phi_{ss} \simeq 1.1$ V. In the other extreme case, the polysilicon gate is heavily *n*-type doped and the substrate is highly *p*-doped so that $\phi_{ss} \simeq -1.1$ V. The threshold voltage of a MOS structure with polysilicon gate differs from the threshold voltage of a device with aluminum gate by an amount

$$\Delta V_T = \phi_{ss} - \phi_{ms} = 0.6 + \phi_{FG}. \tag{19}$$

ϕ_{FG} of a heavily doped *n*-type polysilicon gate is about -0.5 V; therefore, for a *p*-type substrate (*n*-channel device), $\Delta V_T \simeq 0.1$ V. A *p*-type gate with $\phi_{FG} = +0.5$ V and an *n*-type substrate (*p*-channel device) will result in $\Delta V_T = 1.1$ V. Silicon gate threshold voltages as a function of bulk (substrate) concentration are shown in Figs. 9 and 10.

1.5 Threshold Voltage under Nonequilibrium Conditions

It is possible to apply a reverse voltage V between the channel and the substrate (also called bulk) by fabricating a MOS diode (Fig. 11a) with an additional n^+-p junction. We shall see in Chapter 12, in the case of a MOS transistor, that such a junction may constitute the so-called source (a p^+-n diode would be used in the case of a *p*-type channel device). The application of a potential external to the diode makes it operate under nonequilibrium conditions [4] with the quasi-Fermi levels in the *n*- and *p*-type regions separated by qV if V is the reverse junction voltage. At the surface then

$$\phi_n = \phi_F + V, \tag{20}$$

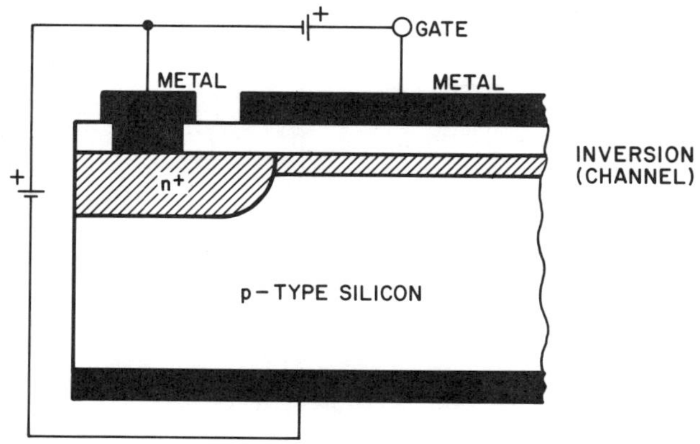

(a)

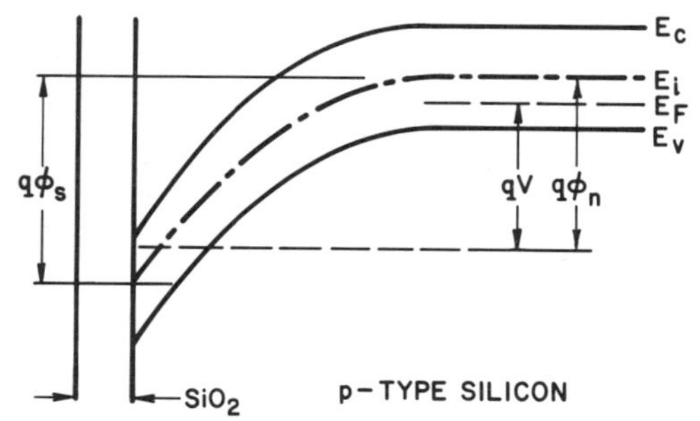

(b)

Fig. 11. (a) MOS diode with an additional reverse biased junction. (b) The energy band diagram for the silicon–silicon dioxide system for the reversed biased junction. (From Grove *et al.* [4].)

where ϕ_n is the quasi-Fermi potential in the n-channel. In the bulk itself, $\phi_p = \phi_F$; i.e., the quasi-Fermi level is the same as the Fermi level. The energy band diagram will be like the one in Fig. 11b [4]. Under conditions of strong inversion, the surface potential now becomes

$$\phi_s = 2\phi_F + V, \tag{21}$$

and expression (9) for the charge in the surface depletion region should be modified according to [4] as

$$Q'_{\mathrm{dmax}} = -[2\epsilon_s N_A q(2\phi_F + V)]^{1/2}. \tag{22}$$

The depletion region width (8) should be modified in the same way. The application of the reverse potential to the channel increases the depletion region maximum width and its charge.

The total charge per unit area in the semiconductor is

$$Q_s = -\epsilon_s E_s = Q_n + Q'_{dmax}. \tag{23}$$

The threshold voltage V'_T in nonequilibrium can be obtained from Eqs. (15) and (21) as

$$V'_T = \phi_{ms} + V + 2\phi_F - Q_{ss}/C_i - Q'_{dmax}/C_i \tag{24}$$

or by using (2) as

$$V'_T = V_{FB} - Q'_{dmax}/C_i + V + 2\phi_F. \tag{25}$$

The threshold voltage for the n-channel device is increased when a reverse potential V is applied to the source–substrate junction.

The threshold voltage for the n-substrate, p-channel device can be computed from expression (25), remembering, however, that Q'_{dmax} will now be a positive quantity since it consists of positive donors. Here, too, a reverse positive potential applied to the substrate–source junction will increase the absolute value of the threshold voltage.

1.6 Channel Charge under Strong Inversion in Nonequilibrium

When the surface is strongly inverted, the charge on the gate can be expressed as

$$Q_G = -Q_{ss} - Q_s = Q_{ss} - Q'_{dmax} - Q_n \tag{26}$$

since Q_n cannot be neglected now. The charge on the gate can also be expressed as a function of the voltage drop across the oxide as

$$Q_G = C_i V_{ox}. \tag{27}$$

The gate voltage is given by

$$V_G = V_{ox} + \phi_{ms} + \phi_s. \tag{28}$$

From (26) and (27) we obtain for the voltage across the silicon dioxide

$$V_{ox} = Q_G/C_i = -Q_{ss}/C_i - Q_s/C_i. \tag{29}$$

Utilizing (29) in expression (28) gives

$$V \quad - \phi_{ms} + Q_{ss}/C_i = \phi_s - Q_s/C_i \tag{30}$$

or

$$V_G - V_{FB} = \phi_s - Q_s/C_i \tag{31}$$

with $Q_s = Q_n + Q'_{dmax}$. Substituting $\phi_s = 2\phi_F + V$ and Q'_{dmax} from (22), we obtain for the mobile charge in the n-channel per unit area

$$Q_n = -C_i[V_G - V_{FB} - V - 2\phi_F] + [2\epsilon_s q N_A(2\phi_F + V)]^{1/2}, \tag{32}$$

or in view of (25) and (22)

$$Q_n = C_i(V_G - V'_T). \tag{33}$$

A similar expression but with the opposite sign is valid for the p-channel charge under strong inversion in nonequilibrium.

1.7 MOS Capacitance

Upon the application of the gate potential, the charge on the gate is

$$Q_G = -(Q_d + Q_n + Q_{ss}). \tag{34}$$

Each component on the right-hand side of the equation contributes to the total charge and therefore to the total device capacitance, defined as

$$C = dQ_G/dV_G \tag{35}$$

with

$$V_G = V_{ox} + \phi_s. \tag{36}$$

The capacitance due to the charge in the silicon surface is defined as

$$C_{si} = -d(Q_d + Q_n)/d\phi_s, \tag{37}$$

the capacitance due to the charge in the oxide is defined as

$$C_{ss} = -dQ_{ss}/dC_i. \tag{38}$$

C_{ss} and C_{si} are in parallel, and their sum is in series with the oxide capacitance so that

$$\frac{1}{C} = \frac{1}{C_i} + \frac{1}{C_{si} + C_{ss}}. \tag{39}$$

This equation can be derived using expressions (35), (37), and (38). Figure 12 represents the device capacitance normalized to C_i as a function of the applied gate voltage. When a large negative potential is applied to the gate electrode, the surface of the p-type silicon attracts holes, and the surface becomes

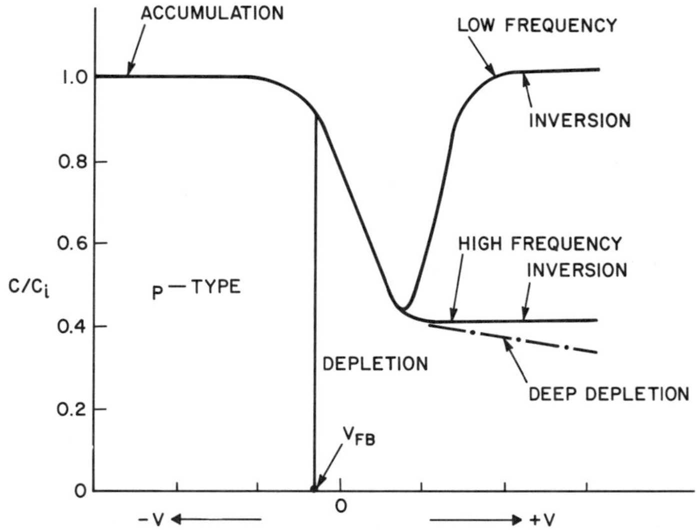

Fig. 12. MOS capacitance C normalized to the oxide capacitance C_i versus applied voltage V.

accumulated. For this situation, $C = C_i$. As the negative potential is decreased in its absolute value, the capacitance becomes smaller since the surface becomes depleted until it reaches a minimum at some positive potential. Beyond the minimum value point the surface is inverted, and the capacitance increases if a very slowly varying incremental voltage is superimposed on the dc bias for the capacitance measurement. A high frequency signal superimposed on the dc bias will show, however, no change in the capacitance value (Fig. 12). The two different branches result from the limited ability of the electron concentration to follow the applied ac signal used for the differential capacitance measurement. When the minority carrier generation rate is fast enough as compared to the ac signal frequency, the capacitance increases. With an n–p junction present, there is no frequency limitation since electrons may be injected rather than generated into the p-type silicon (holes into the n-type silicon). In most cases, only the upper branch of Fig. 12 will be obtained with frequencies from 20 to 100 Hz. The branch marked deep depletion comes about when the gate voltage V_G and the incremental measuring voltage both vary very fast and there are not enough generated minority carriers for inversion, so that the depletion layer widens beyond the x_{dmax} and the capacitance continues decreasing until at some point it merges rapidly with the high frequency branch. Deep depletion may also occur when the gate insulator is a weak conductor and the minority carriers cannot accumulate enough to invert the surface. The depletion layer in the silicon becomes wider than the maximum depletion layer width

x_{dmax}, and the capacitance C decreases. It can be shown [8] that the capacitance in the depletion and in the deep depletion regions follows the expression

$$C/C_i = [1 - (2C_i^2/qN_A\epsilon_s)(V_G - V_{FB})]^{-1/2}. \tag{40}$$

The theoretical capacitance–voltage relationships for ideal MOS capacitors as a function of silicon dioxide thickness were computed by [9]. The ideal diode capacitance curve has a flat-band point at $V = 0$ as there is no band bending in the bias absence. The real-life diodes will have $\phi_{ms} \neq 0$; therefore the entire C–V plot will be shifted as shown in Fig. 12. Any additional charges in the oxide or in the Si–SiO_2 interface will cause further C–V plot shift. The density of the charges located at the fast surface states (Section 1.8.2) is dependent on the gate voltage; consequently, the C–V plot will not only shift, but its slope will change.

1.8 Silicon Surface Charges and States

Several types of electric charges and states exist in the silicon dioxide and the Si–SiO_2 interface. The fixed charges in the oxide in the proximity of the interface and mobile ions in the SiO_2, such as sodium ions or other alkaline ions, are always positive. The energy states at the interface may be donor- or acceptor-like and are known as fast surface states. High energy radiation of electrons, x rays, and gamma rays induces ionized traps in the oxide.

1.8.1 Fixed Charge Density Q_{ss}

Regardless of how clean the silicon oxidation process is, the oxide always contains a certain number of fixed positive ions whose density per unit area Q_{ss}/q under the best conditions is on the order of a few 10^{10} cm^{-2}.

According to [10], the fixed surface charge Q_{ss} is the result of the presence of an excess of positively charged silicon ions left at the Si–SiO_2 interface in an unoxidized state. Q_{ss} depends, therefore, on the final heat treatment temperature and on the ambient gas (Fig. 13 [10]). Q_{ss} obtained after p- or n-type silicon dry or wet thermal oxidation can be reduced by annealing the crystal in the nitrogen or argon atmosphere. The donorlike energy states associated with Q_{ss} are located either close to or above the silicon conduction band and are therefore always at least partially ionized. Q_{ss} depends on the silicon crystal orientation and increases in the order (100) < (110) < (111), approximately in the ratio 1:2:3. By thinning the oxide step by step, it is possible to determine that the charge is located within 20 nm from the silicon–silicon dioxide interface. The positive charge presence shifts the MOS

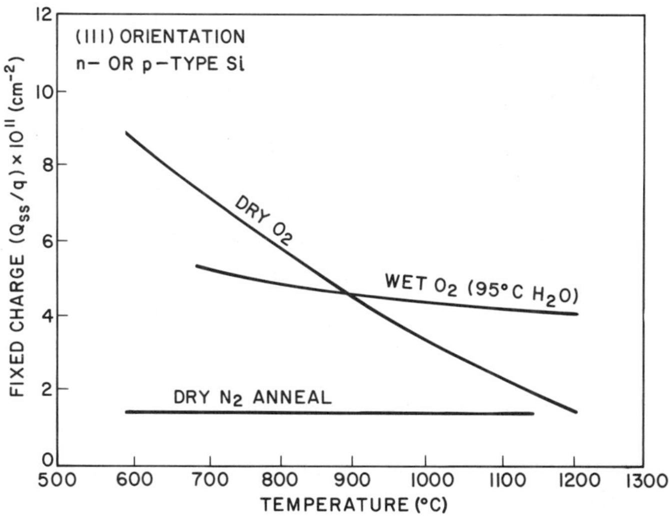

Fig. 13. Fixed positive charge Q_{ss}/q versus temperature. (From Deal *et al.* [10], reprinted by permission of the publisher, The Electrochemical Society, Inc.)

capacitance–voltage plot (Fig. 14) toward the negative potentials for the *p*-type silicon upon the application of the positive electrode (gate) potential. This takes place since the positive charge changes the flat-band voltage by $-Q_{ss}/C_i$.

1.8.2 Fast States (N_{FS})

Fast interface states take their name from the fact that they can exchange charges rapidly with the underlying silicon. It was shown by [11–13] that the fast states distribution, before the heat treatment, has two peaks in the

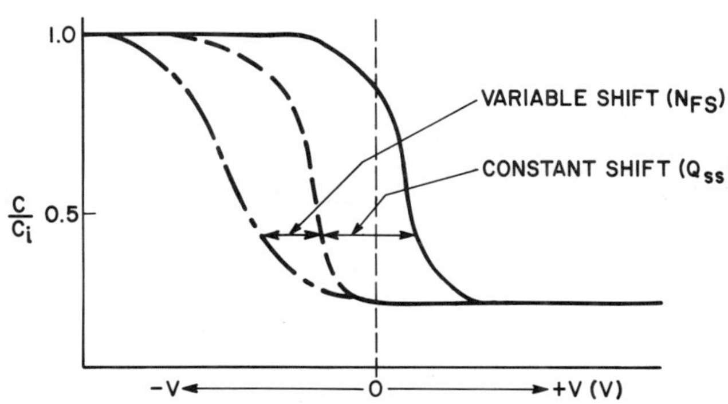

Fig. 14. C–V plot shift due to the interface states and Q_{ss}/q.

silicon energy band gap. One, consisting of the acceptor states, lies near the conduction band edge, and the other, consisting of donor states, resides at the valence band edge. The minimum density of fast states is located at the midgap. The origin of these states is due to the disruption of the silicon crystal periodicity at the surface, where not all valence bands are satisfied. The density of these states is not as high as expected from the number of the atoms at the surface due to the surface oxidation. The observed fast states density is from 10^{10} to 10^{12} cm^{-2}. The heat treatment in hydrogen at around 500°C usually reduces the fast states density to 10^{10} cm^{-2}. The fast states are also known as surface recombination centers since the surface recombination velocity related to surface lifetime is directly proportional to their density. Lowering the number of these states will result in smaller surface generation–recombination (leakage) current of *p–n* junctions. The fast states also reduce the carrier mobility by a scattering mechanism; increase the $1/f$ noise; and by adding a component to the flat-band voltage $-qN_{\mathrm{FS}}/C_i$, for acceptorlike species, shift the C–V plot to the left similarily to the fixed charge Q_{ss}. The magnitude of the shift is dependent on the surface potential and, consequently, on the applied MOS gate potential. N_{FS} depends on the crystal orientation and increases in its magnitude in the order (100) < (110) < (111). Figure 15 [12] illustrates how the interface states density is distributed near

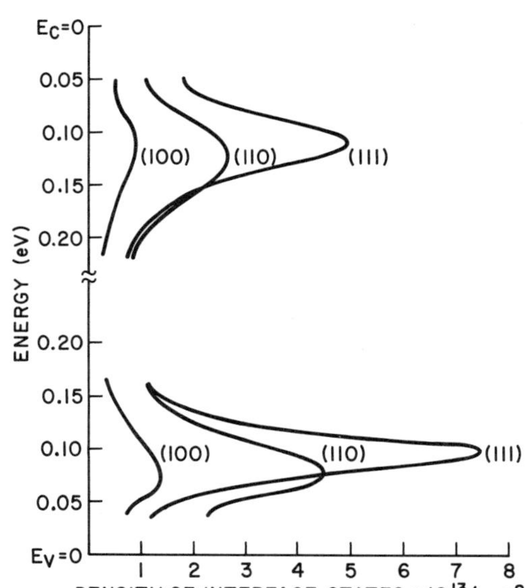

Fig. 15. Interface states energy distribution. (From Gray and Brown [12].)

the band edges for a 1-Ω-cm silicon crystal and an oxide grown with 80 ppm of O_2 at 1000°C. The total number of states is given by the area under the curves.

1.8.3 Mobile Ions

Positive sodium or other aklaline ions are almost always present to some extent in the grown or deposited silicon dioxide. Because of their high mobility, these ions can move easily, particularly at high temperatures (~300°C), so if a positive potential is applied to the MOS capacitor electrode, they will be pushed toward the Si–SiO_2 interface. This will result in a very large change in the system capacitance and in a translation of the C–V plot toward more negative potentials. The shift is not permanent, though, as in the case of Q_{ss}. If the applied voltage polarity is reversed, the ions will be attracted to the negative electrode. This easiness of charge redistribution causes device instability and is perturbing to both MOS and bipolar devices.

To grow practically ion-free oxide, it is necessary to take appropriate processing precautions, and under very stringent clean processing, it is feasible to reduce the ion density to about 10^{10} cm^{-2}. The processing precautions should include a very thorough cleaning of the silicon crystal surfaces, gas purification, furnace walls cleaning, use of extremely clean chemicals, etc. The addition of a small amount of gaseous HCl or chlorine (Cl_2) to the oxidizing atmosphere cleans the furnace quartz walls and drastically reduces the ionic contamination by direct action of Cl on the oxide. The use of chlorine and its compounds in device processing is discussed further in Section 5.5.

Sodium is more soluble in the phosphosilicate glass than in the oxide; so if this glass is deposited on the SiO_2 surface, sodium ions will be segregated into it, leaving the oxide with ion concentration several orders of magnitude lower than the concentration immediately after the steam oxidation [14]. Only thin layers of the phosphosilicate glass are acceptable because the dielectric polarization effects may result in surface instability when a potential is applied for a long period of time. Another dielectric of great importance in protecting against ionic contamination is silicon nitride, Si_3N_4, which should be deposited preferably on the SiO_2 layer rather than directly on the silicon substrate [15]. Sodium ions are not able to penetrate a thin (a few hundred nanometers) Si_3N_4 layer, which seals the underlying silicon dioxide from the outside impurities. The deposition of the nitride directly onto the silicon surface or very thin SiO_2 is not advisable since there exists a large number of traps in the nitride and at the interface that communicate directly by a tunneling effect with the silicon substrate causing hysteresis problems. The $C-V$ plot with the gate potential increasing from 0 to its maximum value is different from the plot obtained by decreasing the potential from the maximum value to 0. The addition of a Si_3N_4 layer changes the threshold voltage of a MOS device since the Si_3N_4 dielectric constant is ~ 12 and that of SiO_2 only ~ 4.0. For the total Si_3N_4 and SiO_2 thickness, the same as that of the oxide alone, the threshold voltage will be lower in absolute value for both the *n*- and the *p*-channel devices.

Al_2O_3 is another dielectric that is impervious to sodium ions [16, 17]. Al_2O_3 is more radiation resistant than thermal oxide.

There is a negative charge located at the Al_2O_3–SiO_2 interface which tends to make the silicon surface more *p*-type.

Hysteresis effects are also observed with the Al_2O_3–SiO_2 system but to a lesser degree than in the silicon nitride–SiO_2 double layer. Both double layer systems accumulate charges at their interfaces due to the electrical conductivity differences. SiO_2 is much less current conducting than the silicon nitride or aluminum oxide. The nitride conductivity can be modified by changing its stoichiometry. The charge accumulation can be eliminated, e.g., by using an oxide layer thicker than the nitride layer and by adjusting the stoichiometry so that the nitride conductivity is reduced.

1.9 Radiation Effects

An exposure of the SiO_2–Si system to ionizing radiation such as high energy electron beams or x or gamma rays, introduces new surface states and charges, modifying both MOS and bipolar device properties. The induced charges are positive and strongly dependent on the applied gate potential of the MOS capacitor [18, 19]. The positive charge generated in SiO_2 for the positively biased electrode may be one order of magnitude higher than that for the negative bias. In the absence of any electric field in the oxide, electrons and holes recombine without leaving any net electric charge. The charges are separated, however, when an electric field is present. A positively biased gate will attract all electrons present, and if there is no electron injection from the silicon substrate, only the slower holes will remain until they are trapped either in the volume of the oxide or at the Si–SiO_2 interface, creating a positive space charge. This charge induces a negative charge in the Si substrate surface. Radiation effects can be eliminated by heat treatment between 150 and 700°C. Photoinjection with uv light accomplishes a similar result.

Radiation-induced states have a degrading effect on silicon properties such as effective lifetime and mobilities, which are important for both bipolar and unipolar transistors. In order to obtain radiation-hardened devices, it is necessary, in some cases, to use other dielectrics such as Al_2O_3 [18] or to expose the SiO_2 film to nitrogen ion implantation [20].

1.10 Impurity Redistribution at the Oxidized Silicon Surface

During the thermal oxidation, the SiO_2–Si interface moves gradually deeper into the underlying silicon. As the silicon converts into the oxide, the doping impurities in the converted silicon have to move either to the oxide or to the silicon. According to [21, 22], *n*-type impurities, such as phosphorus, generally have a tendency to accumulate at the interface, doping the surface more heavily. On the other hand, the *p*-type impurities, such as boron and aluminum, get into the SiO_2 layer depleting the silicon surface of the dopant atoms. The direction of the doping atoms movement is determined largely by the Si–SiO_2 segregation coefficient for a given species and its relative diffusivities in silicon and the oxide. It should be noted also that the thermal oxide occupies a greater volume than the unoxidized silicon. Depletion or accumulation of impurities is more pronounced at lower oxidation temperatures than at higher. The surface impurity concentration is independent of the oxidation time. However, the effects of impurity redistribution are deeper with longer oxidation times. Oxidizing with wet oxygen results in greater impurity segregation than oxidizing with dry oxygen.

At the wet oxidation temperature of about 1000°C [21], the boron concentration at the surface will be about 15% of that in the bulk silicon, but oxidation with dry O_2 will result in the surface concentration of 25% of that in the bulk. Lack of H_2 in the oxidizing atmosphere results in even greater segregation into the oxide, due to the increased boron mobility.

Phosphorus piles up at the interface. At 1000°C, the silicon surface will contain almost 2.7 times more phosphorus than the silicon bulk.

The change in the silicon impurity surface concentration has some effect on the MOS device threshold voltage, particularly when a reverse voltage is applied to the source–substrate junction. In bipolar devices the change in the surface concentration affects the device breakdown and leakage current.

Gold is another impurity particularly important in the fabrication of fast bipolar devices where it is used to reduce the minority carrier lifetime. Gold diffusion from the back surface of the silicon chip piles up and introduces negative charges into the silicon–silicon dioxide interface [23] more or less equal in magnitude to Q_{ss}/q. The C–V plot for the negative charges is shifted in the positive voltage direction for the *p*-type silicon substrate.

1.11 Surface Mobility

The conductance per unit channel width of the n-type inversion layer formed on the p-type bulk silicon is given by [24]

$$g = \frac{q}{L}\int_0^{x_{\text{inv}}} \mu_n(x)n(x)\,dx = \frac{\mu_{\text{eff}}Q_n}{L}, \tag{41}$$

where

$$Q_n = q\int_0^{x_{\text{inv}}} n(x)\,dx$$

is the charge per channel unit area and $\mu_n(x)$, $n(x)$ are the electron mobility and electron concentration at a distance x from the surface, respectively; x_{inv} is the distance below the surface where the semiconductor is just intrinsic. The effective mobility is defined as

$$\mu_{\text{eff}} = \left(q\int_0^{x_{\text{inv}}} \mu(x)n(x)\,dx\right)/Q_n. \tag{42}$$

The mobility $\mu_n(x)$ at the surface differs appreciably from the bulk mobility, which is only limited by the crystal lattice and doping impurities. At the surface there are additional scattering mechanisms modifying appreciably both electron and hole mobilities. The mobility at the surface decreases with the increase of the gate voltage and the electric field in the oxide film. In the n-channel device a higher positive gate attracts electrons closer to the Si–SiO_2 interface, increasing the scattering process [25, 26], which is attributed to the unevenness of the surface and the localized oxide charges. Another proposed mechanism [27] consists of carrier scattering from the so-called misfit dislocations present at the silicon–dioxide interface and negatively charged by electrons in the n-type channel case. Maximum surface mobility occurs at the threshold voltage. For the subthreshold gate bias, there are only a very few free carriers, and the scattering effect due to the randomly distributed oxide charges is very sensibly augmented.

The experimentally observed effective surface mobility can be represented reasonably well as a function of the surface field [28] by

$$\mu_{\text{eff}} = \mu_{\text{eff}}(0)(6\cdot 10^4/\bar{E}_s)^r, \tag{43}$$

where $\mu_{\text{eff}}(0)$ is the low field mobility for the average surface field $\bar{E}_s \leqslant 6\times 10^4\ \text{V cm}^{-1}$. The average surface field $\bar{E}_s = -(V_G - V_T - 0.5V_D)/x_i$, where x_i is the oxide thickness and r is an empirical constant equal to about 0.12 for the n-type channel and about 0.20 for the p-channel, respectively. The electron surface mobility for n-channel devices fabricated on (100) oriented *p*-type silicon is on the order of 500–600 $cm^2\ V^{-1}\ s^{-1}$ and is about one-half of the bulk silicon mobility at the same carrier concentration. Surface electron mobility for *n*-channels on (111) oriented crystal is somewhat lower. The opposite is true for holes in the *p*-channels on *n*-type substrates [29]. Hole mobilities values are about one-third of those of electrons.

1.11.1 Saturated Drift Velocity at Silicon Surface

As the electron velocity in silicon under the applied field approaches the random thermal velocity, an intensified scattering process takes place in which the charge carriers start losing energy with an increased rate. As a result, electrons in silicon reach the limiting velocity of about $10^7\ cm\ s^{-1}$ at 300°K when the electric field reaches the value of $\sim 2 \times 10^4\ V\ cm^{-1}$. A similar effect takes place for holes that reach saturation velocity of about $7 \times 10^6\ cm\ s^{-1}$ at 300°K when the field reaches about $5 \times 10^4\ V\ cm^{-1}$. At the silicon surface the carrier mobility is affected not only by the lattice and impurity scattering but also by the charged interface surface states and oxide charges in the vicinity of the silicon surface. The carrier velocity in the inverted channels saturates as in the silicon bulk at about the same electric field intensities; however, the saturation velocities are significantly smaller [26] and are found to depend on the surface orientation with $v_s(100) > v_s(111) > v_s(110)$ for all temperatures. The limiting velocities for electrons at 300°K are about 6.5×10^6, 5.5×10^6, and $4 \times 10^6\ cm\ s^{-1}$, respectively.

REFERENCES

1. B. E. Deal, E. H. Snow, and C. A. Mead, *J. Phys. Chem. Solids* **27,** 1873 (1966).
2. R. M. Kingston and J. F. Newstadter, *J. Appl. Phys.* **26,** 718 (1955).
3. A. S. Grove, B. E. Deal, E. H. Snow, and C. T. Sah, *Solid-State Electron.* **8,** 145 (1965).
4. A. S. Grove and D. J. Fitzgerald, *Solid-State Electron.* **9,** 78 (1966).
5. P. Richman, *Solid-State Electron.* **11,** 869 (1968).
6. L. L. Vadash, A. S. Grove, T. A. Rowe, and G. E. Moore, *IEEE Spectrom* **6,** 28 (1969).
7. F. Faggin and T. Klein, *Solid-State Electron.* **13,** 1125 (1970).
8. A. S. Grove, "Physics and Technology of Semiconductor Devices." Wiley, New York, 1967.
9. A. Goetzberger, *Bell System Tech. J.* **45,** 1097 (1966).
10. B. E. Deal, M. Sklar, A. S. Grove, and E. M. Snow, *J. Electrochem. Soc.* **114,** 266 (1967).
11. E. H. Nicollia and A. Goetzberger, *Bell System. Tech. J.* **46,** 1055 (1967).
12. P. B. Gray and D. M. Brown, *Appl. Phys. Lett.* **8,** 31 (1966).
13. D. M. Brown and P. V. Gray, *J. Electrochem. Soc.* **119,** 388 (1972).
14. D. S. Kerr, J. S. Logan, P. K. Burkhardt, and W. A. Pliskin, *IBM J. Res. Develop.* **8,** 376 (1964).

15. R. R. Verderber, G. A. Gruber, J. W. Ostroski, and J. E. Johnson, *IEEE Trans. Electron Devices* **ED-17,** 797 (1970).
16. S. K. Tung and R. C. Caffrey, *J. Electrochem. Soc.* **114,** 275 C (1962).
17. A. Waxman and K. H. Zaininger, *Appl. Phys. Lett.* **12,** 109 (1968).
18. K. H. Zaininger and A. G. Holmes-Siedle, *RCA Rev.* **28,** 208 (1967).
19. K. H. Zaininger and A. S. Waxman, *IEEE Trans. Electron Devices* **ED-16,** 333 (1969).
20. R. P. Donovan and M. Simons, *Proc. Gov. Microwave Applic. Conf., Washington, D.C.* (1969).
21. A. S. Grove, O. Leistico, Jr., and C. T. Sah, *J. Appl. Phys.* **35,** 2696 (1964).
22. B. E. Deal, A. S. Grove, E. H. Snow, and C. T. Sah, *J. Electrochem. Soc.* **122,** 308 (1965).
23. S. F. Cagnina and E. H. Snow, *J. Electrochem. Soc.* **114,** 1165 (1967).
24. O. Leistico, A. S. Grove, and C. T. Sah, *IEEE Trans. Electron Devices* **ED-12,** 248 (1965).
25. J. R. Schrieffer, *Phys. Rev.* **97,** 641 (1955).
26. F. F. Fang and A. B. Fowler, *J. Appl. Phys.* **41,** 1825 (1970).
27. G. F. Neumark, *Phys. Rev. Lett.* **21,** 1252 (1968).
28. J. E. Meyer, *RCA Rev.* **32,** 42 (1971).
29. D. Coleman, R. T. Berte, and J. P. Mize, *J. Appl. Phys.* **39,** 1923 (1968).

CHAPTER 2

SEMICONDUCTOR PROPERTIES AT HIGH CARRIER CONCENTRATIONS

Some of the semiconductor properties are greatly modified by high doping levels and by the minority carrier currents with carrier densities exceeding the concentration of the ionized semiconductor impurities. High current densities affect, for example, carrier mobilities and diffusivities and minority carriers' lifetime. High doping levels also have an effect on the silicon energy gap.

2.1 Ambipolar Mobility and Diffusivity

When the injected minority carrier density exceeds the majority carrier concentration in the semiconductor, the flow of the junction injected holes can be visualized as taking place in an electron cloud, and the flow of electrons as in a cloud of holes. Holes and electrons tend to move in the presence of an electric field and/or diffusion gradient in two opposite directions. On the other hand, charge neutrality requires that the electron and hole concentrations be approximately equal so that the electron and hole distributions tend to be held together. The electrostatic attraction of the opposite charge carriers provides the necessary holding force. This effect becomes evident in the case of very large injection levels for which $n' = p'$, $dn'/dx = dp'/dx$, and the recombination rates $n'/\tau = p'/\tau$ are the same (n' and p' are excess carrier concentrations and τ is the lifetime). At very high injection levels, combined steady state hole and electron transport and current continuity equations for a uniformly doped semiconductor yield the so-called ambipolar transport equation

$$-\frac{n'}{\tau} + D_{\mathrm{a}}\frac{d^2n'}{dx} - \mu_{\mathrm{a}}E\frac{dn'}{dx} = 0, \tag{1}$$

where E is the electric field intensity and τ is the minority carrier lifetime, the same for holes and electrons at high injection levels; D_{a} and μ_{a} are defined by

$$D_{\mathrm{a}} = (n + p)D_nD_p/(nD_n + pD_p) \tag{2a}$$

and

$$\mu_a = (n - p)\mu_n\mu_p/(n\mu_n + p\mu_p). \tag{2b}$$

D_{a} and μ_{a} are the so-called ambipolar diffusivity and mobility, respectively, expressed in terms of carrier diffusivities and drift mobilities and the electron and hole *total* concentrations n and p. The concept of the ambipolar transport parameters often simplifies the device analysis since a single equation, valid for both carriers, can be used. Moreover, the presence of majority carriers is automatically taken into account. It should be noted, however, that Einstein's relationship $D = \mu(kT/q)$ does not apply to the ambipolar mobilities and diffusivities expressed by (2b) and (2a).

For very large injection levels $n \simeq p$, the total concentrations of electrons and holes are almost equal so that the ambipolar diffusion constant becomes

$$D_{\mathrm{a}} = 2D_nD_p/(D_n + D_p). \tag{3a}$$

A quantity

$$\mu_{\mathrm{A}} = 2\mu_n\mu_p/(\mu_n + \mu_p), \tag{3b}$$

related to the ambipolar diffusion constant D_{a} [Eq. (3a)] by Einstein's relationship is often also referred to as the ambipolar mobility.

2.2 Carrier–Carrier Scattering

The minority carrier mobility μ at low injection levels depends essentially on two mechanisms: the crystal-lattice scattering and doping impurity scattering, so that the low level mobility can be expressed as

$$\mu^{-1} = \mu_{\mathrm{latt}}^{-1} + \mu_{\mathrm{imp}}^{-1}. \tag{4}$$

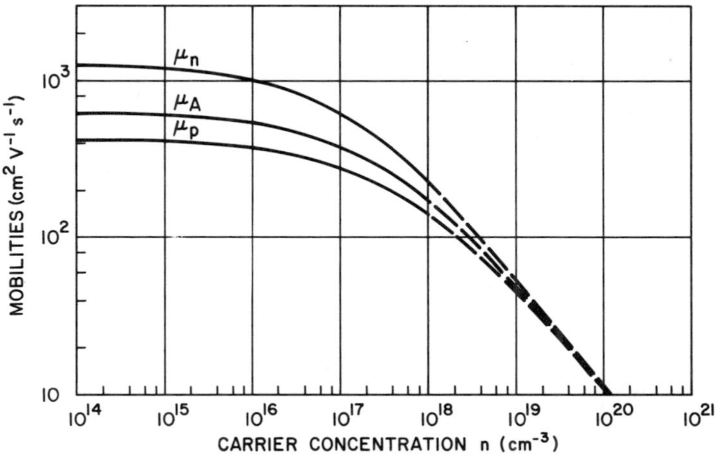

Fig. 1. Electron (μ_n), hole (μ_p), and ambipolar (μ_A) mobilities as a function of carrier concentration. (From J. Burtscher *et al.* [1].)

μ_{latt} and μ_{imp} are the lattice and impurity scattered mobilities, respectively. At high current densities, there is an additional effect of the carrier–carrier scattering, arising from the interaction between current carriers, which contributes another term μ_{cc}^{-1} to Eq. (4), which therefore becomes

$$\mu^{-1} = \mu_{latt}^{-1} + \mu_{imp}^{-1} + \mu_{cc}^{-1} \tag{5}$$

Figure 1 [1] shows the electron, hole, and ambipolar mobilities of silicon as a function of the injected carrier concentration. All three mobilities rapidly fall when the carrier density exceeds about $5 \times 10^{16}\ cm^{-3}$.

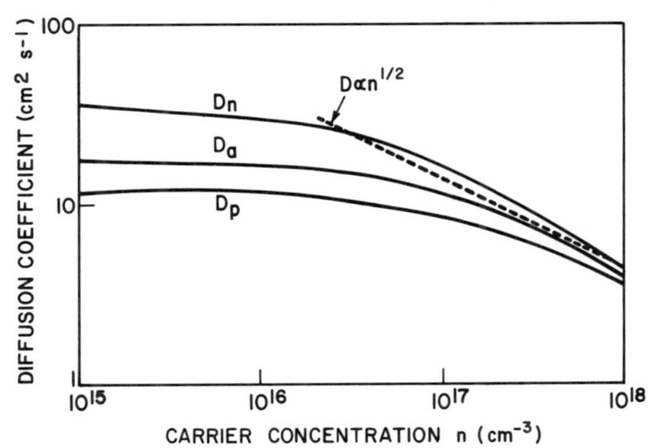

Fig. 2. Electron (D_n), hole (D_p), and ambipolar (D_a) diffusivities as a function of carrier concentration. (From Howard and Johnson [3].)

The ambipolar mobility can be expressed approximately by

$$\mu_A = (C_1 n + C_2)^{-1} \tag{6}$$

for carrier densities greater than about 10^{17} cm^{-3} [2]. Similar expressions hold for the hole and electron mobilities, with different constants however.

The effect of the carrier–carrier scattering is also taken into account in the plot of the diffusivities versus injected carrier concentration [3] of Fig. 2. As in the case of the mobilities, the diffusivities decrease with the increasing carrier concentration and approach the $D \propto n^{-1/2}$ dependence. At the highest injection levels, all diffusion constants are equal to $D = D_n = D_p$.

2.3 Mobility versus Impurity Concentration

An empirical formula based on the analysis of experimental data relates in a simple manner the electron mobility μ_n of silicon, germanium, and GaAs to the impurity concentration N_D by [4]

$$\mu_n = \frac{\mu_0}{1 + (N_D/10^{17})^{1/2}}, \tag{7}$$

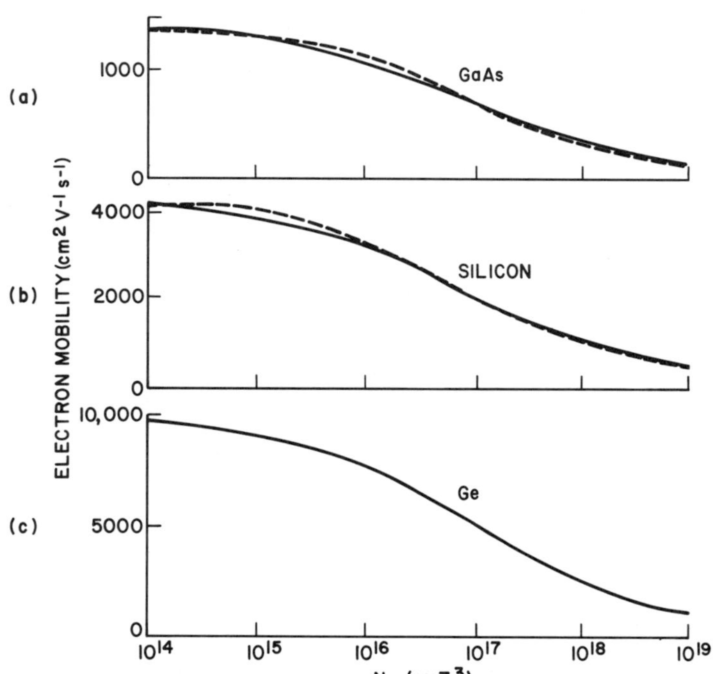

Fig. 3. (a) Gallium arsenide, (b) silicon, and (c) germanium electron mobilities as a function of the n-type impurity concentration. (From Hilsum [4], British Crown copyright reserved.)

where μ_0 is the mobility of the undoped semiconductor at 300°K (about 1400 $cm^2 V^{-1} s^{-1}$ for silicon, about 10,000 $cm^2 V^{-1} s^{-1}$ for GaAs, and about 4200 $cm^2 V^{-1} s^{-1}$ for germanium). Thus $N_D = 10^{17}$ cm^{-3} is the concentration at which the mobility drops to one-half of the original value for all three semiconductors. The formula agrees well with experimental data, but does lead to an underestimation of about 20% at the very high impurity concentrations. Nevertheless, it provides an extremely simple and a fast way of evaluating the mobility values as a function of the impurity concentration for approximate calculations. Figure 3 shows the dependence of electron mobility on donor concentration for GaAs, Si, and Ge. The solid lines represent the plot of the expression (7), and the dashed lines represent the experimental data.

2.4 Carrier Lifetime at High Injection Levels

The rate of the injected carrier recombination can be represented by

$$U = -an' - bn'^2 - cn'^3, \tag{8}$$

where n' is the excess carrier density per cubic centimeter. The first term on the right-hand side of equation (8) is due to the carrier recombination via recombination centers [5], and the coefficient a is the reciprocal of the lifetime at moderate injection levels. The second term is due to the direct band-to-band transition and is normally negligible for silicon with its indirect energy gap. The constant $b \simeq 2 \times 10^{-15}$ $cm^3 s^{-1}$. The third term is the result of the Auger recombination, important at very high injection levels. The constant $c \simeq 2 \times 10^{-31}$ $cm^6 s^{-1}$ for n-type silicon as in [6] or $c = 2.9 \times 10^{-31}$ $cm^6 s^{-1}$ as in [7]. For p-type silicon, $c = 1.2 \times 10^{-31}$ $cm^6 s^{-1}$ [6] at room temperature.

2.4.1 Recombination through Recombination Centers

At low-level injection, when the concentration of the excess minority carriers is much smaller than the silicon equilibrium majority carrier concentration, the carrier lifetime is [5]

$$\tau_{p0} = 1/\sigma_p v_s N_t, \tag{9}$$

$$\tau_{n0} = 1/\sigma_n v_s N_t, \tag{10}$$

where τ_{p0} and τ_{n0} are the low-level hole and electron lifetimes in n- and p-type silicon, respectively; σ_p and σ_n are the hole and electron capture-cross-sections, respectively; $v_s \simeq 10^7$ cm s^{-1} is the thermal velocity; and N_t the concentration of silicon recombination–generation centers per unit volume.

At high injection levels, the electrons and hole lifetimes are the same and given by

$$\tau_p = \tau_n = \tau_{p0} + \tau_{n0}. \tag{11}$$

Relationships (9)–(11) are true when the energy level of the recombination center E_t is at or very close to the intrinsic energy level E_i. This situation exists, for example, for gold in silicon. In the reversed biased depletion region of a p–n junction, the carrier lifetime [8] for recombination centers lying at the energy gap center is

$$\tau_0 = \tfrac{1}{2}(\tau_{p0} + \tau_{n0}). \tag{12}$$

2.4.2 Auger Recombination at High Doping Concentrations

Auger recombination occurs when the energy liberated by the carrier recombination is imparted to a third carrier, which becomes, therefore, more energetic. It is a three-carrier process involving two holes and one electron or two electrons and one hole. Auger recombination also takes place at the low injection levels when the silicon doping level is greater than $\sim 10^{17}$ cm^{-3}. The last recombination rate term of (8) is then replaced, for the n-type silicon, by $-cn^2p'$ with $c \simeq 2 \times 10^{-31}$ cm^6 s^{-1}, n equal to the electron concentration, and p' equal to the excess hole concentration [7].

2.4.3 Effective Carrier Lifetime

Since the band-to-band recombination in silicon can be neglected, Eq. (8) takes the form

$$U = -an' - cn'^3. \tag{13}$$

The effective lifetime at high injection levels is defined by

$$U = -n'/\tau_{\text{eff}}, \tag{14}$$

which together with (13) yields the effective lifetime value as

$$\tau_{\text{eff}} = 1/a + cn'^2. \tag{15}$$

At the moderate injection level of $n' = 10^{16}$ cm^{-3}, the low-level lifetime τ in silicon may be, for example, 10 μs, and the constant $a = 1/\tau = 10^5$ s^{-1}. The effective lifetime at the injected carrier density $n' = 10^{16}\ cm^{-3}$ will hardly be changed, but at $n' = 10^{18}\ \mathrm{cm}^{-3}$, τ_{eff} will be equal to 3.5 μs and will be considerably lower than the low-level lifetime.

2.5 Carrier Concentration Effect on Silicon Energy Band Gap Narrowing

Several effects observed particularly in bipolar transistors may be explained on the basis of the energy band gap E_g narrowing due to the high concentration of majority carriers. The temperature dependence of the current gain and the emitter efficiency belong to this category of effects.

As the ionized impurity concentration increases beyond $10^{17}\ \mathrm{cm}^{-3}$, the current carriers become so numerous that their presence has a screening effect on the periodic potential of the fixed crystal ions. The periodic potential is also modified by the concentration dependent random impurity atom distribution. As a result, the semiconductor band structure undergoes a change in that the band edges are no more sharply defined.

2.5.1 Models of Band Gap Narrowing

At impurity concentrations above about $10^{17}\ \mathrm{cm}^{-3}$, the atoms interact with each other due to their increasing proximity. As a result, the wavefunctions [8] of their associated electrons overlap, and at high concentrations above $10^{18}\ \mathrm{cm}^{-3}$, the impurity energy levels split, forming an impurity band in place of one single level [9]. The second effect is due to the statistically random distribution of the impurity atoms which disturb the crystal periodicity. The effect is very pronounced for concentrations higher than $10^{20}\ \mathrm{cm}^{-3}$ and for highly compensated material. The third effect is due to the interaction of the free charge carriers at concentrations above $10^{20}\ \mathrm{cm}^{-3}$. As a result of these effects, the well-defined band edges at low concentrations will exhibit band tails at the high concentrations [10, 11]. The broadening of the impurity band and the formation of the band edge tails reduce the energy band gap and modify the density of energy states and their shape, which ideally would show parabolic dependence on energy. Figure 4 [12] shows the density of states versus energy for very highly doped crystals with E_{c0} and E_{v0} being the conduction and valence bands edges of the undoped crystal, respectively. The density of states function is concentration dependent and is considered to be the sum of the impurity band states by Kleppinger and Lindholm [13], but is assumed to be equal to the envelope

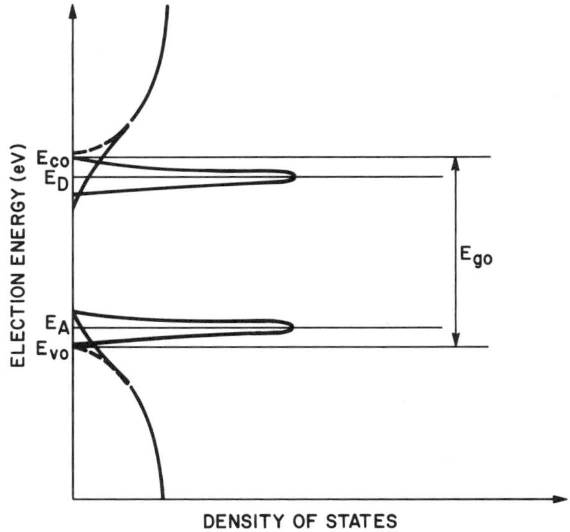

Fig. 4. Silicon density of states distribution as a function of energy. (From R. J. Van Overstraeten *et al.* [12].)

of the conduction (or valence) and the impurity bands by Van Overstraeten [12] and is thus more applicable to the heavily compensated semiconductors. Another model of band gap reduction in silicon, mainly applicable to concentrations of less than 10^{20} cm^{-3}, was proposed by Lanyon and Tuft [14]. Their model takes into account the existence of the electrostatic energy of a minority carrier surrounded by a cloud of the majority carriers that are attracted to it and cause a local departure from the charge neutrality. This energy reduces the activation energy necessary to create an electron–hole pair. This is equivalent from the carrier generation point of view to the energy gap reduction. At some point at distance r from the carrier, the electrostatic potential V of the screened minority carrier can be found from Poisson's equation

$$\nabla^2 V = q[n(r) - n]/\epsilon_s, \tag{16}$$

in which ϵ_s is the silicon permittivity, q is the electron charge, n is the uniform majority carrier concentration, and $n(r)$ represents the majority carrier density at point r. $n(r)$ deviates from the unperturbed distribution due to the minority carrier electrostatic attraction of the majority carriers. The magnitude of the unscreened field is given by

$$E_0(r) = q/4\pi\epsilon_s r^2, \tag{17}$$

and of the screened field by

$$E(r) = \left(\frac{1}{r} + \frac{1}{a_s}\right) \frac{q}{4\pi\epsilon_s r} \exp\left(\frac{-r}{a_s}\right), \tag{18}$$

where a_s is the screening radius (space constant). For nondegenerate material,

$$a_s = (q^2 n/\epsilon_s kT)^{1/2}, \tag{19}$$

where n is the free carrier density. In the strongly degenerate case,

$$a_s = (3\sqrt{\pi}/8)^{1/3}(\epsilon_s h^2/3\pi q^2 m^*)^{n^{-1/6}}, \tag{20}$$

where m^* is the electron's mass effective mass and h is Planck's constant. At the intermediate doping ranges,

$$a_s = (Cq^2 n/\epsilon_s kT)^{-1/2} \tag{21}$$

with constant C also being less than or equal to unity. The electrostatic energy is proportional to the square of the electrostatic field; therefore, the carrier pair self-energy can be expressed as

$$\Delta E_g = \frac{\epsilon_s}{2} \int_{\text{all space}} [E_0^2(r) - E^2(r)]\, d\tau, \tag{22}$$

where τ is the space increment. Substituting (17) and (18) for $E_0(r)$ and $E(r)$, respectively, and integrating (22) with limits $r = 0$ and $r = \infty$ yields the final expression [14]

$$\Delta E_g = 3q^2/16\pi\epsilon_s a_s. \tag{23}$$

The band gap reduction in a nondegenerate silicon is

$$\Delta E_g = (3q^2/16\pi\epsilon_s)(q^2 n/\epsilon_s kT)^{1/2} \tag{24a}$$

and depends on the square root of the majority carrier concentration. At room temperature, (24a) becomes

$$\Delta E_g = 22.5(n/10^{18})^{1/2} \quad \text{meV}. \tag{24b}$$

In degenerate silicon,

$$\Delta E_g = 162(n/10^{20})^{1/6} \quad \text{meV}. \tag{25}$$

The energy gap reduction in this case is temperature independent. For $n = 5 \times 10^{19}$ cm^{-3} (degenerate silicon),

$$\Delta E_g \simeq 144 \quad \text{meV},$$

but in the nondegenerate case of $n = 5 \times 10^{17}\ cm^{-3}$,

$$\Delta E_g = 7.1 \quad \text{meV}.$$

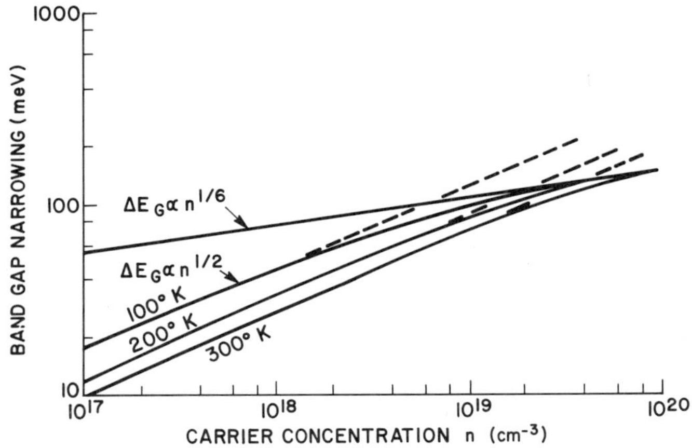

Fig. 5. Silicon band gap narrowing as a function of carrier concentration. (From Lanyon and Tuft [14].)

The empirical relationship developed by [15]

$$\Delta E_g = 9\left[\ln \frac{n}{10^{17}} + \left(\ln^2 \frac{n}{10^{17}} + 0.5\right)^{1/2}\right] \quad \text{meV} \tag{26}$$

yields 112 meV for $n = 5 \times 10^{19}$ and about 6.4 for $n = 10^{17}$.

The theoretical dependence of the band gap narrowing through the pair energy mechanism is shown in Fig. 5 [14] as a function of carrier concentration with temperature as a parameter. The agreement between the theory and the experimental data of most of the other workers is very good.

The band gap narrowing is a consequence of the screening of the minority carriers regardless of their origin; therefore, in the conductivity modulated mode or in an MOS transistor in strong inversion, the injected or induced carrier concentration determines the extent of the energy gap narrowing, rather than the doping level. In the junction depletion region where there are no free carriers, the energy gap is the same as that of the intrinsic silicon.

2.6 Intrinsic Carrier Concentration at High Doping Levels

At low doping levels, the silicon intrinsic carriers concentration is

$$n_i^2 = \text{const} \cdot T^3 \exp(-E_g/kT). \tag{27}$$

Above 250°K, the band gap E_g can be approximated by a linear function of the absolute temperature T as

$$E_g/q = E_{g0}/q - \alpha T, \tag{28}$$

where α is the temperature coefficient and E_{g0} is the energy gap extrapolated to $T = 0°\text{K}$. Substitution of (28) into (27) gives for the intrinsic carrier concentration case

$$n_i^2 = C_1 T^3 \exp(-E_{g0}/kT). \tag{29}$$

According to [16], $C_1 = 9.6 \times 10^{32}$, and $E_{g0} = 1.205$ eV. This is, of course, not the value of the energy gap at 0°K, which for silicon is 1.18 eV. n_i^2 for silicon at 300°K equals 2×10^{20} cm^{-6}. For high impurity concentrations above 10^{17} cm^{-3}, it is necessary to take into account the change in the band gap width ΔE_g. It was shown by [15] that the effective intrinsic carrier concentration can be expressed as

$$n_{ie}^2 = n_i^2 \exp(\Delta E_{g0}/kT) \tag{30}$$

for temperatures above 250°K. n_{ie} is therefore always larger than n_i as can be expected for a smaller energy gap. Expression (30) is valid in the concentration range of up to 3.5×10^{19} cm^{-3}. For $\Delta E_{g0} = 100$ meV we find from (30) that the effective intrinsic concentration at 300°K $n_{ie} = 6.8 n_i$.

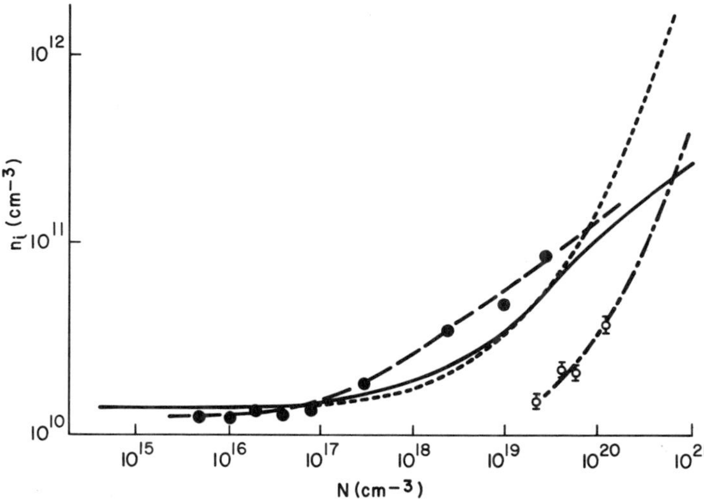

Fig. 6. Silicon intrinsic carrier concentration n_i as a function of impurity concentration N. --Ō-- indicates experimental value of Vol'fson and Subashiev [17]; —— indicates calculated values of Mock [18]; --- indicates calculated values of Van Overstraeten *et al.* [12]; --•- indicates experimental values of Slotboom and deGraaf [15]. (From Slotboom and deGraaf [15].)

Effective intrinsic concentration calculations and experimental values are compared with earlier experiments and calculations of [12, 13, 17, 18] in the plots of n_{ie} versus net impurity concentrations of Fig. 6 [15]. It can be seen that in the silicon doping range between 10^{17} cm^{-3} and 5×10^{19} cm^3 the data of [15] show the highest values. It should be pointed out that the knowledge of accurate n_{ie} values is of great importance in the evaluation of transistor emitter efficiency, which is a very sensitive function of the effective intrinsic concentration.

2.6.1 The Generation–Recombination Current

The space-charge generated current density for reverse biased junction is given approximately by

$$J_{gen} \simeq q|U|W, \tag{31}$$

where U is the carrier generation rate and W is the space-charge region (depletion region) width. The generation rate

$$U = -n_i/2\tau_0, \tag{32}$$

where τ_0 is the carrier effective lifetime in the space-charge region. The space-charge recombination current of the forward biased junction is given by

$$J_{rec} = -\tfrac{1}{2}U \exp(qV_F/2kT) \tag{33}$$

with V_F being the forward voltage. The generation and recombination current expressions remain unaltered at high doping levels since in the depletion region even at high currents, there are not enough free carriers and the band gap remains that of intrinsic silicon [14].

REFERENCES

1. J. Burtscher, F. Dannhäuser, and J. Krausse, *Solid-State Electron.* **18,** 35 (1975).
2. M. Lietz, *Electron. Lett.* **8,** 275 (1965).
3. N. R. Howard and G. N. Johnson, *Solid-State Electron.* **8,** 275 (1965).
4. C. Hilsum, *Electron. Lett.* **10,** 259 (1974).
5. W. Shockley and W. T. Read, *Phys. Rev.* **87,** 835 (1952) and R. N. Hall, *Phys. Rev.* **87,** 387 (1952).
6. N. G. Nilsson and K. G. Svantenson, *Solid State Commun.* **11,** 155 (1972).
7. J. D. Beck and R. Conradt, *Solid State Commun.* **13,** 93 (1973).
8. C. T. Sah, R. N. Noyce, and W. Shockley, *Proc. IRE* **45,** 1228 (1957).
9. T. N. Morgan, *Phys. Rev. A.* **139,** 343 (1965).

10. E. O. Kane, *Phys. Rev.* **131,** 79 (1963).
11. V. L. Bonch-Bruevich, *Soviet Phys.—Solid State* **4,** 1953 (1963).
12. R. J. Van Overstraeten, H. J. DeMan and R. P. Mertens, *IEEE Trans. Electron Devices* **ED-20,** 290 (1973).
13. D. D. Kleppinger and F. A. Lindholm, *Solid-State Electron.* **14,** 407 (1971).
14. H. P. D. Lanyon and R. A. Tuft, *IEEE Trans. Electron Devices* **ED-26,** 1014 (1979).
15. J. W. Slotboom and H. C. deGraaf, *Solid-State Electron.* **19,** 857 (1976).
16. E. M. Putley and W. H. Mitchell, *Proc. Phys. Soc. Lond. A* **72,** 193 (1958).
17. A. A. Vol'fson and V. K. Subashiev, *Soviet Phys.—Semicond.* **1,** 327 (1967).
18. M. S. Mock, *Solid-State Electron.* **16,** 1251 (1973) and M. S. Mock, *Solid-State Electron.* **17,** 819 (1974).

CHAPTER 3

AVALANCHE BREAKDOWN

At high fields encountered in the high voltage p–n junctions, the electrical breakdown takes place by impact ionization, also known as avalanche breakdown. Free carriers always present in a semiconductor crystal may acquire enough energy under the effect of the applied field to break the covalent bonds and to liberate free holes and electrons, which, in turn, create more free carriers. This process becomes an avalanche phenomenon at sufficiently high fields on the order of 10^5 to about 10^6 V cm^{-1}, depending on the silicon-doping level. Higher critical field values are applicable to higher silicon concentrations in the neighborhood of 10^{18} cm^{-3}.

3.1 Avalanche Breakdown Voltage Calculations

Accurate calculation of the p–n junction avalanche breakdown requires the solution of the ionization integral

$$I_p^* = \int_0^{x_d} \alpha_p \exp\left[- \int_0^{x} (\alpha_p - \alpha_n)\, dx' \right] dx = 1 \qquad (1)$$

for hole initiated breakdown or

$$I_n^* = \int_0^{x_d} \alpha_n \exp\left[- \int_0^{x_d} (\alpha_n - \alpha_p)\, dx' \right] dx = 1 \qquad (2)$$

for electron initiated breakdown. In both expressions, x_d is the depletion layer width (Fig. 1) at breakdown and α_n and α_p are electron and hole field dependent ionization rates for silicon, expressed by the empirical relations

$$\alpha_n = A_n \exp(-B_n E^{-1}), \tag{3}$$

$$\alpha_p = A_p \exp(-B_p E^{-1}), \tag{4}$$

where A_n, A_p, B_n, and B_p are constants and E is the absolute value of the electric field at x. The values of A and B were measured by several authors with a wide discrepancy between the obtained values. According to Van Overstraeten and DeMan [1], $A_n = 1.6 \times 10^6\ \text{cm}^{-1}$, $A_p = 5.5 \times 10^5\ \text{cm}^{-1}$, and $B_n = B_p = 1.65 \times 10^6$. According to Lee *et al.* [2], $A_n = 3.8 \times 10^6\ \text{cm}^{-1}$, $A_p = 2.25 \times 10^7\ \text{cm}^{-1}$, $B_n = 1.75 \times 10^6\ \text{V cm}^{-1}$, and $B_p = 3.26 \times 10^6\ \text{V cm}^{-1}$. Both sets of data are valid for peak fields $1.8 \times 10^5\ \text{V cm}^{-1} < E_m < 5 \times 10^5\ \text{V cm}^{-1}$. The data of Moll and Van Overstraeten [3], taking into account the threshold energy influence, are $A_n = 7.03 \times 10^5\ \text{cm}^{-1}$ and $B_n = 1.23 \times 10^6\ \text{V cm}^{-1}$ for electrons and $A_p = 1.58 \times 10^6\ \text{cm}^{-1}$ and $B_p = 2.036 \times 10^6\ \text{V cm}^{-1}$ for holes. These data are valid for the field range of $1.75 \times 10^5\ \text{V cm}^{-1} < E_m < 7 \times 10^5\ \text{V cm}^{-1}$ for electrons and $1.75\ \text{V cm}^{-1} < E_m < 6 \times 10^5\ \text{V cm}^{-1}$ for holes. In many instances it is satisfactory to assume the equality of the ionization coefficients $\alpha_n = \alpha_p = \alpha_{\text{eff}}$ to obtain approximate breakdown expressions as was done by [4, 5, 6]. When α_{eff} is used, the ionization integral becomes

$$\int_0^{x_d} \alpha_{\text{eff}}\, dx = 1 \tag{5}$$

with

$$\alpha_{\text{eff}} = A \exp(-B_n E^{-1}) = 1.07 \times 10^6 \exp(-1.65 \times 10^5 E^{-1}) \tag{6}$$

according to [6] or $\alpha_{\text{eff}} = 1.8 \times 10^{-35} E^7$ as in [5]. For some compound semiconductors, such as GaAs or GaP, $\alpha_n = \alpha_p$ so that expression (5) is accurate and not an approximation.

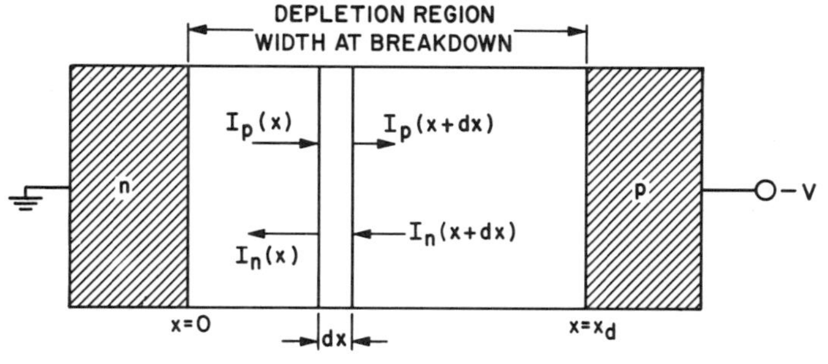

Fig. 1. Carrier multiplication in the depletion region.

The carrier multiplication factor M, i.e., the ratio of the output current to the current entering the avalanche multiplication region, can be expressed as

$$M = (1 - I^*)^{-1}. \tag{7}$$

Breakdown occurs when $I^* = 1$ and M becomes infinite. The multiplication factors for electrons and holes are related to the reverse biased junction voltage V_R by the empirical relationship [7]

$$M = \frac{1}{1 - (V_R/BV)^n} \tag{8}$$

where BV is the silicon or germanium breakdown voltage. For electrons $n = 4$, and for holes $n = 6$. Expression (8) is valid for the collectors of the n–p–n and p–n–p transistors with appropriate values of n. It was originally conceived by [7] for germanium, where $\alpha_p/\alpha_n = 2.2 \pm 02$. A more accurate variation of this empirical relationship for silicon, where this ratio is much different than in germanium, is discussed further in section 8.3 on collector breakdown.

3.2 *p–i–n* Diode Avalanche Breakdown

The simplified version of the ionization integral (5) can be applied very easily, for example, to a p–i–n diode breakdown calculation. When the device's intrinsic region is completely depleted, the electric charge in it is zero so the field is constant as depicted by Fig. 2. The avalanche breakdown takes place when the ionization integral (5) becomes equal to unity, i.e., when

$$\int_0^{x_d} Ae^{-B_n/E}\, dx = 1. \tag{9}$$

Since the field is constant, integration of (9) yields for its value at breakdown

$$E = B_n/\ln Ax_d. \tag{10a}$$

The field is related to the voltage by

$$V = Ex_d; \tag{10b}$$

therefore, the p–i–n breakdown voltage is

$$BV = B_n x_d/\ln Ax_d \tag{11}$$

The electric field at breakdown in the p–i–n structure is equal to about 18–20 V cm^{-1}.

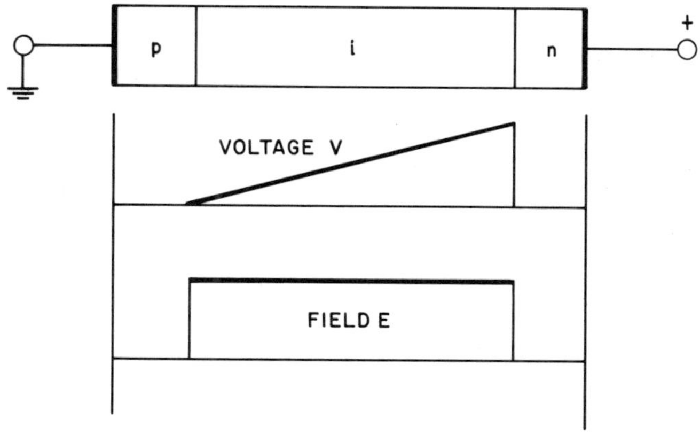

Fig. 2. Voltage and field distribution in the p–i–n diode.

3.3 Plane (One-Dimensional) Junction Breakdown

The avalanche breakdown voltage of a one-sided abrupt silicon junction is approximately given by

$$BV \simeq K_1 10^{13} N^{-3/4}, \tag{12}$$

where N is the impurity concentration of the lightly doped region and the constant $K_1 = 6$ according to [4] or 5.3 according to [5]. The depletion layer width at breakdown for one-sided abrupt junction is

$$x_{\text{dbreakdown}} = (2BV\epsilon_s/qN)^{1/2} \tag{13}$$

or by substituting for N its value computed from (12) and using $K_1 = 5.3$,

$$x_{\text{dbreakdown}} = 2.5 \times 10^{-2} BV^{7/6} \quad \mu\text{m}. \tag{14}$$

The abrupt junction breakdown as a function of the impurity concentration N and the depletion layer width as a function of breakdown are plotted in Fig. 3.

The approximate breakdown voltage of a double sided p–n junction [8] can be obtained by using the effective doping concentration defined by

$$\frac{1}{N_{\text{eff}}} = \frac{1}{N_A} + \frac{1}{N_D}, \tag{15}$$

where N_A and N_D are the acceptor and donor concentrations of the p- and n-side of the junction, respectively.

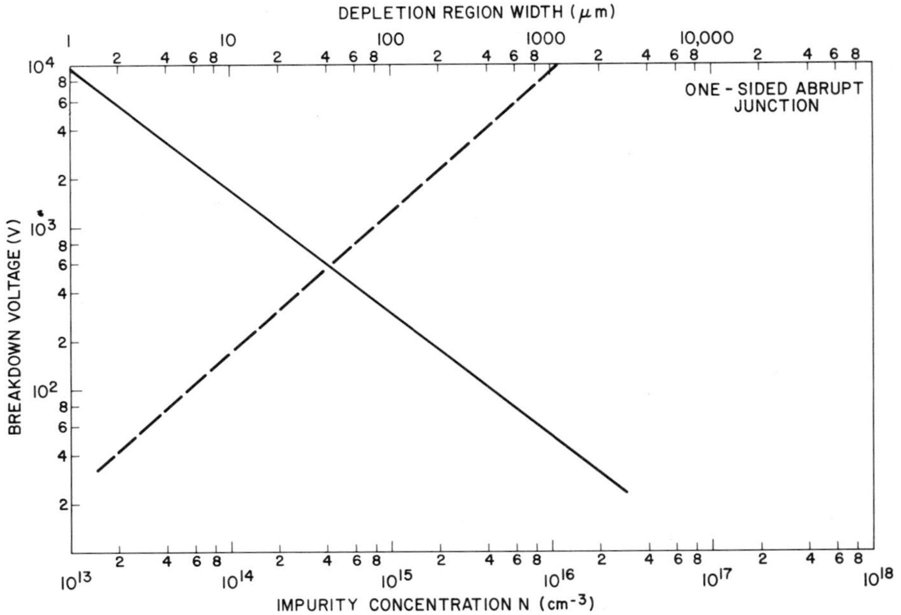

Fig. 3. Solid line is breakdown voltage for one-sided abrupt junctions versus concentration. Dashed line is depletion region width at breakdown.

The breakdown voltage of a linearly graded silicon junction is given by

$$BV = K_2 10^9 a^{-2/5}, \tag{16}$$

where K_2 is equal to 9.3 according to [4] or to 8.93 according to [5], a is the impurity gradient per quadrupled centimeter. The total depletion layer width of a linearly graded junction can be obtained from the expression

$$x_d = (12BV\epsilon_s/qa)^{1/3}. \tag{17}$$

Eliminating a by using (16) gives

$$x_d = 2.15\ 10^{-2} BV^{7/6} \quad \mu\text{m}. \tag{18}$$

The breakdown voltage of the linearly graded junction as a function of the grade constant a and the depletion layer width as a function of breakdown are plotted in Fig. 4.

The breakdown voltage of a diffused junction falls in between the high limit of the linearly graded junction and the low limit of a one-sided abrupt junction. The depletion region width at breakdown for the diffused junction can be determined approximately by averaging the depletion widths of the abrupt and linearly graded junctions.

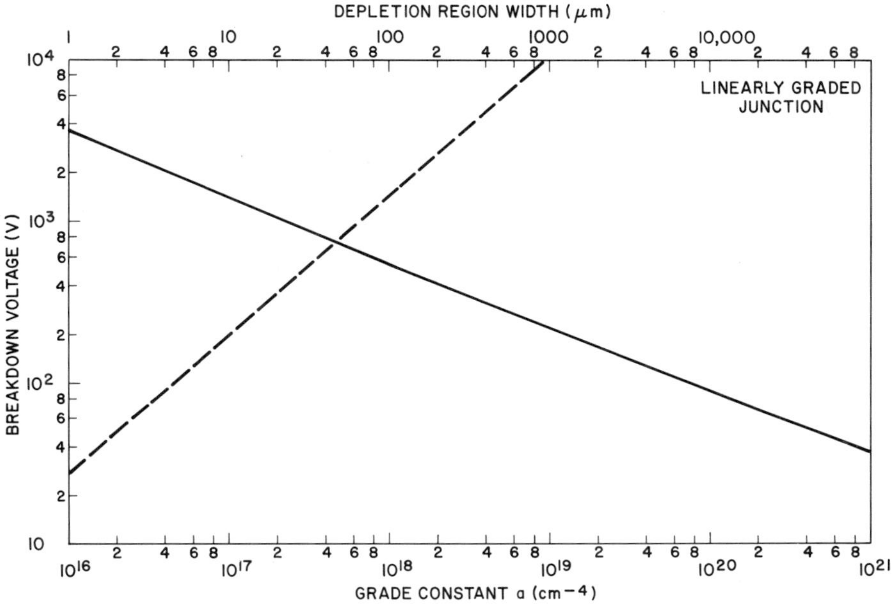

Fig. 4. Solid line is breakdown voltage for linearly graded junctions versus grade constants. Dashed line is depletion width at breakdown.

An empirical *general* relationship between the breakdown voltage and the depletion width at breakdown valid for all three types was more recently developed by [9] as

$$BV = (5.8\ 10^4)x_d^{0.84}. \tag{19}$$

The avalanche breakdown computation for complementary erfc diffusion profiles was performed by [6] using ionization coefficients of [3]. Figures 5 and 6 [6] give the breakdown voltage as a function of the net impurity density of the uniformly doped substrate for junction depths varying between 10 and 100 μm. The surface concentration for Fig. 5 is $C_0 = 10^{18}$ cm^{-3}, and for Fig. 6, 10^{19} cm^{-3}.

The avalanche breakdown voltage depends, of course, on the impurity gradient *a* near the junction. For a certain range of surface and substrate concentrations the gradients for the erfc and Gaussian distribution are practically the same. It was found by [10] that if the ratio of the surface concentration to the substrate concentration C_0/C_B lies between 10^2 and 10^7, then the complementary error function curves can be used to determine the breakdown voltage for the Gaussian impurity distribution if the C_0/C_B ratios of the erfc curves are decreased 2.5 time.

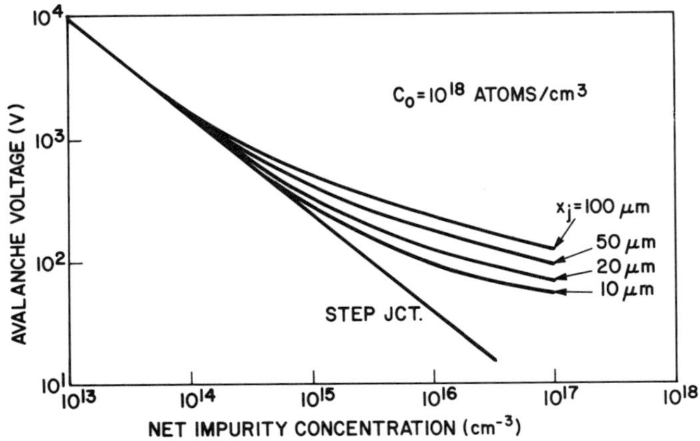

Fig. 5. Avalanche breakdown voltage for a diffused junction as a function of the uniformly doped substrate impurity concentration with junction depth x_j as a parameter. Surface concentration $C_0 = 10^{18}$ cm^{-3}. (From Kokosa and Davies [6].)

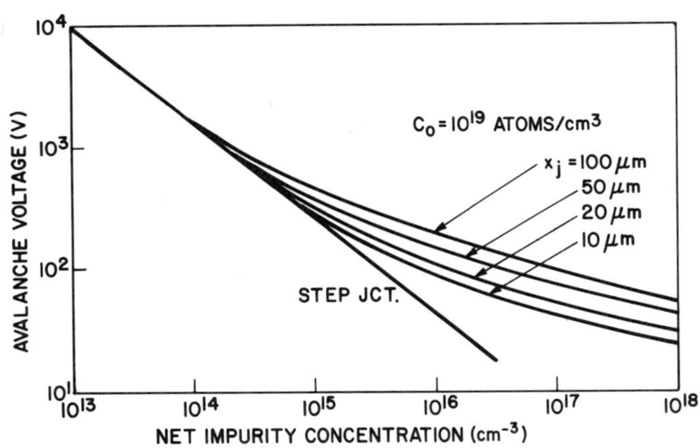

Fig. 6. Avalanche breakdown voltage for a diffused junction as a function of a uniformly doped substrate impurity concentration with junction depth x_j as a parameter. Surface concentration $C_0 = 10^{19}$ cm^{-3}. (From Kokosa and Davies [6].)

3.4 Planar Junction Breakdown

The planar junction breakdown is determined by the impurity concentration profile and, to a large extent, by the curvatures of the junction edges and corners (Fig. 7). Planar junctions, when formed by diffusion through a rectangular mask, include a central plane portion, cylindrical edge regions,

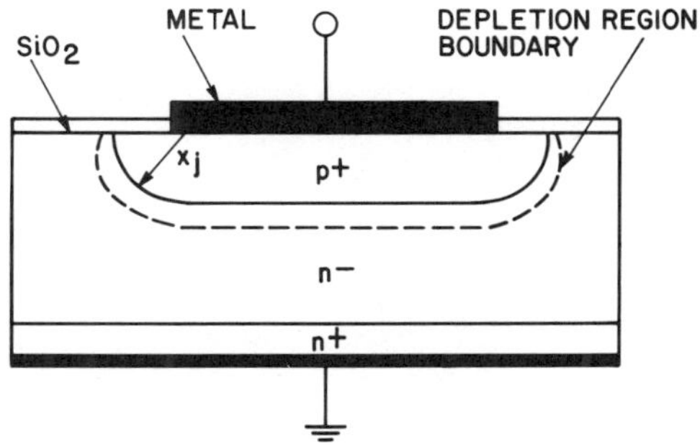

Fig. 7. Cylindrical planar junction with the depletion region boundary.

and spherical regions occurring at the mask corners. Figure 8 [4] shows computer-obtained curves for the planar junction breakdown voltage as a function of the junction depth x_j for the cylindrical and spherical one-sided abrupt junctions. For comparison, the breakdown voltage of a plane- (one-dimensional) type junction is also plotted as a curve with $x_j \to \infty$. The planar junction breakdown is always smaller than that of the plane junction. The silicon dioxide on the device surface usually contains some amount of positive charges; hence the n-silicon surface, in the case of a p^+-n junction, may become accumulated. Consequently, the depletion layer narrows at the surface, as shown by the dotted line of Fig. 7, so that the electric field is increased and the breakdown additionally lowered.

The junction curvature effect on the avalanche breakdown is also shown

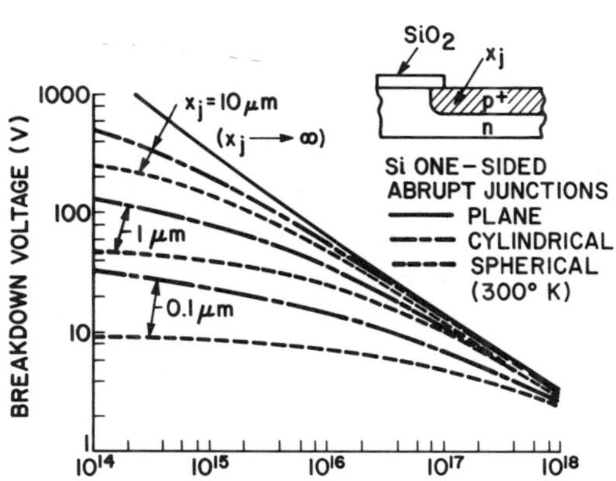

Fig. 8. Breakdown voltage as a function of the substrate concentration for the plane, cylindrical, and spherical junctions with junction depth x_j as a parameter. (From Sze and Gibbons [4].)

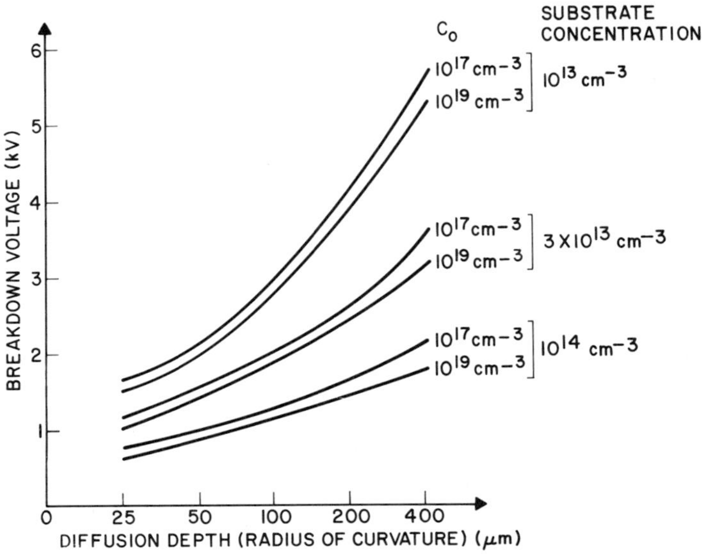

Fig. 9. Breakdown voltage of a diffused junction as a function of the radius of curvature for surface concentrations $C_0 = 10^{17}$ cm^{-3} and 10^{19} cm^{-3}, and substrate concentrations $C = 10^{13}$ cm^{-3}, 3×10^{13} cm^{-3}, and 10^{14} cm^{-3}. (From Temple and Adler [11].)

by Fig. 9 [11]. The curves are plotted for erfc diffusions with *p*-surface concentrations of 10^{17} and 10^{19} cm^{-3}. The *n*-type background concentrations were $C_B = 10^{13}$ cm^{-3}, 3×10^{13} cm^{-3}, and 10^{14} cm^{-3}. The curves demonstrate the large effect of the uniform substrate doping level on the device breakdown. With $x_j = 100\ \mu$m the breakdown voltage for the surface concentration of 10^{19} cm^{-3} and $C_B = 10^{13}$ cm^{-3} is about 2800 V.

A junction with the same parameters, but with $C_B = 10^{14}$ cm^{-3}, has a breakdown of about 1200 V. The dependence on the surface concentration becomes more significant as the diffusion depth increases. The change of the radius of curvature from 100 to 200 μm changes the breakdown voltage from about 1200 to about 1350 V for $C_0 = 10^{19}$ cm^{-3} and $C_B = 10^{14}$ cm^{-3}. Using the same substrate dopings and surface concentrations as in Fig. 9, it was possible, thanks to the normalization method, to present in single curves the maximum bulk electric fields, the maximum surface fields, and the fields of the one-dimensional (flat) portion of the planar junction (Fig. 10 [12]). The curves show that the maximum bulk fields occur at the maximum junction curvatures. The maximum surface fields, on the other hand, are at the point where the junction outline crosses the surface. The figure demon-

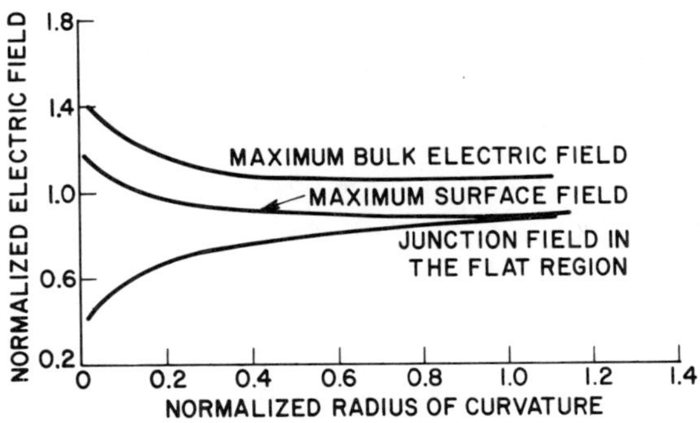

Fig. 10. Normalized electric fields as a function of the planar junction normalized radius of curvature. (From Adler and Temple [12].)

strates that the bulk electric field of a planar junction at breakdown is always greater than that of the electric field of the one-dimensional flat junction and that it always increases with the decreasing radius of curvature. As the radius becomes smaller, the size of the depletion region is reduced at the point of the maximum field, and this field has to increase in order to produce sufficient avalanche multiplication over a shorter distance. The same is true for all cases that cause the depletion region to shrink. These include higher substrate doping and appropriate junction shaping.

One single curve may be sufficient to show the breakdown dependence on the radius of a curvature if these quantities are normalized as proposed by [11]. Figure 11 shows the planar junction avalanche breakdown voltage as

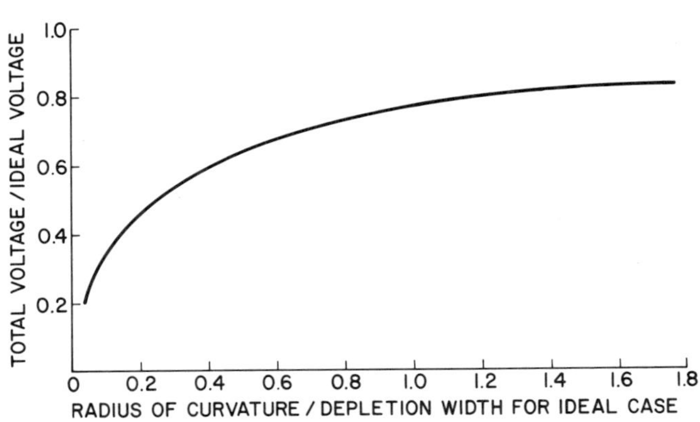

Fig. 11. Planar junction normalized avalanche breakdown as a function of the normalized depletion region width. (From Adler and Temple [12].)

normalized to the ideal one-dimensional (plane junction) breakdown voltage as a function of the radius of curvature normalized to the plane depletion region width at breakdown. The use of Figs. 9 and 11 requires the knowledge of the ideal breakdown voltage and the depletion layer width at breakdown for the ideal one-dimensional plane geometry. Approximate formulas and graphs may be used for this purpose. Reference [11] contains tabulated values for the breakdowns and space-charge widths. The breakdown values quoted by [11] are somewhat lower than in other publications since the breakdown definition occurs when the multiplication factor $M = 2$ rather than infinity. In the work of [11], avalanche ionization coefficients of [1] were used, giving somewhat lower breakdown voltages than expected from the data of [4].

3.4.1 Expression for Planar Junction Breakdown

The breakdown voltage of the one-sided abrupt cylindrical or the spherical junction is given for silicon in volts as [4]

$$BV \simeq 6 \times 10^{13} N^{-3/4} \{[(n + 1 + \gamma)\gamma^n]^{1/(n+1)} - \gamma\} \tag{20}$$

with $\gamma = x_j/x_d$; $n = 1$ and 2 for the cylindrical and spherical junctions, respectively; and x_d equal to the depletion layer width at breakdown in microns for one-sided plane abrupt junctions.

3.5 Avalanche Breakdown Voltage Temperature Dependence

Figure 12 shows how the avalanche breakdown of silicon and germanium vary with temperature [13]. Breakdown increases with increasing temperature, particularly for higher resistivity silicon and germanium crystals. Figure 12 predicts BV(T°K)/BV(300°K) ratios for Ge and Si for abrupt and linearly graded junctions which have the same room temperature breakdown as one-sided abrupt junctions with 10^{15} cm^{-3} doping on the lightly doped side. According to [13], the predicted percentage change in BV is about the same for Si, Ge, GaAs, and GaP. For BV(400°K) the percentage change for a graded silicon junction is about 28%.

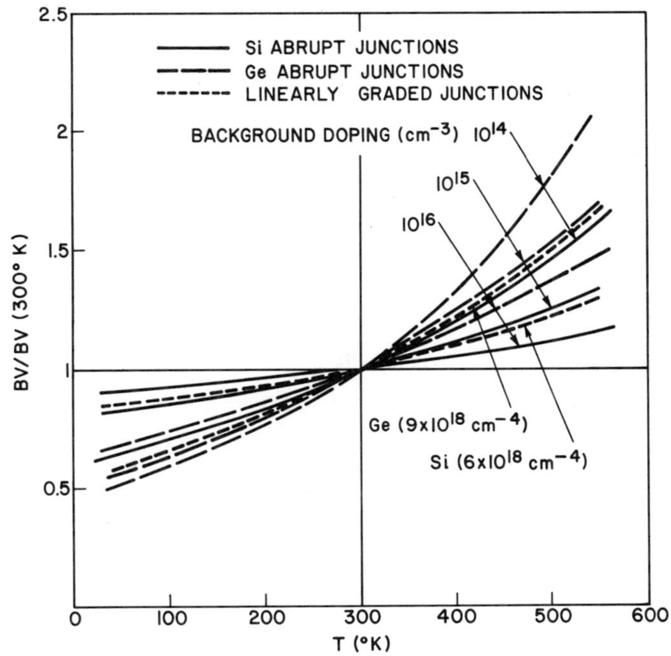

Fig. 12. Breakdown voltage normalized to 300°K breakdown voltage as a function of the absolute temperature. (From Crowell and Sze [13].)

REFERENCES

1. R. Van Overstraeten and H. DeMan, *Solid-State Electron.* **13,** 583 (1970).
2. C. A. Lee, R. A. Logan, R. L. Batdorf, J. J. Kleimack, and W. W. Wiegman, *Phys. Rev. A.* **134,** 761 (1964).
3. J. L. Moll and R. Van Overstraeten, *Solid-State Electron.* **6,** 147 (1963).
4. S. M. Sze and G. Gibbons, *Solid-State Electron.* **9,** 831 (1966).
5. W. Fulop, *Solid-State Electron.* **10,** 39 (1967).
6. R. A. Kokosa and R. L. Davies, *IEEE Trans. Electron Devices* **ED-13,** 874 (1966).
7. S. L. Miller, *Phys. Rev.* **99,** 1334 (1955).
8. P. Brook, *IEEE Trans. Electron Devices* **ED-21,** 730 (1974).
9. R. M. Warner, *Solid-State Electron.* **15,** 1303 (1972).
10. P. R. Wilson, *Solid-State Electron.* **16,** 99 (1973).
11. V. A. K. Temple and M. S. Adler, *IEEE Trans. Electron Devices* **ED-22,** 910 (1975).
12. M. S. Adler and V. A. K. Temple, *IEEE Trans. Electron Devices* **ED-25,** 1266 (1978).
13. C. R. Crowell and S. M. Sze, *Appl. Phys. Lett.* **9,** 242 (1966).

CHAPTER 4

AVALANCHE BREAKDOWN IMPROVEMENT METHODS

The avalanche breakdown voltage of p–n junctions can be substantially enhanced by such techniques as field plates, guard rings, and junction contouring. Their purpose is to shift the possibility of the electrical breakdown from the junction surface to the interior of the semiconductor and to decrease the electric field to a minimum in order to utilize fully the semiconductor capabilities.

4.1 Field Plate

The field plate technique remains one of the most useful approaches to the planar avalanche junction breakdown improvement. It consists of extending the junction contact metallization beyond the n^+–p junction as shown in Fig. 1. The reverse potential applied to the junction makes the field plate positive in respect to the p-type region. When a sufficiently thick SiO_2 layer is used under the field plate, the surface of the p-type silicon region becomes depleted of holes upon the application of the positive potential. A too small oxide thickness and/or high voltage may invert the

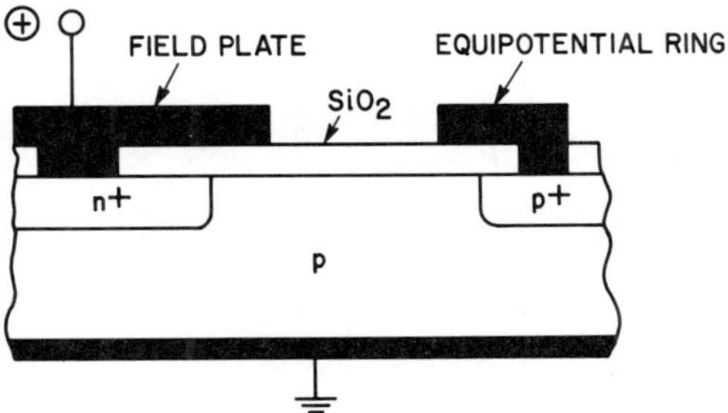

Fig. 1. The n^+–p junction with a field plate and an equipotential ring.

p-type surface causing the appearance of an n-type channel increasing appreciably the total n^+-region area and possibly causing low junction breakdown if extended to the p^+-region. The well-designed field plate with sufficiently thick oxide will not invert the surface, but will deeply deplete the p-region. The field plate is usually used together with another electrode, the equipotential ring, surrounding the field plate and electrically connected to the positively biased p-substrate (Fig. 1). The negatively biased ring attracts holes to the Si–SiO_2 interface counteracting the field plate inverting effect and also gives a path to ground the positive charges accumulated on the SiO_2 surface. The p^+-region improves the equipotential ring contact to the p-type region and, at the same time, serves as a channel stopper since the p^+-region is much more difficult to invert than the p-type region. The presence of the equipotential ring has some negative effect on the junction breakdown as will be shown in Section 4.3. Similar considerations are applicable to the p^+–n-type junctions (Fig. 2[1]).

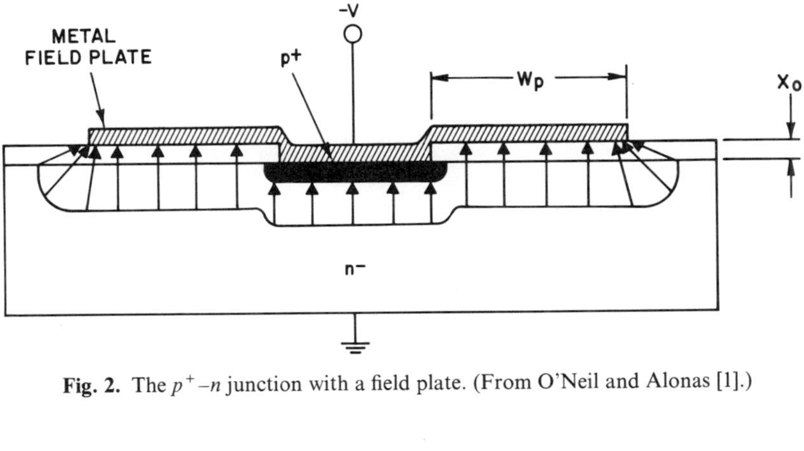

Fig. 2. The p^+–n junction with a field plate. (From O'Neil and Alonas [1].)

4.1.1 FIELD PLATE OPTIMIZATION

A sufficiently thick field plate oxide prevents inversion and permits depletion of the n-region under the field plate (Fig. 2). If the oxide is too thick, however, the n-region will not be depleted enough, however. At the point where the field plate terminates, i.e., at the plate's edge, the field is highly concentrated (Fig. 2).

In order to take into account the edge effect [2] in the breakdown computation, it is necessary to solve a set of two-dimensional Laplace and Poisson equations for the oxide–field plate system. The Laplace equation applies to the SiO_2 layer, assumed to be charge free, i.e.,

$$\frac{\partial^2 V}{\partial x^2} + \frac{\partial^2 V}{\partial y^2} = 0. \tag{1}$$

In silicon the Poisson equation for the fully depleted region is given by

$$\frac{\partial^2 V}{\partial x^2} + \frac{\partial^2 V}{\partial y^2} = -\frac{qC_B}{\epsilon_s}, \tag{2}$$

where C_B is the substrate concentration, and ϵ_s is the silicon permittivity. Along the field plate's AB line (Fig. 3a,b [2]), the potential $V = V_G' = V_G - V_{FB}$, where V_G is the field plate voltage and V_{FB} is the flat-band voltage. Along the BD line, $\partial V/\partial x = 0$, and along the $DFGH$ line, considered to be in the neutral region, $V = 0$. The boundary condition at the HI line in silicon is $V = \phi_s[1 - (x - x_o)/x_d]^2$, where x_o is the oxide thickness and x_d is the junction depletion region thickness; ϕ_s is the surface potential in the central planar portion of the device and is equal to $\phi_s = qC_B x_d^2/\epsilon_s$, where ϵ_s is the permittivity of silicon. At the IA line in the silicon dioxide, the boundary condition is

$$V = V_G' - (x/x_o)(V_G' - \phi_s). \tag{3}$$

The computer solution of this set of equations permitted the calculation of the electric field and potential distributions, which, in turn, allowed the calculation of the ionization integral and the avalanche breakdown. The MN field line of the maximum ionization integral was defined as starting at the maximum field point M, determined for each pair of C_B and x_o values, and following the direction of the electric field. The results of the numerical calculations, plotted in Fig. 4, show that at small oxide thicknesses below 0.1 μm and substrate concentrations of less than 10^{17} cm^{-3}, the avalanche breakdown takes place beneath the edge of the metal plate at B', (Fig. 3 [2]), and that the breakdown voltage increases as the substrate concentration decreases. The oxide voltage drop is small for this range, and the

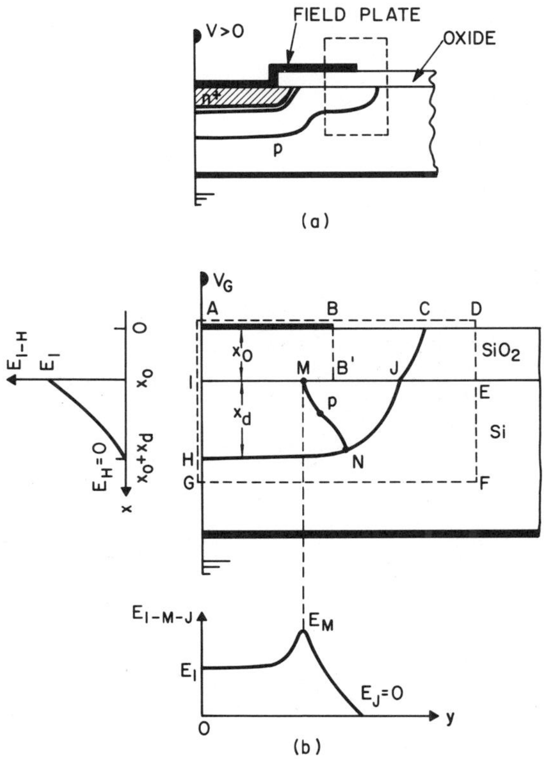

Fig. 3. Two-dimensional model for the field plate calculations. (a) Metal overlapped planar diode (b) Region of interest and space-charge field shown at left. Electric field at the interface shown at the bottom. (From Rusu and Bulucea [2].)

voltage breakdown takes place in the field-induced space-charge region in silicon and behaves like the conventional p–n junction breakdown. However, it is smaller than that of the plane junction because of the field concentration at the field plate edge.

At large oxide thicknesses, i.e., $x_o \geqslant 1\ \mu\text{m}$ and $C_B \geqslant 10^{16}\ \text{cm}^{-3}$, the breakdown voltage occurs at the central plane portion of the device, and it increases as the substrate concentration goes from about 10^{16} up to $10^{18}\ \text{cm}^{-3}$. This is attributed to the field concentrating in the oxide rather than in silicon. For this range of x_o and C_B the voltage drop in the semiconductor is negligibly small compared to the oxide voltage drop V_{ox}. If the critical field for the semiconductor breakdown is E_{crit}, then the field in the oxide E_{oxb} at silicon breakdown is given by

$$E_{oxb} = (\epsilon_s/\epsilon_i)E_{crit}. \tag{4}$$

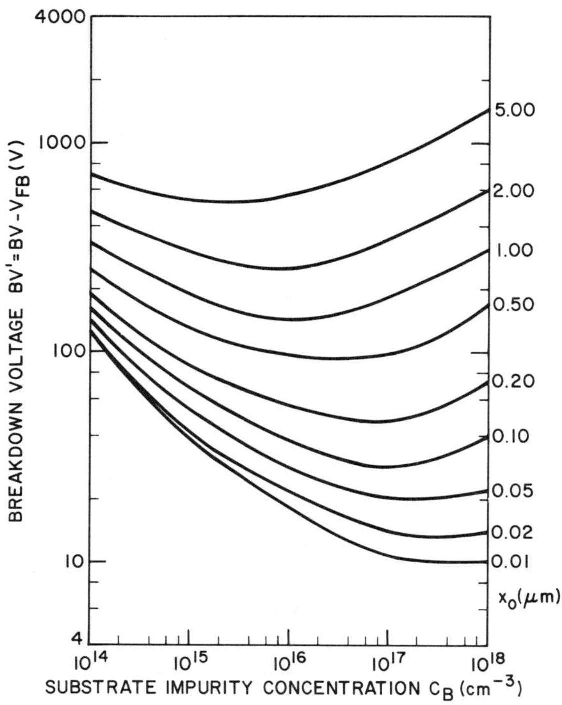

Fig. 4. Breakdown voltage versus substrate impurity concentration with SiO_2 thickness as a parameter. (From Rusu and Bulucea [2].)

The oxide voltage drop at silicon breakdown is therefore

$$V_{oxb} \simeq BV' = (\epsilon_s/\epsilon_i)x_o E_{crit}, \tag{5}$$

where $BV' = BV - V_{FB}$. Since the voltage drop in silicon is negligibly small, $BV' \simeq V_{oxb}E_{crit}$ increases with the background doping, for example, $E_{crit} \simeq 1.5 \times 10^5$ V cm^{-1} at $C_B = 10^{14}$ cm^{-3}, but equals 6×10^5 V cm^{-1} at $C_B = 10^{17}$ cm^{-3}. As a result, the breakdown voltage BV' increases rather unexpectedly with the background concentration. Equation (5) also shows that BV' varies linearly with the oxide thickness for the x_o and C_B range where the approximation of the negligibly small depletion region drop is valid [2], i.e., for $C_B \gg 10^{16}$.

The curves of Fig. 4 show that for a given oxide thickness there is a substrate concentration that gives a minimum breakdown voltage. To the left of the minima the edge breakdown dominates; to the right the breakdown due to the plane (central) portion of the depletion region prevails. The electric field stays uniform at any substrate doping if the condition $x_o/x_{dmaxp} \geqslant 0.3$ is satisfied; x_{dmaxp} is the maximum depletion layer width of the plane junction. The computer-generated results of Rusu and Bulucea [2] agree reasonably well with the results of O'Neil and Alonas [1], who came up with a closed form analytical expression for the breakdown voltage of a planar junction provided with a field plate. The work of [1] also showed that making the oxide thicker reduced the field crowding effect at the field plate edge and raised the breakdown voltage similarily to the increase in junction depth. The increase of the oxide thickness has a greater effect on the field reduction in the silicon substrate than an equal increment in the junction depth x_j. The improvement can be expressed as

$$\Delta x_{j(\text{equiv})} = (\epsilon_s/\epsilon_i)x_o. \tag{6}$$

From Eqs. (3.12) and (3.20) we obtain for the cylindrical planar junction breakdown

$$BV_{\text{planar}} = BV_{\text{plane}}\{[(2 + \gamma)\gamma]^{1/2} - \gamma\}, \tag{7}$$

where $\gamma = x_j/x_d$, x_d being the depletion width for a one-sided abrup junction with the breakdown voltage BV_{plane}. In view of (6), the breakdown voltage is given by the same expression as (7), but with γ replaced by

$$\gamma' = (\epsilon_s/\epsilon_i)(x_o/x_d) \simeq 3(x_o/x_d). \tag{8}$$

The analytical expression for the planar junction breakdown voltage with a field plate is therefore given by

$$BV_{\text{FP}} = 6 \times 10^{13} N_D^{-3/4}\{[(2 + \gamma')\gamma']^{1/2} - \gamma'\}. \tag{9}$$

This relationship is valid for Q_{ss}/q values much below 10^{12} cm^{-2} (for good oxides Q_{ss}/q is about 10^{10} cm^{-2}). An excessively high positive surface charge increases the breakdown voltage of the n^+–p junction, but degrades the breakdown of the p^+–n junction. Figure 5 represents a plot of this expression for the breakdown of a planar junction provided with a field plate versus oxide thickness with $C_B = N_D = 1.1 \times 10^{14}$ cm^{-3}. Also plotted are the experimental data obtained for planar junctions with radii $x_j = 4$ μm and $x_j = 9$ μm. The agreement is very good. The formula is applicable to the n^+–p junctions by just changing the concentration N_D to N_A for the p-type substrate. An example, breakdown voltage obtained by [1] for $N_D = 1.1 \times 10^{14}$ cm^{-3} with $x_o = 1.0$ μm is about 370 V. About the same breakdown can be obtained from the plot of Fig. 5 [2] for the same parameters and with $V_{FB} \simeq -5.0$ V.

According to Grove *et al.* [2a], the breakdown voltage of a p–n junction depends linearly on the field-plate voltage V_G *as*

$$BV = mV_G + \text{const},$$

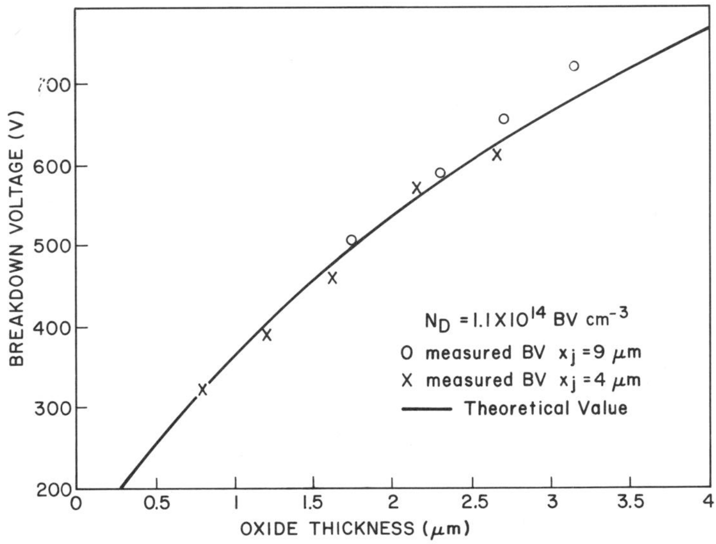

Fig. 5. Planar junction avalanche breakdown voltage as a function of SiO_2 oxide thickness x_o. (From O'Neil and Alonas [1].)

with $m \simeq 1$. However, as V_G is increased (more negative for p^+–n junctions), a point is reached when the breakdown voltage suddenly collapses. A small change in V_G results in a drastic change of the breakdown voltage from a few hundred volts, for example, to a few tens of volts. The origin of this effect was attributed by Grove *et al.* [2a] to the breakdown of the field-plate-induced junction (inversion). The recent study of Rusu *et al.* [2b] explains the collapse as a result of a spatial switching of the avalanche breakdown within the device structure. It may take place even in the defect-free crystal. The understanding of the breakdown collapse mechanism makes Rusu *et al.* believe that it is possible to extend the diode breakdown up to the plane junction breakdown without incurring the breakdown fall-down. To achieve this, the oxide under the field plate should be made thicker than some minimum value x_{ob}, determined by the substrate concentration C_B. For example, x_{ob} at $C_B = 2 \times 10^{14}$ cm^{-3} equals ~ 10 μm; at $C_B = 10^{15}$ cm^{-3}, it equals 2.2 μm. At $C_B = 10^{16}$ cm^{-6}, the critical thickness drops down to only 0.2 μm.

It should be pointed out that the field-plate-controlled p^+–n diodes, which had been previously avalanched at high reverse voltages, show higher breakdowns due to the injection of positive charges (holes) into the oxide. Positive oxide charging appears to be equivalent in silicon to an increase in the oxide thickness.

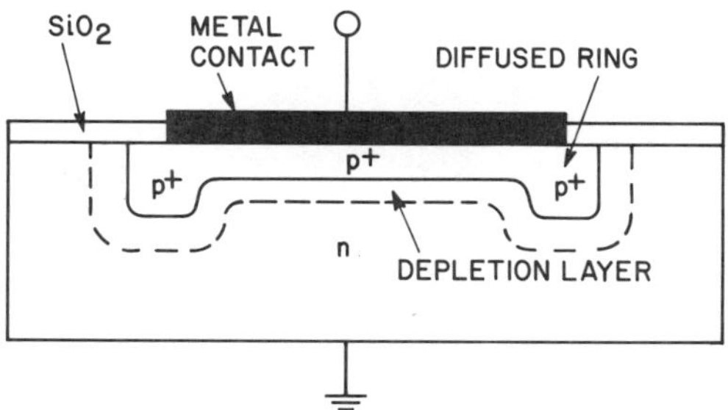

Fig. 6. Diffused guard ring.

4.2 Diffused Guard Ring

A shallow diffused planar junction has a low breakdown voltage due to its small radius of curvature x_j. In many instances it is possible to surround the planar junction at its perimeter by diffusing (Fig. 6) a guard ring of the same polarity but with considerably higher penetration and larger x_j. The deep diffusion also results in a shallower grade constant a and, consequently, improves the device breakdown.

4.3 Equipotential Ring and Channel Stopper

The purpose of the equipotential metallic ring is to prevent the silicon surface inversion in its vicinity. It is normally deposited on the top of a highly doped region (p^+ in the case of Fig. 1), and it remains at the substrate potential. The presence of the equipotential ring also contributes to the silicon surface stabilization by collecting the surface contaminating ions. The purpose of the p^+-diffused region is to prevent the surface from inverting and to assure good ring contact to silicon. The presence of the equipotential ring has a somewhat harmful effect on the junction breakdown voltage due to the increased field intensity near the edge of the ring [3]. Junction field plate combined with the equipotential ring and channel stopper was experimentally studied by Clark and Zoroglu ([4, 5]). It was found that the breakdown voltage increases with the increasing field plate and equipotential ring spacing W_i. The breakdown voltage of the experimental device with 200 μm spacing was 1575 V. An increase of this spacing by 1 μm resulted in an increase of the breakdown voltage by about 2.5 V. The breakdown voltage increases more or less linearly with the increased spacing, but with a rate specific to the particular geometry of the field-plate–equipotential-ring system and the passivant between them.

4.4 Resistive Field Plate

The nonuniform fields appearing at the metal field plate edge and the equipotential ring can be minimized by using a resistive polycrystalline film [5] bridging the space between the metal field plate and the annular ring. The presence of the film makes the electric field more uniform, improves the breakdown, and also suppresses surface arcing. The resistive film used by [5] consisted of undoped polycrystalline silicon 200 nm thick with a sheet resistance of $5 \times 10^{10}\ \Omega$ per square. The metal field plate and the metal annular ring had a square geometry as shown in Fig. 7. The figure demonstrates the effect of the electrode spacing W_i on the breakdown voltage of the polysilicon passivated diode with junction depth $x_j = 20\ \mu$m. The dimensions of the diode and the equipotential ring remained unchanged so as the field plate size was increased, the spacing W_i was decreased and the lateral radius r_0 of the field plate was increased. For the small W_i spacings the slope of the breakdown voltage versus the W_i curve is about 17 V μm^{-1}. When W_i is

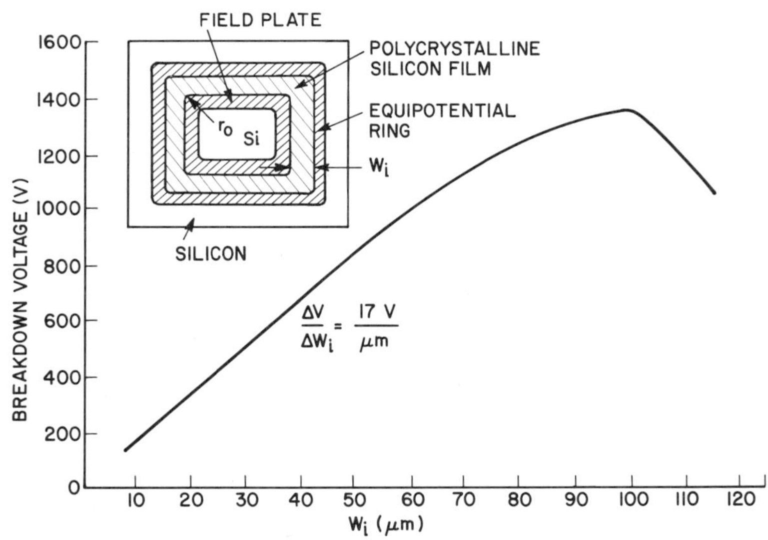

Fig. 7. Planar junction breakdown voltage versus field plate and equipotential ring spacing W_i. (From Clark and Zoroglu [5].)

about 100 μm, the breakdown voltage starts falling because of the decrease in the plate corner radius and subsequent increase in the electric field in this region. A considerably improved version of the simple polycrystalline film is the Semi-insulating polycrystalline silicon (Sipos) described in the next section.

4.4.1 Sipos Passivating Resistive Field Plate

Semi-insulating polycrystalline silicon (Sipos) has been used extensively as a resistive field plate and a surface passivant since first introduced by Aoki *et al.* [6] and Matsushita *et al.* [7] to passivate planar devices. In the planar devices the fixed positive charges in the SiO_2 layer cause silicon surface inversion that results in low breakdown voltages. The positive mobile charges in SiO_2, due to sodium and other contaminants, lead to the device instability. Another problem arises when an SiO_2 passivated planar transistor is subjected to an avalanche breakdown after which carrier recombination increases at the silicon surface in the vicinity of the emitter–base metallurgical junction [8] permanently lowering the device gain. During the avalanche breakdown, hot carriers injected into SiO_2 modulate the surface conductivity and cause the so-called breakdown walk-out phenomena. These problems are largely remedied by the use of Sipos thanks to several factors. The very slightly conducting oxygen-doped Sipos layer allows the flow of the very small current which makes the silicon surface field more uniform. The conductivity of Sipos is so low, however, that the leakage current through the film is either comparable or smaller than the leakage in the silicon bulk or at the silicon surface. The reverse potential applied to the transistor collector is also applied to the deposited Sipos film, which, due to its conductivity, transfers it like a field shield to the surface of the base region (Fig. 8 [7]). In the case of *p–n–p* transistors, electrons near the surface of the *n*-base will be depleted since the collector of the *p–n–p* transistor is negative. The surface will therefore become more resistive, the surface field will be lowered, and the breakdown voltage will increase. If the Sipos film is made too resistive by excessive oxygen doping, then the breakdown approaches that of the thermal oxide. Figure 9 [7] shows the breakdown voltage behavior as a function of the Sipos oxygen concentration for a planar p^+–n junction 25 μm deep with n^--region doped at 10^{14} cm^{-3}. When the oxygen concentration reaches 66.7 at. %, the junction breakdown is the same as that of a planar junction passivated by thermal oxide. The Sipos layer also shields the device surface from the external charges and prevents the accumulation of ions at the silicon–Sipos interface. Charges induced at the silicon surface by the contaminating mobile ions, such as sodium, present on the top of the

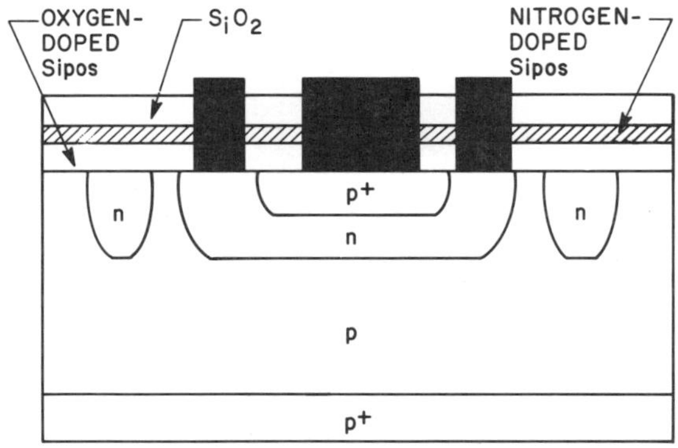

Fig. 8. *p–n–p* transistor with oxygen- and nitrogen-doped Sipos. (From Matsushita *et al.* [7].)

passivating film can drift into the film. As a result, either charge neutralization occurs, or a space-charge region is formed inside the Sipos film, resulting in an electrostatic shielding effect. Hot carriers, due to the film conductivity, cannot be stored for an appreciable time so that the walk-out effect is remedied. The Sipos passivation technique can be applied both to the *n–p–n* and to *p–n–p* transistors. Figure 8 [7] shows a cross section of a *p–n–p* transistor with field-limiting rings and with a triple layer passivation system consisting of a 0.5 μm thick oxygen-doped Sipos film, a 0.15 μm thick nitrogen-doped Sipos film, and a 1 μm thick SiO_2 film, all chemical-vapor deposited on an oxide-free silicon surface. The oxygen-doped silicon polycrystalline film containing about 20 at. % oxygen was obtained at 650°C from the SiH_4–N_2O–N_2 mixture; the nitrogen-doped film with 52 at. % nitrogen was derived at 650°C from SiH_4–NH_3–N_2 system. The silicon

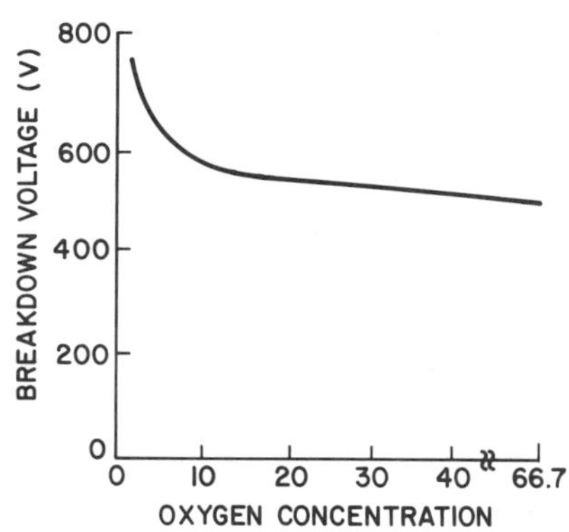

Fig. 9. p^+–n planar junction breakdown as a function of Sipos oxygen concentration. (From Matsushita *et al.* [7].)

nitride film acts as a sealing agent against the penetration of moisture or highly mobile ions such as sodium. The oxide film prevents the surface breakdown of the nitride. Good stability and breakdown voltages of several thousand volts are achievable with Sipos.

4.4.1.1 Sipos Conduction Mechanism and Resistivity

Sipos films consist of crystalline Si grains separated and surrounded by thin sheets of amorphous grain boundaries. Shallow donor and/or acceptor states can exist in the silicon grains due to such impurities as oxygen which contributes donor and acceptor states at 0.16 and 0.39 eV, below the conduction band edge. At the grain boundaries the surface concentration is $N_S = 10^{13}$–10^{15} cm^{-2}. At equilibrium the Fermi levels in the grain and grain boundaries are aligned, causing charge redistribution and band bending. The band structure of Sipos can consequently be considered as a succession of pairs of symmetrical Schottky barriers connected back to back [9] and always forward biased when either an external positive or negative potential is applied. The band gap of the oxygen-rich polycrystalline silicon was found to be similar to that of single crystalline silicon with electrons as majority carriers. A net donor state density of 2.3×10^{19} cm^{-3} existed at 0.12 eV below the conduction edge band in the samples tested by [9]. These donors, located in the Si grains, were completely ionized in the presence of large grain boundary surface states. The surface charge density at the grain boundaries was about -2.3×10^{13} cm^{-2}, and the band bending was 0.44 eV. Tang [9] showed that below 373°K current carriers move by tunneling and that above this temperature they move by thermionic emission. The resistivity of Sipos [10] may be varied by changing the N_2O/SiH_4 ratio γ_N (Fig. 10). For the devices with 1500 V breakdown, the optimum resistivity is about 10^8 Ω cm, obtained with $\gamma_N = 0.23$. The leakage current of the silicon nitride at this voltage is about 15 μA, and the room temperature junction leakage is less than 1 μA at 1500 V and 1 mA at 150°C.

Using Maxwell's equations and appropriate boundary conditions for a Sipos and Si_3N_4 passivated high voltage silicon *p–n* junction, it is possible to obtain electric field distributions [11] for this passivation method. It was found that there is a high-field region approaching the dielectric breakdown of Si_3N_4 in the case of thin nitride layers. As a consequence of the conductivity and dielectric differences between the Sipos and the nitride, an interface charge density on the order of $5 \times 10^{12} q$ cm^{-2} is induced at the voltages near the nitride breakdown. Since the relaxation time of these charges is comparatively long, on the order of milliseconds, the effect of this charge on the transient junction behavior may be substantial. Moreover, the field ceases to be uniformly distributed as expected, but reaches the maximum

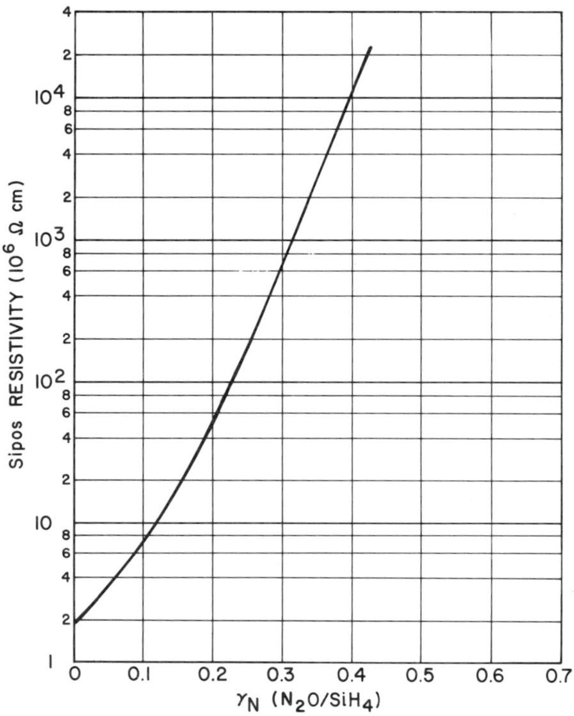

Fig. 10. Sipos resistivity as a function of N_2O/SiH_4 ratio (γ_N). (From Goodman [10], as quoted in reference [24] of Chapter 5.)

value near the surface depletion region. A thin nitride film on the Sipos surface prohibits the surface field relaxation. Much more uniform surface fields are achievable by using thick dielectrics, such as glass, in place of the nitride film. It is advantageous to use a silicon dioxide layer between the Sipos and the glass. The glass composition should be chosen to provide a good thermal expansion match to the silicon. A thin SiO_2 layer is usually applied over the glass to protect the glass during metal etching.

4.5 Field Limiting Ring

The field limiting ring consists of a diffused region (Fig. 11 [12]) of the same polarity as the highly doped region of the reverse biased junction but electrically floating. The distance between the ring and the main junction is small enough to permit the depletion regions of the two junctions to merge prior to avalanche breakdown. This limits the electric field to a value lower

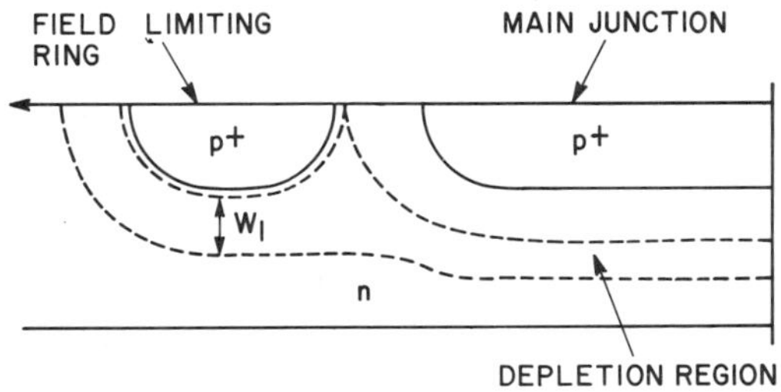

Fig. 11. Field limiting ring. (From Kao and Wolley [12].)

than that of the critical breakdown field. As the junction potential is further increased, there is a complete merger of the two space-charge regions (Fig. 11). The field limiting ring acts, to some extent, as a voltage divider and helps in an appreciable reduction of the curvature of the main junction depletion region.

Field intensity calculations with floating field ring were performed by Adler *et* al. [13], who solved numerically the two- and three-dimensional Poisson equations for a reverse-biased junction. Figure 12 shows the depletion layer contours for the usual case in which the positive SiO_2 charge Q_{ss} causes surface inversion and leads to the presence of an undepleted triangular region. The increase of the junction potential beyond the punch-through voltage results in a depletion region with a much larger radius of curvature than that of the main planar junction already noted in connection with Fig. 11.

The two- and three-dimensional analysis established a lack of validity

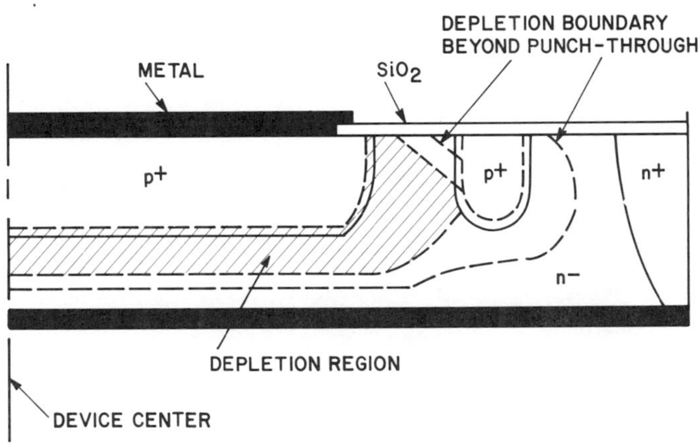

Fig. 12. Depletion layer contours with field limiting ring for positive oxide charge. (From Adler *et al.* [13].)

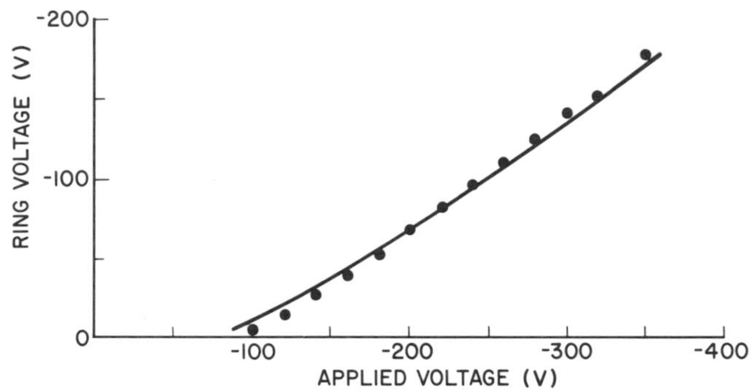

Fig. 13. Field limiting ring voltage as a function of junction voltage. Dots denote experimental values; the line denotes theoretical values. (From Adler *et al.* [13].)

for the original one-dimensional concept [12] that the voltage drop between the field ring and the main junction remains unaltered for voltages higher than those needed for the merger of the two depletion regions. The one-dimensional analysis predicts a slope of unity for the plot of the ring voltage as a function of the main junction voltage. The multidimensional analysis and experimentation of Adler *et al.* [13] show that the slope is about 0.65 (Fig. 13) and that the potential drop is increasing with the rise of the junction potential. This means that, due to a larger voltage drop, a smaller potential appears at the ring and that the ring divides the voltage less efficiently than expected from the one-dimensional analysis. The breakdown usually occurs, not at the surface, but at the point of maximum curvature. At that point the field ring action is inferior to the expected perfect voltage divider. The potential difference between the main p^+-region and the ring may vary from 0 to 45% of the applied voltage. Figure 14 shows how much an optimally designed field limiting ring improves the main junction breakdown voltage. This voltage is normalized with respect to the ideal *plane* (one-dimensional)

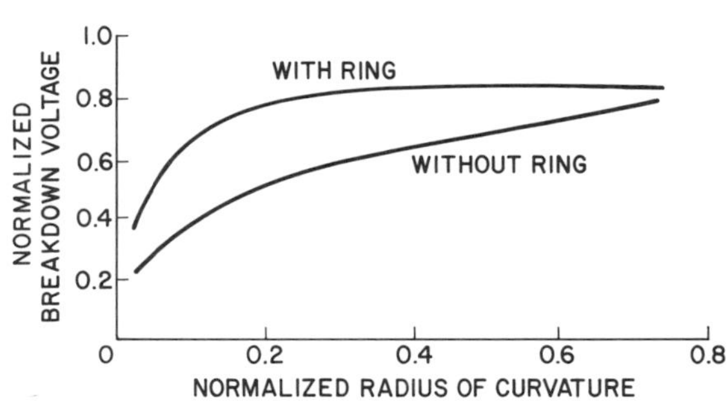

Fig. 14. Main junction normalized breakdown as a function normalized junction radius of curvature. (From Adler *et al.* [13].)

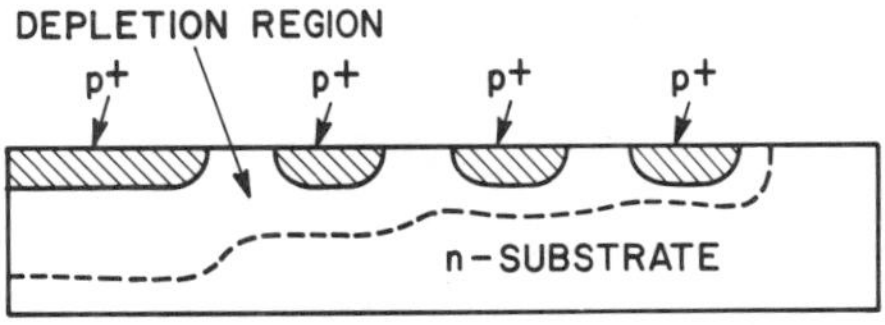

Fig. 15. Planar junction with multiple field limiting ring. (From Kao and Wolley [12] reprinted by permission of the publisher.)

junction breakdown voltage, and the radius of curvature is normalized to the *plane* junction depletion layer width at breakdown. This way the curves are valid for all substrate dopings and junction depths.

The spacing between the main junction and the ring has a pronounced effect on the ring's voltage dividing capability, and there is always an optimum distance at which the effect is the greatest. Q_{ss} and other positive charges induce channels in the *p*-type silicon surface and make the field limiting rings less effective.

Multiple field limiting rings may be used to further improve the breakdown. Figure 15 [12] schematically shows a device with such multiple concentric rings. The results of the Kao and Wolley [12] study of the multiple ring structure show that the largest part of the applied potential is supported by the main junction. Each succeeding ring is supporting a smaller part of the potential applied to the main junction. Breakdown of 2000 V was achieved by [12] for a p^+–n junction with 80-Ω-cm substrate. The results with the n^+–p diodes generally were less satisfactory because of the *n*-channel appearance at the *p*-type surface that resulted in excessive leakage currents. The multiple ring structure, although electrically very satisfactory, requires the availability of a large surface area for high voltages, thus making the device expensive.

4.6 Junction Beveling

For devices with a few hundred volts operating capability, deep moat etching usually achieves satisfactory breakdowns (Fig. 16). High voltages, those above 400–500 V, however, require a less haphazard approach by using techniques that can consistently reproduce the same device contour on a production scale.

A junction shaping method which combines an excellent reproducibility with great breakdown improvement consists of mechanical beveling of the device as described by [14]. A device can be beveled with a positive or a

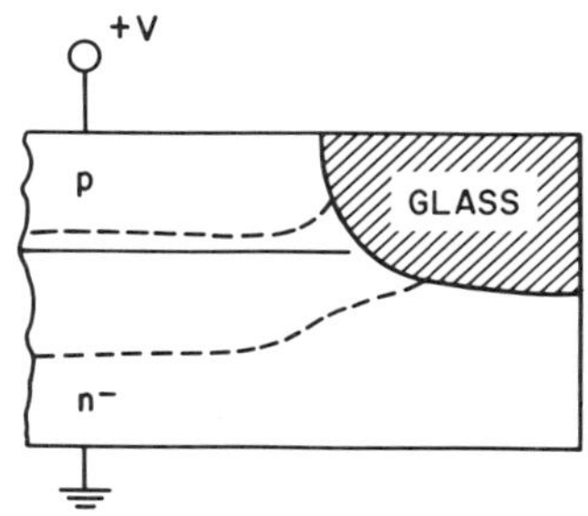

Fig. 16. Deep moat etched junction with glass passivation.

negative angle according to the convention of Fig. 17. If the p–n junction has a decreasing area when going from the more to the less doped side, then the bevel angle is considered to be positive (Fig. 17b). If the area is increasing, the bevel is considered to be negative (Fig. 17a). There are advantages and disadvantages of either polarity. The analysis of [14–18] provide an excellent insight into the behavior of the surface and bulk electric fields for both beveling polarities. The theoretical approach to the field computation for devices with either the Gaussian or complementary error function impurity distribution consists [16] of approximating the diffusion profile by an exponential function, warranted for deep, high voltage junctions. Hence, for a diffusion taking place into the z direction (Fig. 18), the net impurity concentration as a function of the distance z from the metallurgical junction is given by

$$N(z) = N_D[\exp(-z/\lambda) - 1], \tag{10}$$

where N_D is the uniform substrate doping and λ is a space constant. Since all practical devices have some degree of symmetry, it is usually enough to consider just two two-dimensional Poisson equations for the field determination

$$\nabla E = \frac{q}{K\epsilon_0}\rho(x, y), \tag{11}$$

where E is the electric field, $\rho(x, y)$ is the net electric charge density, K is the

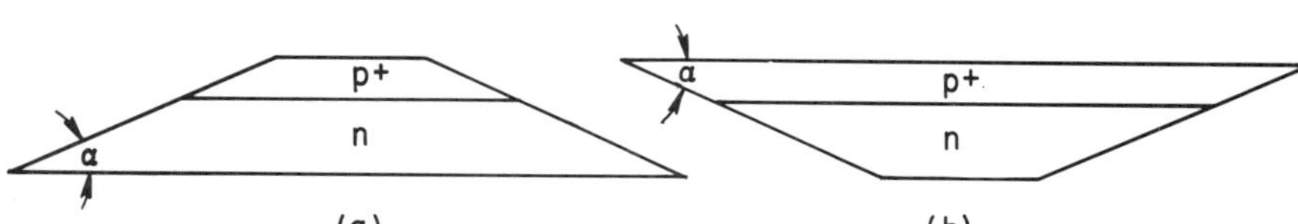

Fig. 17. (a) Negatively and (b) positively beveled junctions.

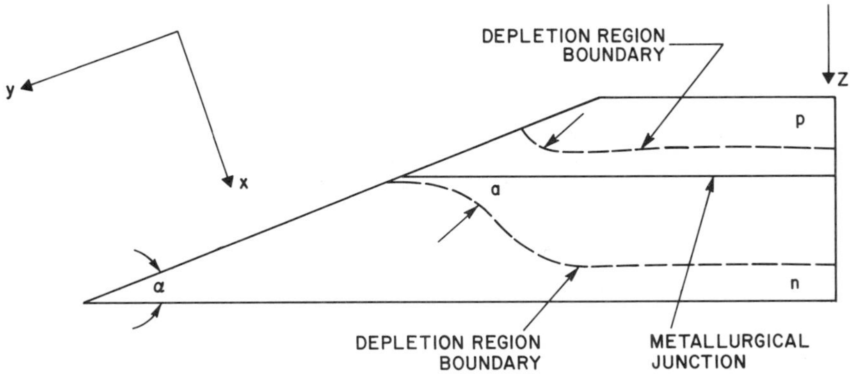

Fig. 18. Depletion region boundaries of a negatively beveled junction. (From Cornu [16].)

dielectric constant, and ϵ_0 is the free space permittivity. In the space-charge region, ρ is equal to the net impurity charge density at each point of coordinates x and y. In the regions such as dielectrics, the charge may be equal to zero and Eq. (11) becomes Laplace's equation

$$\nabla E = 0. \tag{12}$$

For the reverse-biased junction, it is often necessary for simplicity to assume that the reverse current is negligibly small so that the magnitude of the electric field at the depletion region boundaries and beyond is also zero. Inside the space-charge region the charge distribution can be determined by knowing the impurity distribution [Eq. (10)] and by first determining by consecutive approximations the depletion region boundaries. The results of the analytical and experimental studies for both bevel polarities show that the positive angle bevel is much more effective in the breakdown improvement than the negative angle.

4.6.1 Junction with Positive Bevel Angle

Figure 19 [16] shows an abrupt p^+–n junction with a positive bevel angle α. The experimental investigation of [14] and [17] agrees with theoretical predictions that the positively beveled junctions are capable of achieving maximum obtainable breakdown. However, positive angle beveling requires mechanical device shaping and is therefore expensive. Due to the positive bevel, the depletion layer boundary in the highly doped p^+-region comes closer to the junction so that the space-charge region narrows. On the high resistivity n^--side the depletion layer thickness increases, and the space-charge region expands. Consequently, the surface field is reduced much

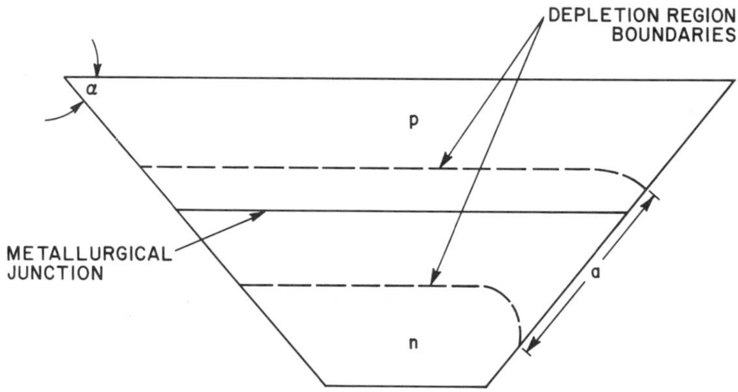

Fig. 19. Depletion region boundaries of a positively beveled junction. (From Cornu [16].)

below the bulk value since the applied voltage appears across a longer distance a. The peak surface field does not occur where the junction meets the surface, but is shifted at the surface toward the center of the lightly doped side (Fig. 20 [19]). The smaller the bevel angle, the larger the shift is and the lower the maximum surface field is. The bulk field is always higher than the surface field so that the breakdown always happens in the bulk [19]. For instance, a device with a 45° positive bevel angle will have a peak surface field equal to ~45% of the maximum bulk field at avalanche breakdown, and it will be shifted from the place where the junction meets the surface by about $0.7x_d$. (x_d is the depletion layer width at breakdown.) As can be seen

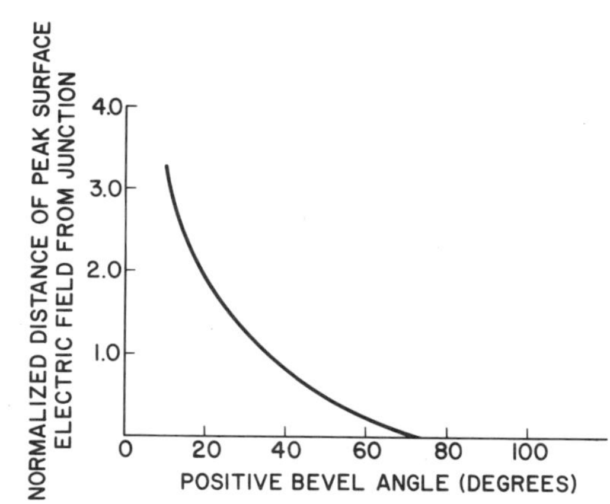

Fig. 20. Normalized distance of the peak surface electric field from the junction as a function of positive bevel angle. (From Adler and Temple [19].)

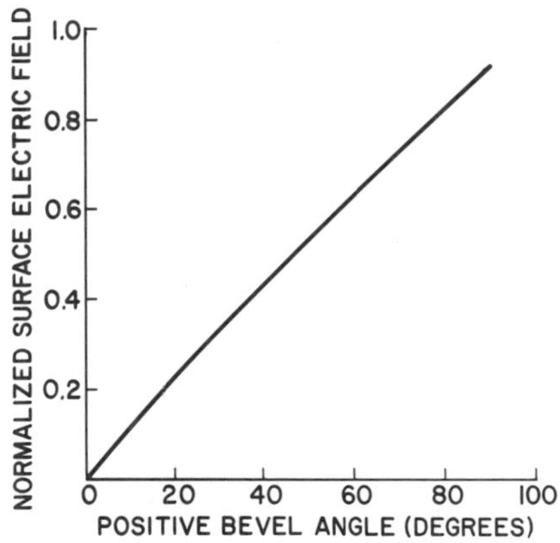

Fig. 21. Normalized surface electric field as a function of positive bevel angle. (From Adler and Temple [19].)

from Fig. 21 [19], the surface field, as normalized to the ideal one-dimensional value, is linearly reduced with the decreasing bevel angle. Thanks to the normalization method, the curves apply to substrates with all doping levels.

4.6.2 Junction with a Negative Bevel Angle

The computerized solution of the two-dimensional Poisson equation for the diffusion profile approximated by the exponential expression (10) yields the space-charge boundaries shaped as shown in Fig. 18 [16]. It can be seen that the minimum distance between the two depletion region boundaries of the negatively beveled junction occurs in the silicon bulk below the device surface. The maximum field is thus shifted from the surface to the device interior. The point of maximum field is always located in the more heavily doped side of the junction and away from the surface as illustrated by Fig. 22 for three negative bevel angles [16]. The investigated device diffusion profile was approximated by an exponential impurity distribution with a space constant $\lambda = 10\ \mu m$. The substrate was uniformly doped with donor concentration $N_D = 10^{14}\ cm^{-3}$, and the achieved breakdown was about 2000 V. As the figure shows, the bulk electric field was higher than the surface field. Deep in the bulk the field value becomes constant and equal to $2.08 \times 10^5\ V\ cm^{-1}$. The surface and bulk electric fields decrease as the absolute value of the beveling angle decreases. The absolute maximum field value was located for the investigated device at about 25 μm away from the device side surface,

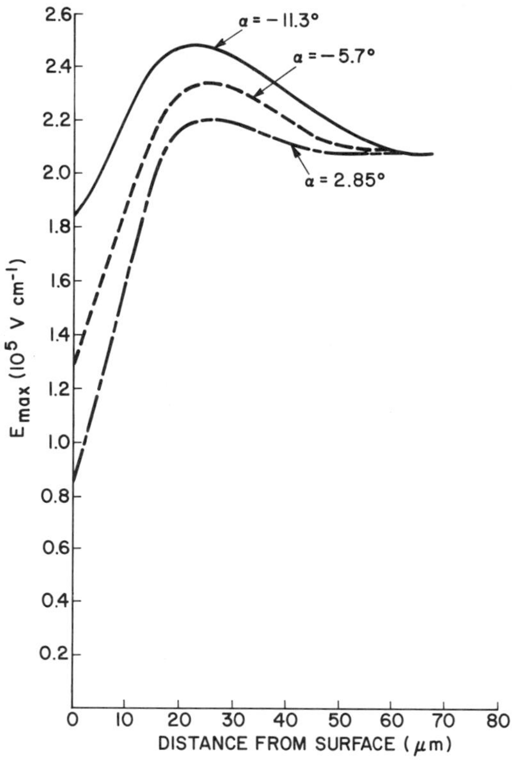

Fig. 22. Maximum electric field as a function of the distance from the surface of negatively beveled junction. (From Cornu [16].)

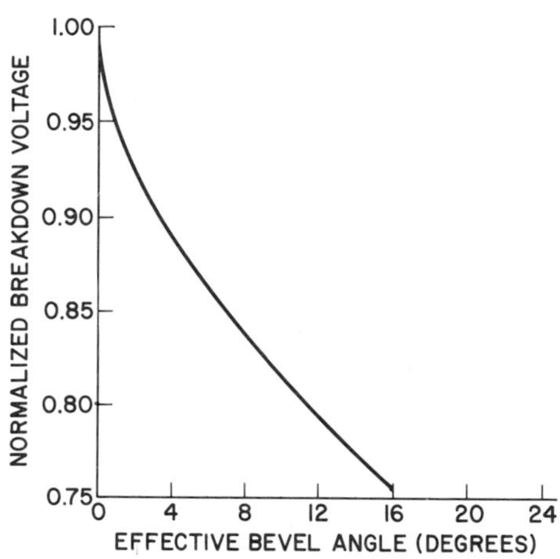

Fig. 23. Normalized breakdown voltage as a function of effective negative bevel angle. (From Adler and Temple [19].)

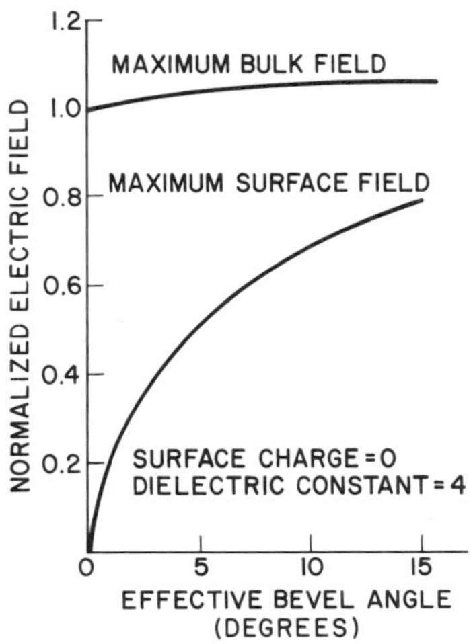

Fig. 24. Maximum normalized bulk and surface fields as a function of effective negative bevel angle. (From Adler and Temple [19].)

and avalanche breakdown took place at this location. Figure 23 shows the fraction of the ideal breakdown voltage for a negatively beveled device as a function of the effective bevel angle [19] defined as

$$\theta = 4 \times 10^{-2}(W_L/W_H)^2\alpha, \tag{13}$$

where α is the actual bevel angle and W_H and W_L are the depletion region widths of the one-dimensional device on the heavily doped and lightly doped sides, respectively. The advantage of using the effective angle concept is that the data for many devices may be reduced to just one curve. Figure 24 [19] illustrates the behavior of the normalized surface and bulk fields as a function of the negative bevel angle. The bulk field increases rather slowly with the increasing bevel angle in contrast to the surface field, which varies rapidly with θ_{eff} and reaches 80% of the bulk value at $\theta_{eff} = 15°$. Figure 25 [19] shows the position of the maximum field as the normalized distance from the junction as a function of the effective angle. The point of the maximum field is displaced very rapidly as the effective angle becomes smaller than 2°.

4.6.3 Dielectric Constant and Surface Charge Effect (Negative Bevel)

The theoretical and experimental studies of [20] and [17] show that it is impossible to use a single correction factor to account for the effects of either

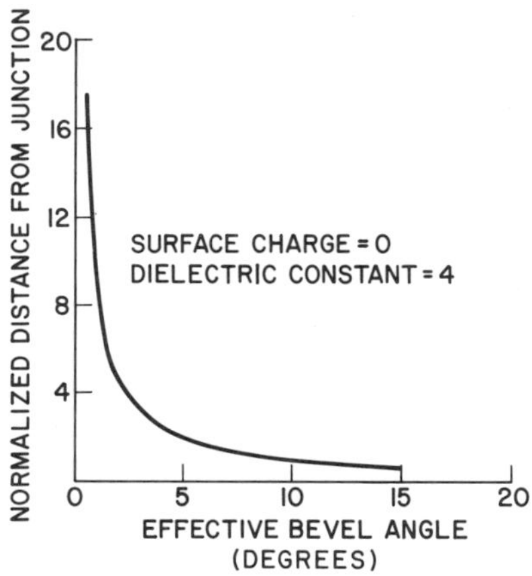

Fig. 25. Normalized distance from the junction of the maximum field as a function of effective negative bevel angle. (From Adler and Temple [19].)

the passivant's dielectric constant or of the existent surface charges on the device's electric fields and breakdown.

Figure 26 [19] represents the fraction of the ideal breakdown voltage of a p^+-n diffused junction as a function of a few variables. The bevel angle dependence is shown just for comparison to the dielectric constant and surface charge effects. The plots were obtained for a device whose lightly

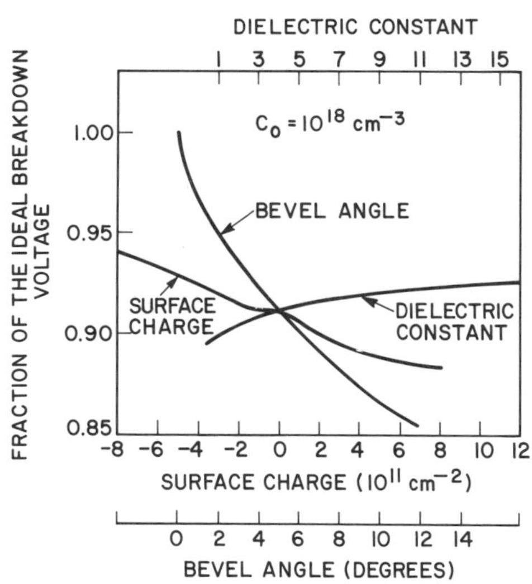

Fig. 26. Normalized ideal breakdown voltage as a function of negative bevel angle of surface charge, and of dielectric constant with surface concentration $C_0 = 10^{18}$ cm^{-3}. (From Adler and Temple [20].)

doped n-region had a donor concentration $N_D = 2 \times 10^{14}$ cm^{-3}. The surface charge and the dielectric constant curves were plotted for an effective bevel angle of about 3° and an actual angle of 5°. Similar plots are shown in Fig. 27 [20] for N_D also equal to 3×10^{14} cm^{-3}, but with the p-region surface concentration of 10^{19} cm^{-3} and about 50 μm junction depth. The dielectric constant and surface charge curves are plotted here for an effective negative angle of about 13° corresponding to the actual bevel angle of 6°.

As another illustration of the surface charge effects, Fig. 28 [17] shows the maximum bulk field as a function of the negative bevel angle α with Q_{SS}/q (the number of the fixed surface states) as a parameter. The typical range for the fixed surface charges lies between 10^{10} and 10^{12} cm^{-2}. The plot of Fig. 28 shows the field for $Q_{ss}/q = 0$ and $Q_{ss}/q = 10^{10}$ cm^{-2}. The investigated diffused p^+–n diode had a substrate concentration $N_D = 6 \times 10^{13}$ cm^{-3}, and the p^+-region surface concentration was $C_0 = 3 \times 10^{19}$ cm^{-3}. The impurity profile was described by the complementary error function.

The presence of the positive charge on the silicon surface increases the electron concentration in the n-type region. In the p-type region Q_{ss} decreases the hole concentration. As a result, the depletion region shrinks at the

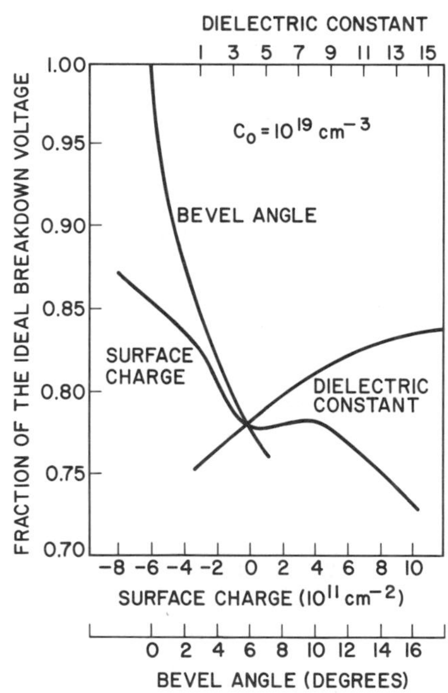

Fig. 27. Normalized ideal breakdown voltage as a function of negative bevel angle, of surface charge, and of dielectric constant with surface concentration $C_0 = 10^{19}$ cm^{-3}. (From Adler and Temple [20].)

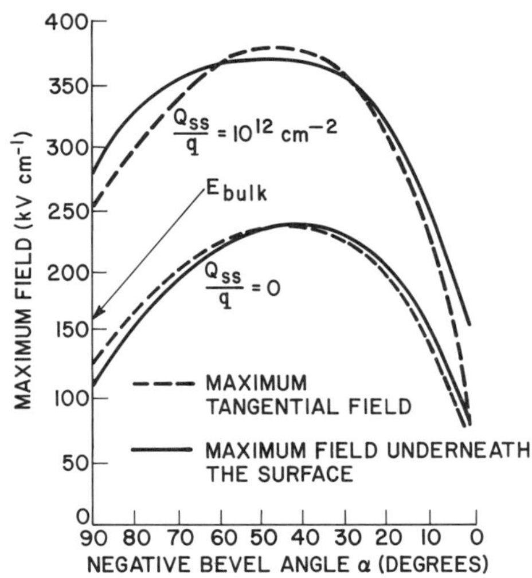

Fig. 28. Maximum electric field as a function of negative bevel angle with Q_{ss}/q as a parameter. (From Bakowski and Lundström [17].)

surface on the n-type side and expands on the p-type side. With $Q_{ss}/q = 10^{12}$ cm^{-2}, the tangential surface field almost doubles for the angles between 30 and 60° because of the presence of this positive surface charge as compared to the ideal case of $Q_{ss}/q = 0$. The figure also demonstrates that positive surface charge effects are considerably smaller for very small negative angles for which the maximum fields are not very much affected by the presence of $Q_{ss}/q = 10^{12}$ cm^{-2}. At bevel angles smaller than about 6°, the depletion region boundary at the surface of the lightly doped side comes very close to the metallurgical junction since the surface becomes highly accumulated in the presence of the fixed positive charge $Q_{ss}/q = 10^{12}$ cm^{-2}. On the p-side, the depletion region boundary moves up until it reaches the highly doped p^+-region since the p-type side has been depleted at the surface. Further positive charge increase cannot, at these small bevel angles, affect the surface field because the depletion region boundaries at the surface became fixed previously.

Negative surface charges are much less common on silicon surfaces, but may occur, for example, with some types of surface passivating glasses. Maximum surface and bulk fields are usually reduced in the negative charge presence. If the negative charge density is high enough, a p-type inversion layer (channel) may be induced in the n-type substrate, leading to an increased leakage current and a lower breakdown voltage. Very precise control of negative charges is therefore of special importance.

4.7 Depletion Etch Method

High avalanche breakdown voltage can be achieved for both plane- and planar-type junctions by chemically etching away a part of the highly doped junction region beyond the metallic contact as shown in Fig. 29a,b [21]. This approach proves to be very effective electrically and economically and therefore represents a major improvement in both planar and plane junction device processing. The purpose of this partial etching technique, analogous to the negative beveling, is to increase the depletion region penetration into the heavily doped portion of the junction. Figure 30a,b [21] compares a chemically etched junction to a negatively beveled one. The *n*-type region was uniformly doped to 5×10^{13} cm^{-3}; the *p*-type region had a surface concentration of about 10^{16} cm^{-3}. The electrical charges etched or beveled away together with silicon are denoted in both cases by Q_R. Q_{H} denotes the negative charge on the more heavily doped *p*-side and is contained in a region that extends above the one-dimensional region boundary that would have been obtained without etching. The effective negative charge Q_{L} in the lightly doped n^--region is not contributing to the total depletion region charge after etching takes place. The purpose of the contouring is to redistribute the charges for breakdown improvement. Since the charge neutrality has to be conserved,

$$Q_R = Q_{\mathrm{H}} + Q_{\mathrm{L}}. \tag{14}$$

Point *J* (Fig. 30b) marks the region where, in the case of the one-dimensional analysis, the field is maximum. The closer Q_{H} is to the junction, the higher its

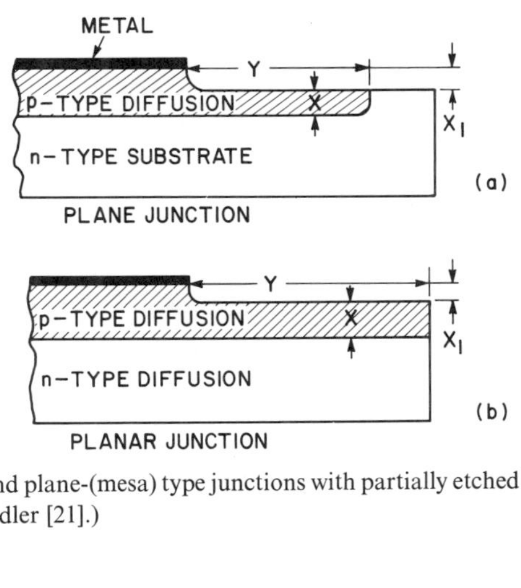

Fig. 29. Planar- and plane-(mesa) type junctions with partially etched highly doped *p*-regions (From Temple and Adler [21].)

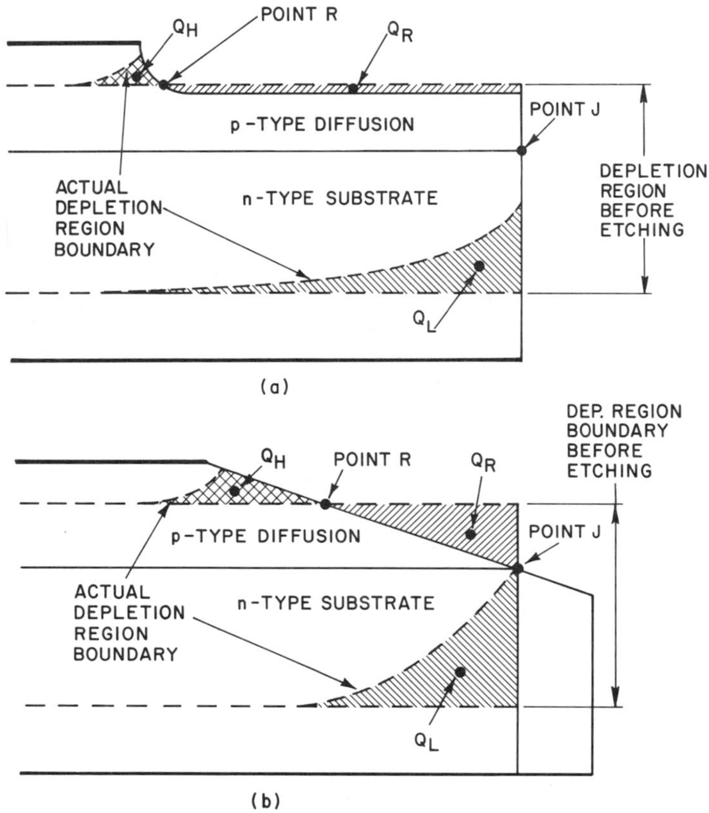

Fig. 30. Charge distributions for (a) the depletion etch and (b) the negative bevel. (From Temple and Adler [21].)

effect on the peak bulk field. To reduce the maximum field, it is desirable to reduce Q_{H} and have it spread as far away from the junction as possible. Obviously, the field at the point R should be minimized as well. It is assumed that a part of Q_R [21] within the distance D from point R is equivalent to Q_{H}, and the part to the right of D is equivalent to charge Q_{L}. If W_{H} and W_{L} are the depletion region widths in the plane (unidimensional) part of the junction for the highly doped and lightly doped junction sides, then the effective charge coupling distance D can be approximated by $(W_{\mathrm{H}} + W_{\mathrm{L}})/2$.

4.7.1 Depletion Etch Method versus Negative Beveling

The etched contour for the *plane* junction of Fig. 29 shows the etching depth as X_1. In the y direction the etch region spreads by a distance Y. If the unetched part X is exactly equal to the depletion layer thickness in the heavily doped silicon, then there is no effect on the electric field at the junction. X has to be reduced to a value somewhat smaller than W_H to avoid avalanche breakdown at the metallurgical junction. At the same time, the extent of the lateral etching, i.e., the distance Y, must also be carefully chosen. Experimental results and exact computer calculations show that best results are obtained for $Y \simeq 2D \simeq W_L$. In the case of an abrupt negatively beveled junction, making the bevel angle smaller increases the discarded depletion region charge Q_R, increases Q_L, and reduces Q_H, which spreads more laterally as well. Consequently, the peak surface and bulk fields close to Q_H are reduced. Mechanical beveling, however, removes too much of Q_R compared to the chemical etching technique so that the field at the point J is zero. To achieve low Q_H, very small angles are needed, making the devices too large to be economical. On the other hand, the etch method permits adequate control of the depth X and the lateral extent Y. Moreover, it makes it possible to find the optimum Q_H/Q_L ratio. By the proper choice of these charge magnitudes it is possible to have the bulk breakdown occur near point R while the surface is breaking down close to the junction (point J). Figure 31 [21] represents the plot of the breakdown voltage versus the etch depth X_1 for both the plane and the planar junctions. The diodes had the same doping profiles as those used for Fig. 30. The substrate concentration was 3–5 × 10^{13} cm^{-3}. The p-type diffused region was 20 μm deep with the surface concentration of 3×10^{16} cm^{-3}. The peak bulk voltages (solid lines) for the two structures differ only by a couple of hundred volts. The surface avalanche breakdowns are very approximately represented by the dashed lines for which the peak surface field was higher than 80% of the peak bulk field. Figure 32 [21] represents the peak surface field of a diffused junction as a

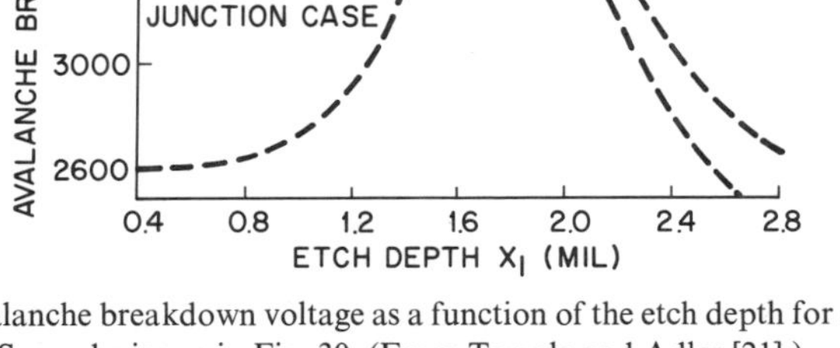

Fig. 31. Avalanche breakdown voltage as a function of the etch depth for a plane and planar junction diode. Same device as in Fig. 30. (From Temple and Adler [21].)

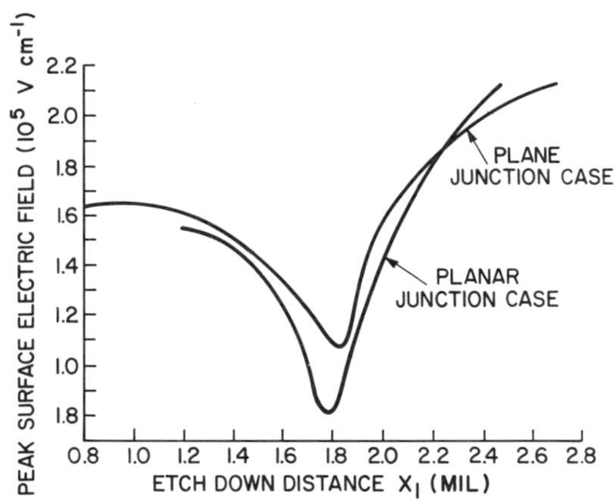

Fig. 32. Maximum surface electric field as a function of the etch depth. Same device as in Figs. 30 and 31. (From Temple and Adler [21].)

function of the etch depth X_1. In the cases studied by [21] the peak bulk and surface fields were both occurring close to the junction (point J of Fig. 30a) since the Q_L charge was small. As soon as the optimum etch depth was exceeded, the two peaks shifted closer to point R at charge Q_H vicinity (Fig. 30a). The etch depths for the lowest surface field differ somewhat from those necessary to obtain the minimum bulk field. Experimental results with plane junctions proved the technique to be superior electrically and economically to the negative beveling. Planar breakdown achieved was not far below the plane junction results.

The depletion etch method appeared initially to be difficult to apply to practical devices because of the rather critical requirements on the depth control. It turned out, however, that a very reasonable control level of about $\pm 20\%$ was possible by diffusing the higher doped p-region more deeply into the n^--region, achieving this way a very shallow grade [22]. The p-base of a transistor fabricated by [21] was diffused to a junction depth of $x_j = 60\ \mu$m with surface concentration $C_0 = 1 \times 10^{16}$ cm^{-3}. The collector-to-emitter sustained breakdown voltages $BV_{CEO(sus)}$ obtained with the n^--region of about 10^{14} cm^{-3} and planar junctions were close to 2000 V.

4.8 Substrate Etch Termination versus Positive Beveling

In the case of the depletion etch method the junction had less silicon left on the *highly doped* junction side than on the lightly doped side and was therefore analogous to the negative bevel junction. It used, however, much less silicon area and did not require device mechanical shaping. Its main drawback is the need for relatively precise etch depth control. Another etch method, known as substrate etch termination [23], is applied to remove a part of the silicon crystal from the *low doped* junction region in analogy to the positive bevel technique. Similar to the depletion etch technique analogous to the negative bevel, the positive analog does not require any mechanical surface shaping. It needs, however, about the same crystal area. Another advantage is the adaptability to both the plane and planar junctions.

In the deep moat etch technique (Fig. 16) that has been in use for many years to shape the mesas of the high voltage transistors, etching takes place from the top silicon wafer surface. In the substrate etch termination, the etching process takes place from the opposite side of the wafer (Fig. 33 [23]), resulting in a very significant (50%) and consistent surface field reduction, whereas the deep moat etching does not give consistent results and does not sufficiently reduce the surface fields. In Fig. 33, schematically representing the device with an etched substrate, X_{E} is the etch depth while Y_{E} is its extent. Since a part of the crystal has been removed from the *n*-substrate, a part of the depletion region charge was lost as well. This lost part is denoted by Q_R. Q^- represents the net reduction in the depletion charge in the more heavily doped *p*-side of the junction, and Q^+ is the additional positive charge which appears on the lighter doped substrate side outside of the originally existing depletion region. Because of the charge neutrality requirement $Q_R = Q^+ + Q^-$. Similar to the situation with the negative bevel angle, the larger Q^- is in its absolute value, the smaller the surface field becomes in the neighborhood of point 1 (Fig. 33). The effect of the newly appeared positive charge Q^+ on the lightly doped side has only a very insignificant effect on the maximum bulk electric field as opposed to the situation with the negative bevel angle. Q^+ for the positive bevel extends over a longer region in the low doped side whose distance from the metallurgical junction is large. Due to this negligible effect, variations in the etch depth X_{E} become much less important in the positive bevel case. The surface fields will be reduced

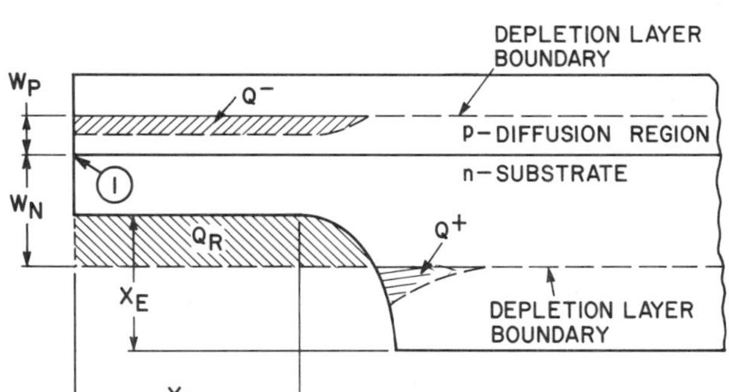

Fig. 33. Charge distribution for the etched substrate. Plane mesa junction. (From Temple and Adler [23].)

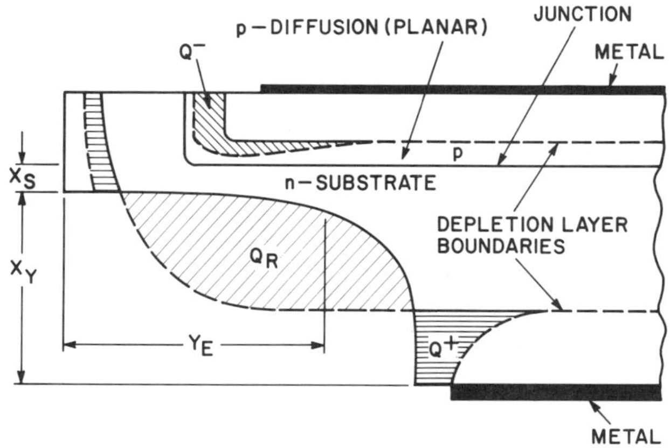

Fig. 34. Charge distribution for the etched substrate planar junction. (From Temple and Adler [23].)

sufficiently, providing the etch extent Y_E is long enough to make Q^- large and maximum surface field low. A large fraction of charge Q_R is located close to the junction region before its removal by etching, and its removal radically reduces the bulk field at the metallurgical junction.

A similar etch technique can be applied to a device with the planar *p*-type diffusion (Fig. 34 [23]). As in the case of the plane junction, the charge Q^- reduces the peak surface and junction fields in the junction termination region. The appearance of Q^+ as the result of the substrate etching has again a negligible effect on the bulk field at the junction as being very far removed. By making the distance Y_E from the junction edge to the scribed surface long enough, i.e., equal to several times the depletion width of the one-dimensional junction, it is possible to reduce to zero the field at the scribed edge and achieve near ideal breakdown voltage.

Calculations and experimentation of [23] show that the extent of the lightly doped region Y_E could be made as much as one-half of the substrate (N-region) depletion width to obtain near ideal breakdown. The etch depth X_E is not critical, but it is reasonable to etch away between 50 and 75% of the depletion layer on the lightly doped side.

4.9 Depletion Region Charge Control by Ion Implantation

The depletion etch technique removes a large portion of the semiconductor edge region with the purpose of reducing the surface electric field without the bulk field increase. The shaping of the depletion region and the subsequent field control can also be achieved by ion implantation rather than by beveling or etching [24]. This approach appears to be less critical and therefore less expensive. Figure 35a,b [24] shows schematically how the plane and planar junctions can be extended by an implant done at some convenient part of the device processing. The experiments performed by [24] showed that the device characteristics are practically unchanged for Y_1 values between ~13 and 130 μm. The X_E dimension was not critical either as long as most of the *p*-type semiconductor was removed at the edges. In the devices studied, the X_E size for the plane junction was about 90 ± 25 μm with no effect on the achieved near ideal breakdown voltage. The junction depth was about 60 μm.

The net activated charge of the implant required to obtain the maximum breakdown voltage is estimated as follows [24]. The depletion region charge per unit area Q_d for an ideal plane junction breakdown condition is given by $Q_d = \epsilon_s E_b$, where E_b is the junction field at breakdown. The implant dose Q_I is some fraction of Q_d with smaller doses giving smaller peak surface fields. Q_I should not be made too low, however, since the maximum bulk field becomes excessively large. Calculations and experiments of [24] show that $Q_I = (0.8\text{–}0.9)Q_d$ gives the best results if the surface passivation does not introduce undesired charges. Under these conditions the junction breakdown achieved exceeds 95% of the ideal breakdown.

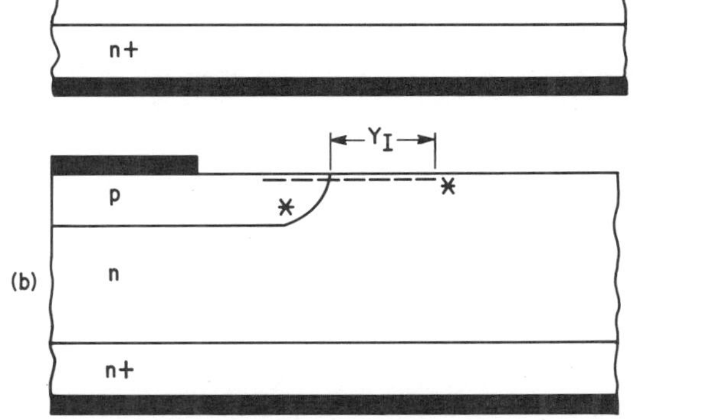

Fig. 35. Ion implanted surface of (a) a plane and (b) a bipolar junction. High field points are shown by asterisks. (From Temple [24].)

REFERENCES

1. V. P. O'Neil and P. G. Alonas, *IEEE Trans. Electron Devices* **ED-26,** 1098 (1979).
2. A. Rusu and C. Bulucea, *IEEE Trans. Electron Devices* **ED-26,** 201 (1979).
2a. A. S. Grove, O. Leistico, Jr., and W. W. Hooper, *IEEE Trans. Electron Devices* **ED-14,** 157 (1967).
2b. A. Rusu, O. Pietrareanu, and C. Bulucea, *Solid-State Electron.* **23,** 473 (1980).
3. F. Conti and M. Conti, *Solid State Electron.* **115,** 95 (1972).
4. D. S. Zoroglu and L. E. Clark, *IEEE Trans. Electron Devices* **ED-19,** 4 (1972).
5. L. E. Clark and D. S. Zoroglu, *Solid-State Electron.* **15,** 653 (1972).
6. A. Aoki, T. Matsushita, Y. Yamoto, H. Hayashi, M. Okayama, and Y. Kawana, *Electrochem. Soc. Spring Meeting Extended Abstracts,* **74-1,** 64 (1974).
7. T. Matsushita, T. Aoki, T. Antsu, H. Yamoto, H. Hayashi, M. Okayama, and Y. Kawana, *IEEE Trans. Electron Devices* **ED-23,** 826 (1976).
8. B. A. McDonald, *IEEE Trans. Electron Devices* **ED-17,** 871 (1970).
9. M. L. Tarng, RCA Labs, Princeton, New Jersey. Private communication (September 1977).
10. A. Goodman, RCA Labs, Princeton, New Jersey. Private communication (November 1977).
11. F. J. Campbell and M. L. Tarng, RCA Labs, Princeton, New Jersey. Private communication (January 1979).
12. Y. C. Kao and E. D. Wolley, *Proc. IEEE* **55,** 1409 (1967).
13. M. S. Adler, V. A. K. Temple, A. P. Ferro, and R. C. Rustay, *IEEE Trans. Electron Devices* **ED-24,** 107 (1977).
14. R. L. Davies and F. E. Gentry, *IEEE Trans. Electron Devices* **ED-11,** 313 (1964).
15. J. Cornu, *Electron. Lett.* **8,** 169 (1972).
16. J. Cornu, *IEEE Trans. Electron Devices* **ED-20,** 347 (1973).
17. M. Bakowski and K. I. Lundström, *IEEE Trans. Electron Devices* **ED-20,** 555 (1973).
18. M. Bakowski and B. Hanson, *Solid-State Electron.* **15,** 651 (1975).
19. M. S. Adler and V. A. K. Temple, *IEEE Trans. Electron Devices* **ED-25,** 1266 (1978).
20. M. S. Adler and V. A. K. Temple, *IEEE Trans. Electron Devices* **ED-23,** 956 (1976).
21. V. A. K. Temple and M. S. Adler, *IEEE Trans. Electron Devices* **ED-23,** 950 (1976).
22. R. Denning and J. White, RCA Technical Communications, Princeton, New Jersey, p. 66 (1979).
23. V. A. K. Temple and M. S. Adler, *IEEE Trans. Electron Devices* **ED-24,** 1077 (1977).
24. V. A. K. Temple, *IEDM Tech. Digest,* p. 423 (1977).

CHAPTER 5

SELECTED FABRICATION TECHNIQUES

Most of the classical semiconductor fabrication methods, such as diffusion and epitaxy used in conjunction either with the Czochralski or with float zone grown silicon substrates for junction formation, remain largely the mainstay of the power transistor technology. Ion implantation, which is of great value in *p–n* junction or MOS channel formation, is a more recent addition to an old arsenal of the available tools that either complement or improve or supplant the older techniques. Passivation of silicon surfaces by doped polycrystalline films became decidedly one of the most significant steps in the achievement of device operational and shelf-life stability and is only now coming of age. Silicon neutron transmutation doping for high voltage device fabrication, dry (plasma) etching of various chip components, and improved oxide and lifetime control techniques belong to the more recent technological advances which are discussed in what follows.

5.1 Neutron Transmutation Doping (NTD)

The quality of the starting silicon crystal in high voltage power transistor processing is of utmost importance. Until recently, the best very high resistivity bulk silicon material with very low contamination levels, i.e., with a minimum of oxygen, carbon, and heavy metals content, was obtained by the float zone technique. The low level of contaminants is essential for the prevention of precipitates formation leading to premature breakdown. The float zone silicon cannot be grown, however, without radial microresistivity variations consisting of big resistivity changes taking place over several micrometers-long radial distances. These striations, due to thermomechanical equipment problems, are present even with the best crystal growing equipment. They can be almost entirely eliminated, though, by the use of the neutron transmutation doping (NTD) [1–3]. Very uniformly phosphorus-doped high-resistivity silicon can be produced by bombarding silicon with thermal (mostly slow) neutrons. Since the silicon isotope ^{30}Si is uniformly distributed in the silicon crystal, its transmutation to phosphorus by neutron irradiation results in a very accurate and uniform doping concentration due to the precise neutron flux control.

Neutrons do not carry any electric charges and therefore can penetrate a silicon layer 1 m thick, easily reaching the positively charged silicon nuclei. The slow moving thermal neutrons even allow for a greater interaction time between the nucleus and the neutron. In other words, the neutron capture cross section representing the probability of the interaction is inversely proportional to the neutron velocity or to the square root of its kinetic energy.

The reaction may be represented by

$$^{30}\mathrm{Si}(n,\gamma) \rightarrow {}^{31}\mathrm{Si}^{2.62\mathrm{hr}} \rightarrow {}^{31}\mathrm{P} + \beta^-, \tag{1}$$

i.e., after the irradiation and absorption of the neutron n, gamma radiation is emitted, and the ^{30}Si isotope is transmuted to the ^{31}Si isotope, which is unstable and decays with a half-life of 2.62 hr emitting beta rays (β^-). The decayed ^{31}Si is converted to the phosphorus isotope ^{31}P.

The requisite irradiation time depends on the required phosphorus concentration, the neutron flux density, and the reactor structure. Two hours were, for instance, necessary to produce 200 Ω cm resistivity silicon with a neutron flux density of 2×10^{13} neutron $cm^{-2}\,s^{-1}$ by [4]. A small portion of the phosphorus isotope also undergoes a transmutation under the neutron bombardment resulting in an undesirable side-effect reaction expressed by

$$^{31}\mathrm{P}(n,\gamma) \rightarrow {}^{32}\mathrm{P}^{14\,\mathrm{day}} + {}^{32}\mathrm{S} + \beta^-. \tag{2}$$

The phosphorus isotope ^{31}P is transformed into the ^{32}P isotope, which, in turn, decays into ^{32}S sulfur with the emission of β^- radiation. This decay, with 14 days half-life, is the main source of the spurious radiation.

Because of the comparatively long decay time of the phosphorus–sulfur transmutation and for safety reasons, the produced doped silicon can be released for use only after a sufficiently long time.

The NTD process damages the silicon crystal lattice by displacing the silicon atoms because of the presence of some fast neutrons, charged particles, and gamma- and beta-ray recoils. Damage increases, of course, with higher neutron flux needed for higher doping levels. The radiation-created lattice defects act as carrier recombination and/or mobility scattering centers. Lattice defects can be removed largely by short thermal annealing, for example, at 750°C for 2 min in argon atmosphere to obtain stable and correct resistivity [2]. Generally, annealing recovers carrier mobility well, but much less completely the minority carrier lifetime. The after-annealing macro- and microuniformity of crystal doping is excellent. The doping accuracy presently achievable with NTD is about 1% at the very high resistivity levels required for power transistor production, and it is economical with doping levels as high as 1×10^{14} cm^{-3}.

5.2 Ion Implantation

In the past ten years ion implantation has become one of the basic tools of silicon technology. The advantages of ion implantation over the thermal deposition and diffusion technique are mainly due to the possibility of the dopant introduction at lower temperatures, at specified places, and with a very exact dosage. Materials such as photoresist, metals, and SiO_2 and SiN_4 may be used to various degrees for masking purposes during implantation. Ion implantation control and uniformity lead to improved performance and lower costs.

The ion implantation consists of bombarding the semiconductor surface with the desired impurity ions by accelerating them to very high velocities by the application of high potentials in an apparatus which is essentially a mass spectrometer. The combination of the magnetic and electric fields assures an excellent separation of various ionic species, hence the high purity of the ions hitting the target. The implanted atoms penetrate the semiconductor, interact with the target electrons, and stop when their energy is entirely dissipated.

The surface orientation for this process is of great importance. In the case of improperly chosen orientation, the ion beam may penetrate deeply into the crystal by "channeling" between the atoms. By tilting the crystal away from any major crystal axis, the silicon lattice will appear to be randomly distributed to the incident beam. The crystal behaves, therefore, in this case, as an amorphous solid, the channeling is avoided, and the impurity is distributed uniformly in a shallow surface layer. The exact dose is assured by integrators that measure the total implanted charge with an accuracy of better than 1%.

The implanted atoms interact with the crystal lattice causing considerable damage to it. Moreover, only a small part of them are electrically active. In order to repair the crystal damage and to reactivate the implanted impurities, it is necessary to heat-treat (anneal) the semiconductor so that the crystal lattice is returned to its original perfection and the impurities get located into the substitutional sites. For instance, boron as an implant requires about 850°C heat treatment in a nonoxidizing atmosphere. Annealing temperatures are normally much lower than the diffusion temperatures so that no significant diffusion occurs during the heat treatment.

Just after implantation, the dopant distribution below the silicon surface follows Gauss's law [5], i.e.,

$$N_I(x) = N_{max} \exp[-(x - R_p)^2/2(\Delta R_p)^2] - N_b, \tag{3}$$

where

$$N_{max} = N_{dose}/(2\pi)^{1/2}\, \Delta R_p, \tag{4}$$

$N_I(x)$ is the implanted concentration per unit area as a function of the depth x measured from the silicon surface; N_{dose} is the total implanted dose per unit area; R_p is the range, i.e., the statistically averaged distance from the surface; and ΔR_p is the standard deviation in range (Fig. 1 (6], also known as straggle. N_b is the background doping concentration.

When a protective silicon oxide layer is present on the silicon surface during implantation, N_{dose} is that portion of the ion dose that resides in the silicon crystal, and R_p, with SiO_2 present, is the distance between the Si–SiO_2 interface and implanted Gaussian peak. Depending on the required doping profile and junction depth, the implanted impurities may be further diffused as in the case of other methods of impurity deposition. The use of a thick

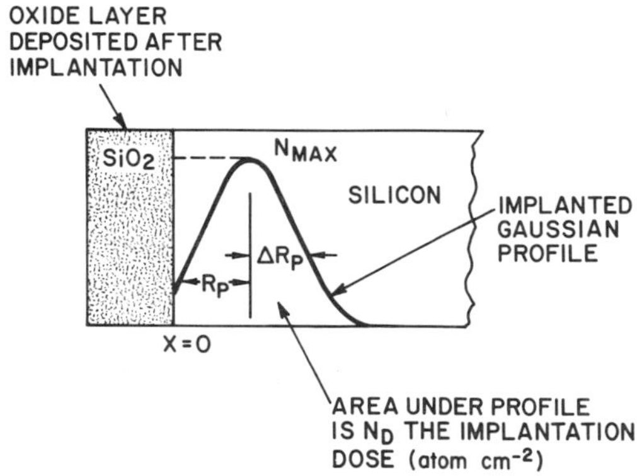

Fig. 1. Ion implantation with initial truncated Gaussian profile. (From Douglas and Dingwall [6].)

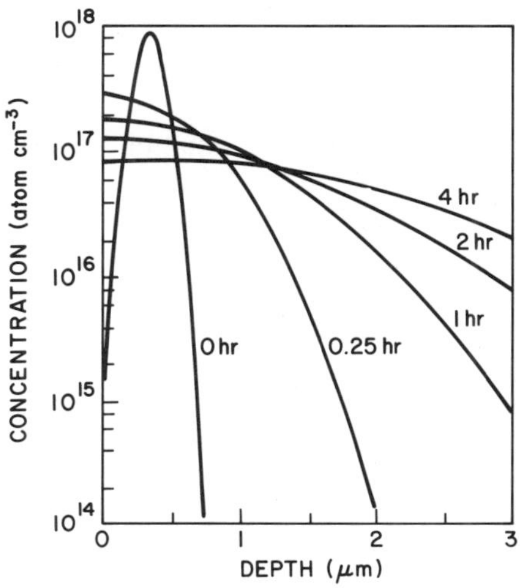

Fig. 2. Calculated diffusion profiles for boron implanted in silicon. (From Douglas and Dingwall [6].)

capping oxide and of nonoxidizing atmosphere during diffusion prevents the loss of the implanted impurities by segregation into a growing oxide. [7]

The range R_p for boron ions in silicon at 100 eV is about 0.3 μm, the standard deviation 7.6×10^{-3} μm, and the junction depth $x_j = 0.65$ μm if N_{dose} per unit area is assumed to be numerically equal to the substrate concentration per unit volume. Tables for R_p, ΔR_p, and x_j are readily available [8]. When diffusion follows ion implantation, the original Gaussian distribution flattens up as shown in Fig. 2 [6], which is a plot of the calculated impurity concentration (boron) versus depth x. The curves are plotted for the diffusion times $t = 0$, 0.25, 1, 2, and 4 hr with a dose of 2×10^{13} cm^{-2}, ion energy of 100 keV, diffusion temperature of 1200°C, and boron diffusion constant of 10^{-12} cm^2 s^{-1}.

5.3 Dry Etching[1]

Etching and cleaning of semiconductors and their components was accomplished exclusively by wet, high purity chemical solutions until about ten years ago. Further miniaturization prompted the development of the dry processing techniques using gas plasma discharge. These techniques, which are essential in highly packed devices, are also useful in the fabrication of

[1] See [9–13].

power devices, such as MOS power transistors with narrow channels or microwave power transistors with highly interdigitated geometries with very narrow line widths and/or small ratio-of-line-width-to-film thickness. Even for wider line width in metals, such as aluminum, sharp delineation is of great advantage.

In contrast to wet etching, dry etching is anisotropic. Wet isotropic etchants tend to undercut photoresist or silicon dioxide.

Dry etching has some disadvantages; the most serious is that of the radiation damage, which may be mostly annealed, however, at comparatively low temperatures (400–450°C) in the forming gas atmosphere. Other disadvantages are lower etching selectivities than those of wet etchants and increased difficulty of photoresist removal.

5.3.1 Radio Frequency Sputter Etching and Milling

For radio frequency (RF) sputter etching, the substrates are placed on a cathode electrode (Fig. 3). The diode's anode is grounded while the cathode is coupled to a RF generator. The diode is placed in a vacuum chamber filled with a rarefied inert gas such as argon. Argon ions formed as the result of discharge are accelerated across the cathode dark space and hit the targets with high energy, causing substrate sputtering. The typical ion energy is equal to about 1000 eV.

Sputtering, an entirely physical process, has the ability to uniformly etch very fine patterns over large target areas. Owing to its many disadvantages, such as electron radiation damage and ion implantation, its use is restricted to noble metal etching, such as gold or platinum.

Ion milling resembles plasma sputtering, except that the ion beam is

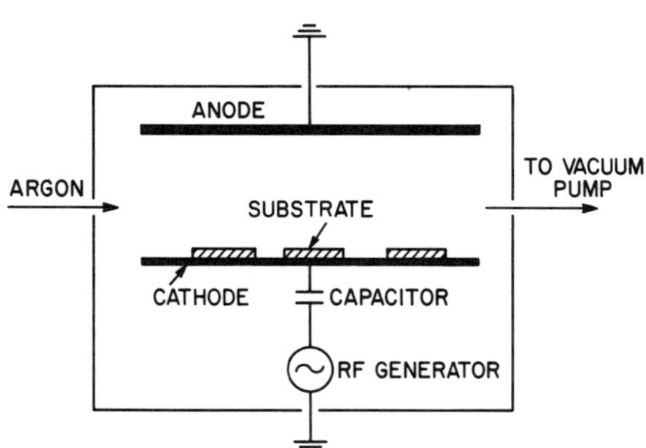

Fig. 3. RF sputter etching.

collimated by the use of special grids. The substrate holder may be rotated in the milling-type equipment for the control of the ion beam impingement angle.

5.3.2 Reactive Ion Etching (RIE)

Inert gases, such as argon, are used in the sputter and milling etching equipment, resulting in a purely physical process.

In reactive ion etching, reactive gases such as fluorocarbons are used. The gas pressure for the reactive ion etching is in the 0.1–1.5 Torr range, whereas for sputtering the gas pressure is maintained in the 10–150 Torr range. The etch rates of RIE are in some cases up to ten times higher than those of sputtering. The RIE is a chemical process with good selectivity and without any appreciable radiation damage. The likely candidates for this type of etching are silicon oxides and nitrides, polycrystalline silicon, and aluminum.

5.3.3 Tunnel and Flat-Bed Plasma Etching

Plasma etching takes place in a reactor when a reactive gas is used without, however, high field ion acceleration toward the substrate surface. The evolved volatile chemical compounds are removed from the chamber by a vacuum pump (Fig. 4).

Two reactor types are available for plasma etching: the tunnel type and the flat-bed (parallel-plate) type. In the tunnel-type reactor the substrates are stacked vertically like silicon wafers in a diffusion furnace. The RF power is

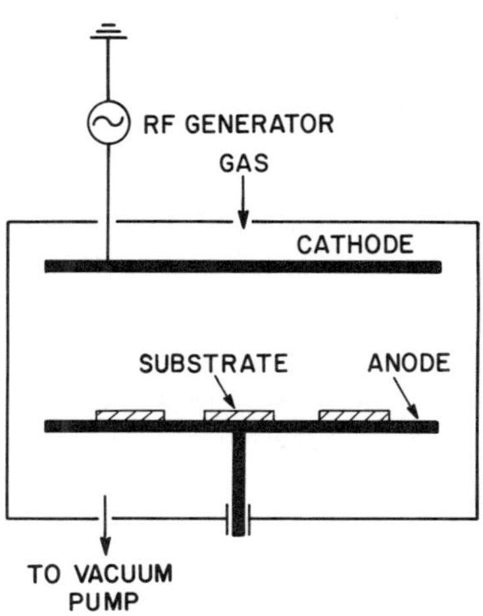

Fig. 4. Flat-bed (parallel-plate) plasma etching.

applied to metal electrodes surrounding the load and placed, for example, on the tunnel wall. The tunnel-type equipment, not much superior to the wet etching, is adequate for less demanding operations, such as photoresist stripping, nitride, and polysilicon etching.

The flat-bed (parallel-plate) etching reactors, on the other hand, are capable of very fine processing. The parallel-plate plasma etching apparatus (Fig. 4) consists of a flat, grounded anode and parallel to it a cathode connected to the RF supply. Substrate to be dry etched is placed on the anode approximately 2 in away from the cathode. In this type of apparatus, different gases attack various materials at different rates, resulting in very good selectivity. For instance, by changing gases it is possible to etch silicon faster or slower than the silicon dioxide. Silicon nitride and oxide, polycrystalline silicon, aluminum, etc. can be economically etched by this technique.

5.4 Minority Carrier Lifetime Control

Several techniques perfected over a period of years are now available to either improve or reduce the minority carrier lifetime of single crystal silicon.

5.4.1 Lifetime Improvement Techniques

In many types of power transistors good minority carrier lifetime is necessary to achieve reasonable current gains and/or leakage currents. It is essential, therefore, to maintain the silicon lifetime at a comparatively high level during device processing. This requires very clean techniques that do not introduce contaminating impurities, which not only lower the minority carrier lifetime but at the same time cause soft breakdowns. The impurities that affect silicon the most are gold, copper, and iron since they introduce deep recombination centers. These types of impurities may sometimes be present in the starting material, but more often are introduced during such high temperature processing as diffusion. All of the most effective lifetime killers are very rapid interstitial diffusants.

Oxygen is another impurity that affects silicon crystal properties and exists even in the so-called float zone silicon and upon high temperature processing, converts into SiO_2 precipates, contributing to soft breakdown. Crystal dislocations provide precipitation sites for most impurities and together with vacancies act as recombination centers as well.

A few excellent techniques are presently available to reduce the amount of these undesirable impurities during processing. The most popular probably is the use of a mixture of O_2 and HCl gas for the gettering of metallic impurities. There are several types of materials, such as phosphosilicate and borosilicate glasses, that are quite effective in gettering many impurities at elevated temperatures. The phosphosilicate glass, for instance, when deposited at high temperature on the silicon surface, is a good gettering material for gold.

5.4.2 Lifetime Reduction Techniques

Fast switching bipolar transistors may require low minority carrier lifetimes for the reduction of the electric charges stored in the base and collector regions. In such cases lifetime is reduced to a reasonably low value by the introduction of the lifetime reducing agents known as "killers" into the device processing.

The most commonly used lifetime "killer" is gold [14], which diffuses very rapidly into silicon with an interstitial diffusion constant $D \simeq 9 \times 10^{-6}$ $cm^2\ s^{-1}$ at 1000°C and $4 \times 10^{-6}\ cm^2\ s^{-1}$ at 800°C. Gold is usually diffused from a very thin evaporated film at temperatures from 800 to 1000°C and rapidly quenched to avoid precipitation. Another metal of great interest for this application is platinum [15, 16], normally diffused from the spin-on paints containing platinum compounds. Both impurities introduce acceptor levels that reduce lifetime in *n*-type silicon. The gold acceptor is 0.54 eV below the silicon conduction band in *n*-type silicon, whereas, the dominant platinum acceptor level is shallower and only 0.26 eV below the conduction band according to [15] or 0.42 eV according to [17]. In the *p*-type silicon, however, the two elements are very similar, introducing a donor level at 0.35 eV above the valence band.

Since the platinum acceptor level in the *n*-type silicon is shallower than that of gold, the space-charge generation rate is lower than that for gold as impurity. As a result, the leakage current of the platinum-doped *n*-type silicon is smaller at low and high temperatures.

The lifetime of the platinum-doped silicon is strongly dependent on the injection level. At high injection levels, lifetime may be three times shorter than that at low injection levels. The use of Pt for lifetime reduction is therefore advisable at high injection level switching.

The recombination tail in the reverse recovery of bipolar devices is predominantly controlled by the low-level lifetime, and for this situation gold is definitely superior to platinum.

A more recent lifetime reduction technique employs electron beam irradiation, which has the advantage over gold or platinum doping of being done after the device fabrication completion. This technique is basically clean, can be performed at room temperature, and is easily controlled. Many

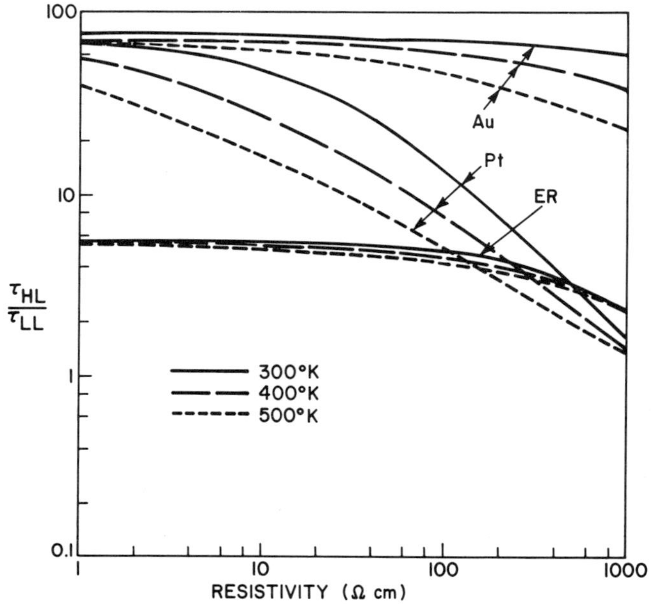

Fig. 5. High-level to low-level lifetime ratio as a function of resistivity and temperature. Curves marked Au, Pt, and ER are for gold doping, platinum doping, and electron irradiation of silicon, respectively. (From Baliga and Sun [16].)

defects caused by irradiation are annealed at temperatures above the treated device's operating temperature. Electron beam irradiation introduces deep level recombination centers, resulting in lifetime reduction. The device lifetime before and after irradiation are related by

$$\tau_{\text{eff}}^{-1} = \tau^{-1} + K\phi, \tag{5}$$

where τ_{eff} and τ are the minority carrier lifetime after and before irradiation, respectively; ϕ is the electron radiation fluence in $e\ \text{cm}^{-2}$; and K is a radiation damage coefficient [18]. There are three deep electron radiation-induced recombination centers left after annealing the irradiated silicon at 300°C [18]. The deepest is 0.36 eV below the conduction band.

Calculations and experimental work of [16] on devices with *n*-type resistivities from 1.0 to 500 Ω-cm indicate that gold-diffused bipolar devices have the best trade-off between forward voltage drop and reverse recovery time and have leakage currents higher than those for platinum doping or electron irradiation. Electron-irradiated devices have a worse forward drop–reverse recovery trade-off curve than those with platinum doping, but the leakages are comparable.

These results are explained on the basis that a large high-level lifetime leads to low forward drop in bipolar diodes or transistors; small low-level lifetime, on the other hand, allows fast device recovery during the turn-off transient. Consequently, a large ratio of high-level to low-level lifetime τ_{HL}/τ_{LL} is desirable to obtain low forward voltage drop at high current levels, but fast recovery when the device is turned off and the turn-off tail is supposed to be as short as possible. A large ratio of space-charge region effective lifetime and high-level lifetime τ_o/τ_{HL} is necessary for a low reverse leakage current. τ_{HL} is the largest for gold and the smallest for electron irradiation (Figs. 5 and 6 [16]).

It should be pointed out that the introduction of impurities into silicon for lifetime-killing changes the resistivity of the treated crystal. On the other hand, the τ_{HL}/τ_{LL} ratio is dependent on the crystal resistivity. It is a more sensitive silicon resistivity function for platinum doping than for gold doping or for electron irradiation.

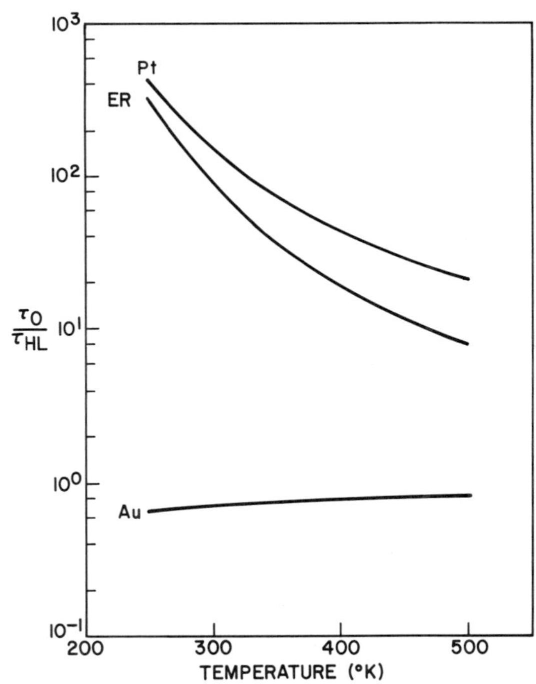

Fig. 6. Space-charge lifetime to high-level lifetime ratio as a function of temperature for gold (Au), platinum (Pt), and electron irradiated silicon (ER). (From Baliga and Sun [16].)

5.5 Surface Stability and Device Passivation

Surface properties of both the bipolar and field-effect transistors determine the device shelf- and operational-life stability. Careful fabrication techniques assure a contamination-free surface, which must be further preserved by proper device passivation and packaging.

5.5.1 Surface Stabilization by Use of Chlorine and Its Compounds

Chlorine (Cl_2) and its compounds, such as hydrochloric acid (HCl) or trichloroethane (TCE), are now widely used in the processing of the bipolar and field-effect transistors. The achieved cleaning effect is probably the most important aspect of this technique.

During device processing, the silicon wafers are usually thermally oxidized, either in steam or O_2, at elevated temperatures at which many impurities contained in the furnace quartz tubing wander into the SiO_2 film formed on the silicon surface. This type of oxide contamination is avoided by precleaning the furnace prior to the oxidation, using a mixture of O_2 and Cl_2 (or its compounds). The most dreaded contaminants are the alkali ions, such as sodium Na^+, which easily move through the silicon dioxide and drastically change the surface properties. Other impurities, such as Au, Cu, and Fe, affect silicon lifetime as discussed in Section 5.4. The addition of chlorine results in the formation of volatile, easily removed, metallic chlorides.

The precleaning of the furnace tube usually is sufficient to avoid silicon contamination. Very often, oxidation itself is also carried out in a chlorine containing atmosphere. Furnace precleaning remains, nevertheless, the essential processing step.

Surface stability by the removal or neutralization of sodium ions can be achieved to the same degree by using, for example, 6% HCl/O_2 or 0.75% TCE/O_2, which have about the same Cl_2 [19] concentration. A significant reduction of oxide charges due to radiation effects can be obtained by processing silicon in HCl-precleaned furnaces. However, high Cl_2 concentrations in the oxidizing atmosphere increase the radiation effects, probably because of chlorine migration in SiO_2.

Another benefit derived from the oxidation in the Cl_2-containing atmosphere is the reduction by one order of magnitude of the interface state density as compared to those of oxides grown without Cl_2. This fast state reduction (Chapter 1) takes place because during the high temperature oxidation both HCl and TCE evolve large quantities of H_2 and OH^- groups, which fill the dangling silicon bonds. The orientation dependence of the interface state density is similar to that of the oxide grown in the absence of chlorine (Section 1.8). HCl, Cl_2, TCE, and HBr [20] also improve, to various degrees, the silicon dioxide integrity and its breakdown by the oxide defect reduction.

The in-process passivation effect of chlorine and its compounds is rather controversial. After oxidation in the chlorine atmosphere, the device undergoes processing steps such as photoresist coating and metallization, which introduce large amounts of sodium and other impurities. For oxides grown at high temperatures and very long time periods, the contaminating sodium may still be neutralized due to the existence of the high chlorine concentration at the silicon–oxide interface. This in-process passivation is limited to a narrow range of HCl concentrations of 3 to 4% for contaminating sodium levels of about 10^{12} ions $cm^{-}2$. Up to this level, the complete neutralization of Na^+ and its immobilization appear to be still possible.

Inclusion of high Cl_2 concentrations in O_2/HCl grown oxides exhibits room temperature bias instability not found in oxides processed without Cl_2. In the C–V plots, the flat band shifts toward the negative gate voltages and is accompanied by C–V curve distortion indicative of the increase of the surface states (Section 1.8). The field intensity and the chlorine content affect the magnitude of these effects [21]. For oxides obtained in Cl_2/O_2 atmosphere flat-band shifts occur for positive and negative polarities of the gate voltage suggesting the presence of charges of both polarities. As for the HCl/O_2 oxides, the interface states are generated at high fields. The use of TCE results in less pronounced instability [19]. According to [19], it is impossible to achieve all improvements by using one type of chlorine compound, and compromises are necessary in all cases. Generally, however, a HCl chlorine source is superior to Cl_2, which is extremely reactive even at room temperature. On the other hand, HCl is very corrosive and roughens the silicon surface. TCE, which was the least studied, appears to be very attractive because of the ease in which it can be handled.

5.5.2 Device Passivation

Surface passivation of semiconductor devices is essential for the preservation of their operational or storage capability. Passivation protects the surface from the undesirable extraneous contaminants and from moisture. Many present-day devices are still passivated by simple technological means, such as thermal oxide and silicon nitride coatings or thermal oxide and phosphosilicate glass deposited at about 400°C. Organic polymer films have been used with variable results since the early days of the semiconductor technology and normally consist of silicon rubbers and epoxies. Thermally grown SiO_2 is not used for discrete *high voltage* devices because of drift effects and high charge density. Silicate glasses or organic polymers are more appropriate for this application.

A passivation system that, at the same time, hermetically seals the device from the outside world and permits devices to be tested and investigated without final packaging has a big cost saving feature. Glass is such a passivant. A very useful glass, for example, is the Innotech-type 760 lead–alumina–borosilicate glass with a thermal expansion coefficient of 4.8×10^{-6} per °C. The alumina–silica glasses are also very suitable for high voltage devices. Glasses may be applied, for example, by techniques such as electrophoresis [22] in which the anode consists of some noble metal and both electrodes are immersed in an electrolyte consisting of the colloidal suspension of glass in, for example, methyl alcohol. This technique is even adequate for deep mesa high voltage junctions. Another simple glass deposition method consists of the application to the surface of the glass slurry in an organic binder. The coating is made thin and uniform by removing the excess by "doctor blading," and the binder is removed by a bake-out at comparatively low temperatures.

Semi-insulating polycrystalline silicon [23] is a comparatively recent addition to the arsenal of planar junction passivants. It can be used for both the *n–p–n* and *p–n–p* transistors with excellent results. Sipos consists essentially of an oxygen-doped polycrystalline silicon film deposited on the junction surface. The polycrystalline film can be obtained, for example, from a SiH_4–N_2O–N_2 mixture. The electrical conductivity of Sipos, due to oxygen doping, can be made extremely low so that the leakage currents are negligibly small. The oxygen-doped Sipos film is normally further covered with a silicon nitride film obtained from the SiH_4–NH_3–N_2 system, which

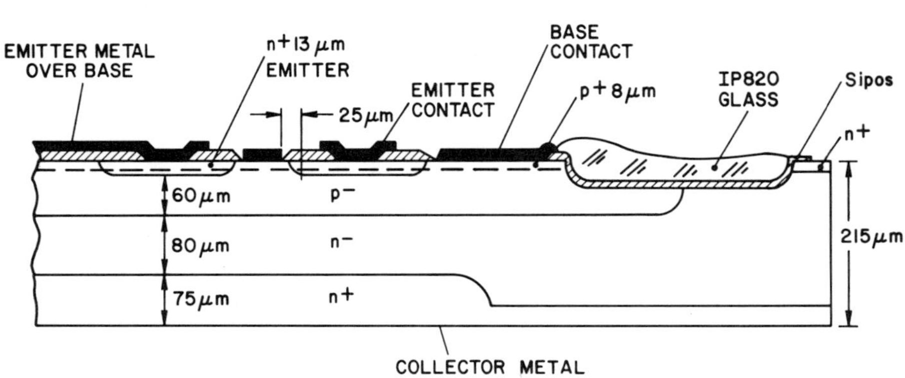

Fig. 7. Collector–base junction passivated by a Sipos–glass system. (From Denning and White [24].)

acts as a sealer against mobile sodium ions and moisture. To prevent the silicon nitride from electric breakdown, a SiO_2 film is deposited on it. A glass layer may be added for further mechanical stability and hermeticity (Fig. 7 [24]). The Sipos technique is described in some detail in Section 4.4.

Another promising passivating film consisting of hydrogenated amorphous silicon was studied by Tarng and Pankove [25]. Such a film can be deposited on silicon crystal, for example, from SiH_4 by glow–discharge decomposition at 350°C. The Si:H conductivity is very small so that there is no appreciable reverse leakage current present, even at elevated test temperatures. Due to the crystal surface hydrogeneration, the interface state density is so much reduced that the leakage current becomes about two orders of magnitude lower than that obtained with thermal oxidation. The hydrogeneration of the crystalline silicon surface takes place during the deposition of the amorphorus Si:H. The hydrogen present in the amorphous Si:H film serves as a source of hydrogen which ties up the crystalline silicon dangling bonds so that the operating and shelf-life stability are assured by the continuous rehydrogenration process.

Amorphous Si:H used in experimentation [25] contained up to 50 at. % of hydrogen. Higher concentrations resulted in porous polycrystalline films.

REFERENCES

1. H. A. Herrman and H. Herzer, *Electrochem. Soc.* **122,** 1568 (1975).
2. H. M. Janus and O. Malmros, *IEEE Trans. Electron Devices* **ED-23,** 797 (1976).
3. M. J. Hill, P. M. VanIsegham, and W. Zimmerman, *IEEE Trans. Electron Devices* **ED-23,** 809 (1976).
4. E. W. Haas and O. Malmos, *IEEE Trans. Electron Devices* **ED-23,** 803 (1976).
5. J. Lindhard and M. Scharf, *Phys. Rev.* **124,** 128 (1961).
6. E. C. Douglas and A. G. F. Dingwall, *IEEE Trans. Electron Devices* **ED-21,** 324 (1974)
7. S. Wagner, *J. Electrochem. Soc.* **119,** 1570 (1972).
8. B. Smith, "Ion Implantation Range Data for Silicon Germanium." Research Studies Press, Inc., Forest Grove, Oregon, 1978.
9. J. A. Bondur, *J. Electrochem. Soc.* **126,** 226 (1979).
10. J. L. Vossen and W. Kern, EDS, "Thin Film Processes." Academic Press, New York, 1974.
11. P. D. Parry and A. F. Rodde, *Solid State Technol.* **22,** 125 (1979).
12. J. W. Coburn and E. Kay, *Solid State Technol.* **22,** 117 (1979).
13. P. S. Burgraaf, *Solid State Technol.* **22,** 49 (1979).
14. W. M. Bullis, *Solid-State Electron.* **9,** 173 (1966).
15. M. D. Miller, *IEEE Trans. Electron Devices* **ED-23,** 1279 (1976).
16. B. J. Baliga and E. Sun, *IEDM Tech. Digest*, p. 495 (1976).
17. K. P. Lisiak and A. G. Milnes, *J. Appl. Phys.* **46,** 5229 (1975).
18. P. Rai-Choudry, J. Bartko, and J. E. Johnson, *IEEE Trans. Electron Devices* **ED-23,** 814 (1976).

19. B. R. Singh and P. Balk, *J. Electrochem. Soc.* **125,** 453 (1978) and *J. Electrochem. Soc.* **126,** 1288 (1980).
20. C. M. Osborn, *J. Electrochem. Soc.* **121,** 809 (1974).
21. Y. J. Van der Meulen, C. M. Osburn, and A. Ziegler, *J. Electrochem. Soc.* **122,** 284 (1975).
22. K. Miwa *et al.*, *Proc. Denki Kagaka* **40,** 1478 (1972).
23. T. Matsushita, T. Aoki, T. Antsu, H. Yamoto, H. Hayashi, M. Okayama, and Y. Kawana, *IEEE Trans. Electron Devices* **ED-23,** 826 (1976).
24. R. Denning and J. White, RCA Technical Communication, Princeton, New Jersey, p. 66 (1979).
25. M. L. Tarng and J. I. Pankove, *IEEE Trans. Electron Devices* **ED-26,** 1728 (1979).

PART II

Bipolar Power Transistors

CHAPTER 6

POWER TRANSISTOR STRUCTURES. BIPOLAR TRANSISTOR MODELS

Bipolar power transistors found applications as low and high frequency amplifiers and oscillators and as power switching devices. One can find them in a wide variety of circuits such as sound amplifiers, car ignition systems, TV horizontal deflection circuits, and power supplies.

The main consideration in the power transistor design is its voltage, current, and power handling capability. Power transistors must be capable of high power dissipation without causing any device degradation. Because they carry large collector currents and support high collector voltages, their structure differs from the simple, small signal n–p–n or p–n–p transistors.

The power transistor electrical models are, at this time, entirely based on the models developed for the low power devices and are applicable, in all accuracy, to one-dimensional transistors. An exact power transistor model that takes into account effects occurring in all three dimensions is still to be developed.

6.1 Power Transistor Structures

The simplest n–p–n transistor structure (Fig. 1) consists of the n^+-emitter and n^+-collector, both diffused simultaneously into the uniformly doped

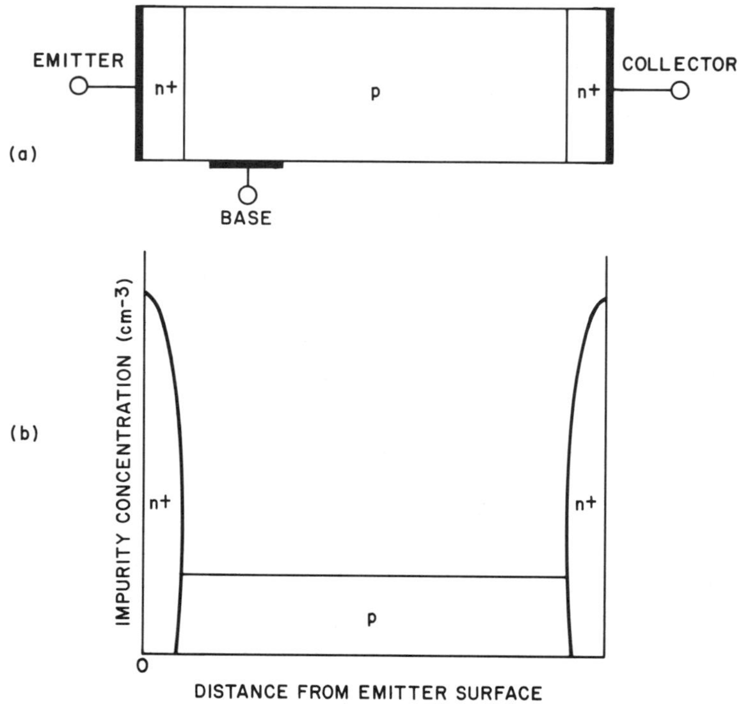

Fig. 1. (a) n–p–n power transistor structure. (b) Single diffused power transistor doping profile.

p-type silicon, hence the name "single diffused transistor." The p-base is lightly doped in order to support the requisite high breakdown voltage of the collector–base junction. The basic disadvantage of this structure is the resulting low current gain because the base width must be made large to avoid punch-through and to support the required collector breakdown voltage.

The device of Fig. 2a represents a considerable improvement over that of Fig. 1a. The p-base is diffused into the n^--collector region with a profile shown by Fig. 2b. The collector depletion region extends for this case mainly into the collector n^--region and only slightly into the more heavily doped p-base. The n^--region, often called the ν-region, is just wide enough to support the breakdown voltage, and the punching through the p-base is entirely avoided because of its diffusion profile. This type of transistor may be fabricated, for instance, by a double diffusion of the p-base, n^+-emitter, and n^+-collector into a high resistivity n^--silicon substrate. A third type of power transistor structure, which contains an n-type collector region in addition to

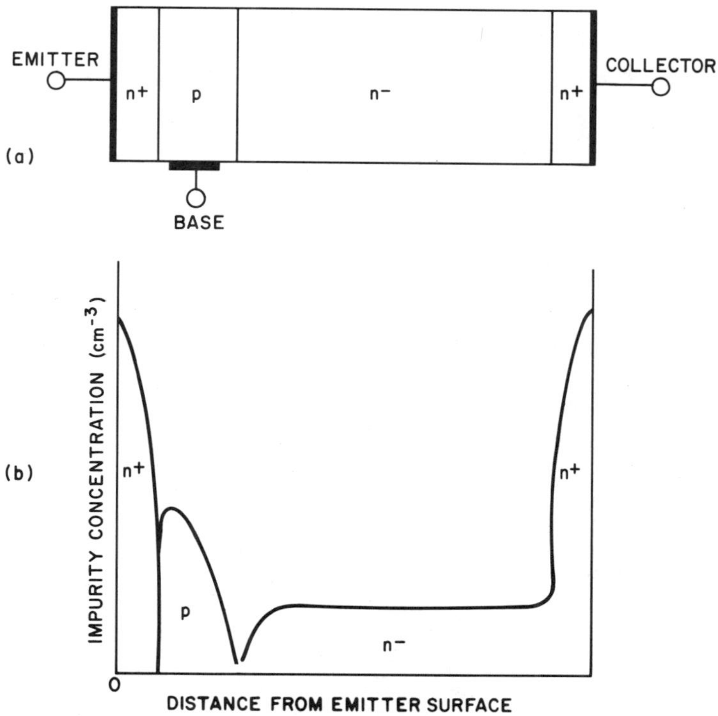

Fig. 2. (a) n^+–p–n–n^+ power transistor structure. (b) Power transistor doping profile with high resistivity n^+-collector region.

the high resistivity n^--region, is shown by Fig. 3. The n-layer average doping concentration is several times higher than that of the n^--region. As will be shown in Chapter 10, the purpose of this n-type buffer zone is to improve power transistor reverse second-breakdown capability. In many cases the n^-- or ν-region is obtained by epitaxial silicon growth on an n^+-substrate.

A transistor with a double epitaxial structure known as n^+–p–π–ν–n^+ device [1, 2] permits the collector depletion regions to spread in the ν-region and also into the very lightly doped π-region of the base (Fig. 4 [1]). For a highly nonsymmetrical one-sided step junction with the p-region doping N_A much larger than the doping concentration N_D of the lightly doped side, if the built-in potential is neglected, the maximum field is

$$E_1(\max) = 2qV_{Ap}N_D/\epsilon_s)^{1/2} \tag{1}$$

when the applied voltage is V_{Ap}. On the other hand, for a symmetrical π–ν junction where $N_D = N_A$, the maximum field is

$$E_2(\max) = E_1(\max)/2^{1/2} \tag{2}$$

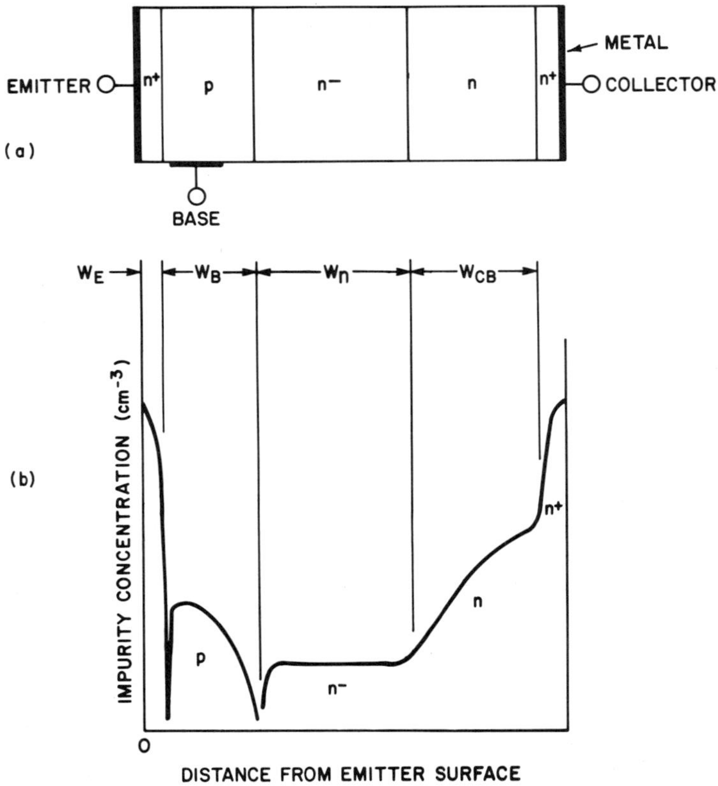

Fig. 3. (a) n^+–p–n^-–n–n^+ power transistor structure. (b) Power transistor doping profile with n-type collector buffer zone.

if the depletion layer can spread freely on both sides of the metallurgical π–v junction. It is possible then to double the applied voltage V_{Ap} for the symmetrical case without changing the maximum field. The π–v structure therefore improves the collector breakdown capability. The π–v structure may also be helpful in improving transistor gain, as discussed in Chapter 7 on current gain. The basic processing steps for the n^+–p–π–v–n^+ structure are shown in Fig. 5.

Power transistors may be integrated with other high power and small signal-type devices. The basic limitation here is the cost effectiveness. In one of the RCA Corporation's exploratory development-type devices an n–p–n power transistor with 10 A and 80 V capability was integrated with a p–n–p power transistor, power Zener diodes, and several other semiconductor components [3]. The power transistors took up about $\frac{2}{3}$ of the chip area, which was on the order of $0.4 \times 0.4\ \text{cm}^2$. The p–n–p power transistor used

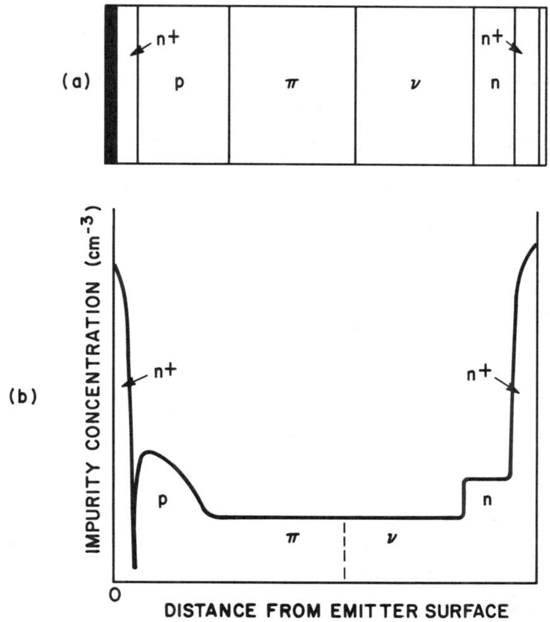

Fig. 4. n^+–p–π–ν–n–n^+ power transistor structure. (b) Power transistor doping profile. (From Blicher [1].)

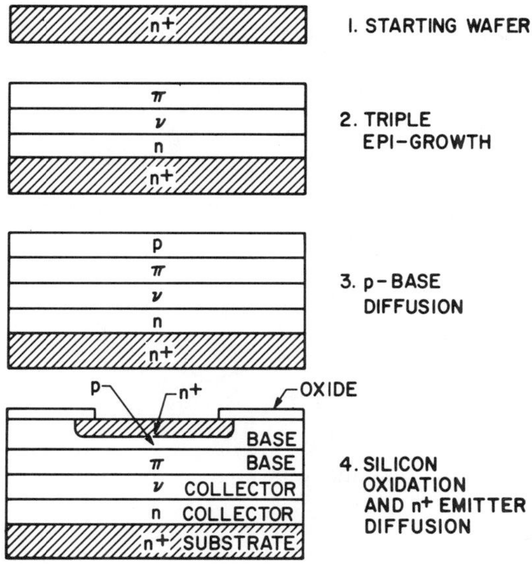

Fig. 5. Basic steps in the fabrication of the n^+–p–π–ν–n–n^+ epitaxial transistor.

for this design was of the so-called lateral type often used in the small signal integrated circuits for compatibility with the *n–p–n* device processing. Junction isolation was used between the various integrated components. The integrated device could be operated, for example, as a high quality audio amplifier or a switching low-loss dc to dc converter.

6.2 Bipolar Transistor Models

6.2.1 Low-Level Injection—Moll–Ross Transistor Model[1]

Consider a bipolar transistor with nonuniform arbitrary base impurity distribution (Fig. 6) and with heavily doped emitter and collector regions. The shallow, diffused, highly concentrated n^+-emitter and -collector regions may be approximated by abrupt junctions. In a transistor with reasonable current gain, the base width W_B is always considerably smaller than the minority carrier diffusion length. It is, therefore, assumed that the carrier recombination in the base can be neglected when computing the collector current. This assumption is, of course, not acceptable for the base current computation since the very existence of the base current is, among other things, the result of the minority carrier recombination. It is further assumed that the carrier mobilities in the base are constant, independent of the base concentration, and therefore of position x. The injection level into the base is assumed to be low, i.e., the injected minority carrier density is smaller than the majority carrier concentration in the base. Due to the low-level injection, there is no appreciable conductivity modulation in the base region.

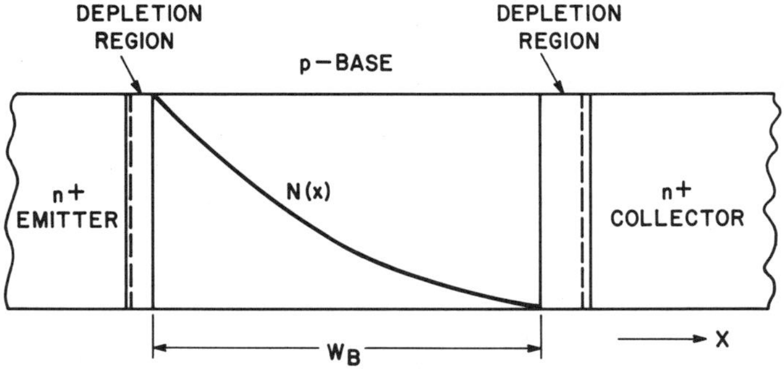

Fig. 6. *n–p–n* transistor with arbitrary base doping $N(x)$.

[1] See [4].

The base electron current density J_n and the hole current J_p are given by the transport equations

$$J_n = q\mu_n nE + qD_n\, dn/dx, \tag{3a}$$

$$J_p = q\mu_p pE - qD_p\, dp/dx, \tag{3b}$$

where μ_n and μ_p, D_n and D_p are the electron and hole mobilities and diffusivites, respectively, and n and p are the electron and hole densities in the base region, respectively. Since the n^+-emitter and -collector regions are heavily doped, there is no appreciable hole flow in the base; consequently, we can assume that $J_p = 0$. This allows us to find the value of the electric field E from Eq. (3b) as

$$E = \frac{kT}{q}\frac{1}{p}\frac{dp}{dx}. \tag{4}$$

By multiplying both sides of (3a) by p and substituting (4) for E in (3a), we obtain

$$J_n = \frac{qD_n}{p}\frac{d(pn)}{dx}. \tag{5}$$

If the cross sectional transistor area A is uniform, we can write for the current

$$I_C = -I_n = -AJ_n \tag{6}$$

since we have assumed that recombination is zero and that the current I_n is independent of position. By separating the variables and integrating (5), we have

$$I_C \int_{x=0}^{x=W_B} \frac{p\,dx}{qD_n} = -A \int_{x=0}^{x=W_B} \frac{d(pn)}{dx}\, dx. \tag{7}$$

The limits of integrations are the emitter and collector edges $x = 0$ and $x = W_B$ (Fig. 6). The integration of the right-hand side of (6) yields

$$I_C \int_0^{W_B} \frac{p\,dx}{qD_n} = -A[p(W_B)n(W_B) - p(0)n(0)], \tag{8}$$

where

$$p(W_B)n(W_B) = n_i^2 \exp(qV_{BC}/kT) \tag{9}$$

and

$$p(0)n(0) = n_i^2 \exp(qV_{BE}/kT). \tag{10}$$

n_i is the intrinsic carrier concentration, V_{BC} and V_{BE} are the collector and

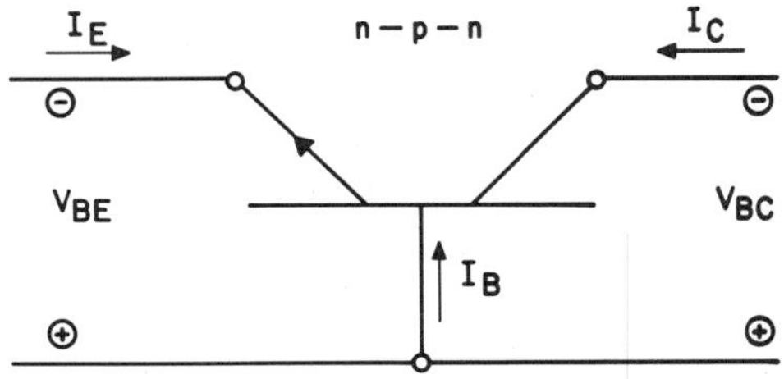

Fig. 7. *n–p –n* transistor current and voltage polarity convention.

emitter voltages, respectively (Fig. 7). For the reverse biased collector, $n(W_B) = 0$ so $p(W_B)n(W_B) = 0$. With this boundary condition Eq. (8) becomes

$$I_C = AD_n q n_i^2 \exp\left(\frac{qV_{BE}}{kT}\right) \Big/ \int_0^{W_B} p\,dx. \tag{11}$$

This is the collector current for the transistor in the active mode of operation (Fig. 8). When the transistor is in the saturated mode, not only the emitter but also the collector is injecting minority carriers into the base. The boundary condition at the collector is therefore $p(W_B)n(W_B) \neq 0$, and the collector current becomes

$$I_C = -AD_n q n_i^2 \left[\exp\left(\frac{qV_{BC}}{kT}\right) - \exp\left(\frac{qV_{BE}}{kT}\right)\right] \Big/ \int_0^{W_B} p\,dx \tag{12}$$

or

$$I_C = -I_S[\exp(qV_{BC}/kT) - \exp(qV_B/kT)] \tag{13}$$

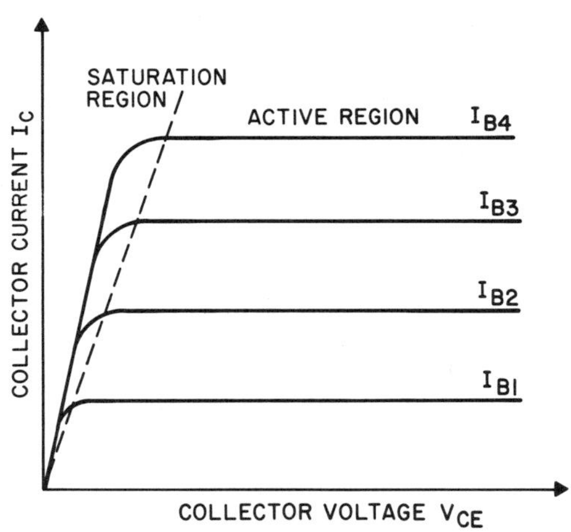

Fig. 8. Collector current as a function of collector voltage characteristic with the base current I_B as a parameter ($I_{B1} < I_{B2} < I_{B3} < I_{B4}$).

with

$$I_S = AD_n q n_i^2 \Bigg/ \int_0^{W_B} p\,dx. \tag{14}$$

Since we have assumed that all the impurities in the base are activated, the hole concentration $p = N_A$. The integral

$$\int_0^{W_B} N_A(dx) = \frac{Q_{BO}}{qA} \tag{15}$$

represents the total number of activated base impurities per unit area. Q_{BO} is the total majority carrier charge in the base [5]. Q_{BO}/qAD_n is known in the literature as the Gummel number. In the derivation of (12) we assumed concentration-independent diffusivity, but Eq. (12) is still valid for the diffusion constant varying with concentration profile if D_n is replaced by the properly averaged value $\bar{D}_n$.

6.2.2 High-Level Injection—Gummel–Poon Transistor Model

The collector current expression (12) was extended by [6] and [7] to the more general case, i.e., when the injected current density is larger than the base majority carrier concentration. In the Gummel–Poon model the collector current, derived without the use of low injection approximation, is given by

$$I_C = -I_S Q_{BO}[\exp(qV_{BC}/kT) - \exp(qV_{BE}/kT)]/Q_B - I_{BC} \tag{16a}$$

and base current by

$$I_B = I_{BE} + I_{BC}. \tag{16b}$$

Q_B is the "base charge" of *all majority carriers* that communicate with the base terminal (holes for the *n–p–n* transistors); Q_{BO} is the zero bias value of the base charge and, from (15), is equal to

$$Q_{BO} = qA \int_0^{W_B} N_A\,dx, \tag{17}$$

and

$$I_S = q^2 A^2 n_i^2 D_n / Q_{BO}. \tag{18}$$

I_{BE} and I_{BC} are the base current components arising from carrier recombination. The dominant component of the integral charge control relations (16a) takes into account only the transport equation of the carriers passing from the emitter to the collector. The base recombination is being taken care of by the base current components. Relation (16a) is a good approximation for *one-dimensional transistors* operating with a large signal and can be adapted for modeling of the high power devices exhibiting base widening (Section 6.2.4).

6.2.3 TRANSISTOR CURRENT GAIN

The collector current I_C as a function of the emitter current I_E is expressed by

$$I_C = \alpha I_E + I_{CBO}, \tag{19}$$

where α is the dc common base current gain and I_{CBO} (or I_{CO}) is the collector leakage current.

The dc current gain in the common base configuration, assuming that the leakage current is negligibly small, is given by

$$\alpha = I_C/I_E \tag{20}$$

and also by

$$\alpha = \gamma\alpha_T M, \tag{21}$$

where α_T is the base transport factor, γ the emitter efficiency, and M the collector avalanche multiplication factor. Far enough from the collector breakdown region,

$$\alpha = \gamma\alpha_T. \tag{22}$$

For the common emitter configuration, the dc current gain is given by

$$h_{FE} = I_C/I_B, \tag{23}$$

where I_B is the transistor base current. Since $I_E = I_C + I_B$,

$$h_{FE} = \alpha/(1 - \alpha). \tag{24}$$

The small signal current gains in both configurations can be obtained by differentiating equation (19) with respect to I_E with the result

$$\alpha_s = \alpha + I_E\, d\alpha/dI_E. \tag{25}$$

The small signal alpha α_s is the sum of the dc alpha and the product of the emitter current and the dc alpha slope. The corresponding small signal current gain h_{fe} in the common emitter configuration becomes

$$h_{fe} = \frac{h_{FE}}{1 - I_B(dh_{FE}/dI_C)}. \tag{26}$$

In first-order theory the dc current gains are emitter current independent. More accurate theory has to take into account few effects which make the gain dependent on the emitter current such as the base conductivity modulation (Webster's effect) and base widening (Kirk's effect). Moreover, the current gain is a function of the collector voltage through Early's effect.

When a transistor operates as a switch and is in the "on" state, i.e., in the saturation mode, the collector becomes forward biased. This reversed transistor is characterized by a reverse common base current gain α_R and reverse common emitter current gain h_{RE}.

6.2.3.1 *Base Transport Factor*

The base transport factor of Eq. (21) is always less than unity because of the minority carrier recombination so that not all emitter injected carriers reach the collector, and it is given by

$$\alpha_T = 1 - (I_r/I_E), \tag{27}$$

where I_r is the recombination current. I_r can be found from

$$I_r = \frac{qA}{\tau}\int_0^{W_B} n'(x)\,dx, \tag{28}$$

where τ is the minority carrier lifetime in the base, n' is the excess minority carrier density in the base, A is the cross sectional area of the transistor, and W_B is the electric base width. The transport factor of uniformly doped base is given by

$$\alpha_T = [\cosh(W_B/L_n)]^{-1}, \tag{29}$$

where L_n is the electron diffusion length in the p-base (L_p would be used in a p–n–p transistor). For base widths much less than the diffusion length, the current gain may be approximated by

$$\alpha_T = 1 - \tfrac{1}{2}(W_B/L_n)^2. \tag{30}$$

A transistor with a base width $W_B = 20$ μm and electron diffusion constant at high current density $D_n = 20$ has a transport factor of 0.98 if $\tau = 5$ μs.

6.2.3.2 *Emitter Injection Efficiency*

The emitter efficiency γ of an n–p–n transistor is defined by

$$\gamma = [1 + |J_p(0)/J_n(0)|]^{-1}, \tag{31}$$

where $J_n(0)$ and $J_p(0)$ are the electron and hole current densities at the emitter–base edge, injected into the base and the emitter, respectively. The emitter efficiency of a transistor in the common emitter configuration is defined, consequently, by

$$\gamma_E = |J_n(0)/J_p(0)| \tag{32}$$

and for a good transistor should be much greater than unity. For the base–emitter voltage $V_{BE} \gg kT/q$, the injected electron current is, from (12),

$$J_n(0) = A^2 q\bar{D}_n n_i^2 \exp(qV_{BE}/kT)/Q_{BO} \tag{33}$$

with

$$Q_{BO} = \int_0^{W_B} qAN_A \, dx$$

and with the assumption that $N_A = p$, i.e., all base acceptors are active. Analogously, the hole current injected from the base into the emitter is given by

$$J_p(0) = -A^2 q\bar{D}_p n_i^2 \exp(qV_{BE}/kT)/Q_{EO} \tag{34}$$

with

$$Q_{EO} = \int_0^{W_E} qAN_D(x) \, dx, \tag{35}$$

where W_E is the effective emitter width equal to or smaller than one hole diffusion length L_p and N_D is the effective emitter donor concentration. Using (33) and (34), the emitter injection efficiency (32) becomes

$$\gamma_E = (\bar{D}_n/\bar{D}_p)(Q_{EO}/Q_{BO}). \tag{36}$$

This expression is valid essentially for lightly doped emitters and low injection levels. A more general expression will be derived in Chapter 7.

6.2.4 Base Width Modulation Effects on the Collector Current

The total majority carrier charge in the base of the n–p–n transistor consists of several components and can be represented by

$$Q_B = Q_{BO} + Q_{TE} + Q_{TC} + Q_F + Q_R. \tag{37}$$

Q_{BO} is the built-in base majority charge, and Q_{TE} and Q_{TC} are the charges on the emitter and collector junction capacitances, respectively. The transition capacitances giving rise to Q_{TE} and Q_{TC} are low current level, incremental, nonlinear capacitances dependent on the applied voltage. The excess majority base charge due to the transistor in the normal operation is given by

$$Q_F = \tau_F I_F, \tag{38}$$

whereas that in the inverted transistor is given by

$$Q_R = \tau_R I_R. \tag{39}$$

τ_F and τ_R are the base transit times for the normal and the inverted transistor, respectively. The forward and reverse currents are given by

$$I_F = I_S(Q_{BO}/Q_B)[\exp(qV_{BE}/kT) - 1] \tag{40}$$

and

$$I_R = I_S(Q_{BO}/Q_B)[\exp(qV_{BC}/kT) - 1]. \tag{41}$$

Under high-level operation a base push-out (widening) (Section 7.2) can be taken care of by noting that the base transit time will be increased by a factor B so that the excess charge due to the forward injection becomes [7]

$$Q_F = B\tau_F I_F. \tag{42}$$

Factor B is a complicated function of several variables and can be found in [7] and [8].

Another type of base width modulation is known as the Early effect: the base width decreases as the collector depletion region width increases with the reverse collector potential. The base width is also affected by the emitter depletion width variation with the applied voltage. It is logical, therefore, to introduce the reverse Early effect [8]. The normal Early effect [9] results in the collector current increase with the reverse collector bias as the emitter injected current is kept constant. This results in the increase of the forward current gain. Increased also is the collector incremental conductance, which in the *active region of an ideal transistor*, is equal to zero. Due to the "normal" Early effect, the collector incremental conductance becomes finite. Due to the finite slope g_c of the transistor active region characteristics, it is possible to extend the I–V lines as shown in Fig. 9 so that they intersect the negative voltage axis at approximately one point, which determines the so-called Early voltage V_A [10]. From the geometry of Fig. 9 we obtain

$$g_c = I_C(0)/|V_A| \tag{43}$$

It was shown by [10] and [8] that

$$|V_A| \simeq Q_{BO}/\bar{C}_{TC}, \tag{44}$$

where $\bar{C}_{TC}$ is the averaged value of the collector junction capacitance. A

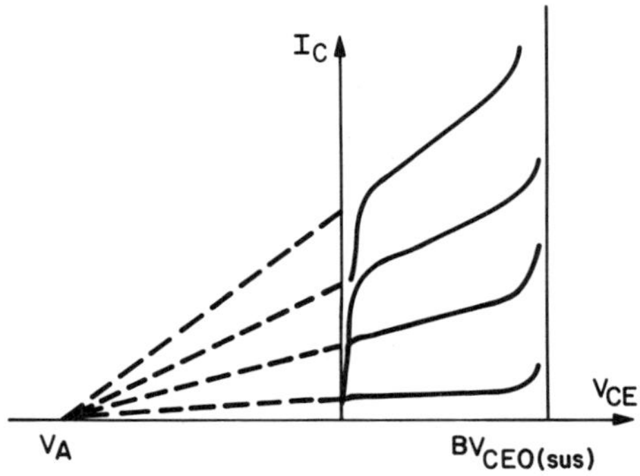

Fig. 9. Early voltage V_A. $BV_{CEO(sus)}$ denotes the collector breakdown in the common-emitter connection.

similar expression is valid for the emitter with the reverse Early voltage V_B, *namely*,

$$|V_B| \simeq Q_{BO}/\bar{C}_{TE}, \tag{45}$$

where $\bar{C}_{TE}$ is the averaged emitter capacitance. The charge on the collector capitance can be expressed, therefore, as

$$Q_{TC} = V_{BC}\bar{C}_{TC} = V_{BC}Q_{BO}/|V_A| \tag{46}$$

and the charge on the emitter capacitance $\bar{C}_{TE}$ as

$$Q_{TE} = V_{BE}\bar{C}_{TE} = V_{BE}Q_{BO}/V_B. \tag{47}$$

These relationships are useful for computer device modeling and calculations and are valid for uniform cross sectional area devices.

At low injection levels, the collector current–emitter voltage relationship follows the equation

$$I_C \propto \exp(qV_{BE}/n_1kT), \tag{48}$$

where the $n_1 \simeq 1.0$ [11]. The base current I_B, highly affected by the space-charge generation, is given by

$$I_B \propto \exp(qV_{BE}/n_2kT). \tag{49}$$

For a one-dimensional transistor, $n_2 = 2$; for the two-dimensional double diffused transistor at high injection levels, $n_2 = 1.72$ [11].

When current density becomes very high, i.e., at $(qV_{BE}/kT) \gg 1$, and when the collector is highly reverse biased, i.e., $|V_{BC}| \gg kT/q$, the base charge, Eq. (37), reduces to the following expression for the negligible depletion region charges:

$$Q_B = (Q_{BO}B\tau_F I_S)^{1/2} \exp(qV_{BE}/kT), \tag{50}$$

and the collector current becomes

$$I_C \simeq -I_F \simeq (I_S Q_{B0}/B\tau_F)^{1/2} \exp(qV_{BE}/2kT). \tag{51}$$

At very high injection levels, the collector current, Eq. (51), starts growing less rapidly with the increasing emitter voltage V_{BE} due to the base conductivity modulation as predicted by [12].

6.2.5 Large Signal Ebers and Moll Transistor Model[2]

In this one-dimensional transistor model the attention is focused on the recombination in the base since at the time when it was developed [13], the base widths were large due to the technological difficulties and it was necessary to introduce into the model the forward and reverse current gains α_F and α_R. The modern, high voltage power transistors still have wide effective bases, but the use of the Ebers–Moll equations is warranted only after such effects as base widening and current gain fall-off with increasing current density are taken into account. The Ebers–Moll terminal current relations depend on the superposition of two current components; one coming from the emitter junction and the second from the collector. Either of these junctions may be forward or reverse biased depending on the mode of operation. In the active mode only the emitter will be forward biased, but in the switching saturation mode the collector will also be forward biased. In the "off" switching condition both junctions will be reverse biased. For the *n–p–n* transistor the Ebers–Moll current voltage relations are

$$I_E = -I_{ES}[\exp(qV_{BE}/kT) - 1] + \alpha_R I_{CS}[\exp(qV_{BC}/kT) - 1], \tag{52}$$

$$I_C = \alpha_F I_{ES}[\exp(qV_{BE}/kT) - 1] - I_{CS}[\exp(qV_{BC}/kT) - 1], \tag{53}$$

$$I_B = -I_E - I_C, \tag{54}$$

where

$$I_{ES} = \alpha_F I_{E0}/(1 - \alpha_F\alpha_R) \tag{55}$$

and

$$I_{CS} = \alpha_R I_{C0}/(1 - \alpha_F\alpha_R). \tag{56}$$

α_F and α_R are the common-base current gains of the normal and the inverted transistor, and I_{C0} and I_{E0} are the saturation currents of the normal and the inverted transistors (collector forward biased and emitter reverse biased).

[2] See [13].

The Ebers–Moll set of equations utilizes transistor property known as reciprocity relationship, i.e.,

$$\alpha_F I_{EO} = \alpha_R I_{CO} \tag{57a}$$

or

$$\alpha_F I_{ES} = \alpha_R I_{ES}. \tag{57b}$$

This relationship is valid for an arbitrarily doped base, but has been proven only for the low-level injection case. It is possible to show that the component of the emitter and collector currents of the Ebers–Moll expressions is the same as the one in the Moll–Ross expression (12). The Ebers–Moll current–voltage relationships were derived assuming that the forward and reverse current gains are current independent and the injection levels are comparatively low so that classical relationship $I \propto \exp(qV/kT)$ for the emitter and collector currents as a function of the applied voltages is valid.

6.2.5.1 Ebers–Moll Equivalent Circuit

The Ebers–Moll equations can be presented in a concise form by

$$I_E = -I_F + \alpha_R I_R, \tag{58}$$

$$I_C = \alpha_F I_F - I_R, \tag{59}$$

where the forward and reverse currents are given by

$$I_F = I_{ES}[\exp(qV_{BE}/kT) - 1], \tag{60}$$

$$I_R = I_{CS}[\exp(qV_{BC}/kT) - 1]. \tag{61}$$

Equations (58) and (59) suggest immediately the equivalent circuit of Fig. 10 in which the diodes represent the emitter–base and base–collector junctions, and I_F and I_R are defined by expressions (60) and (61). I_F represents the total current flowing across the base–emitter junction while I_R is the total current flowing across the base–collector junction.

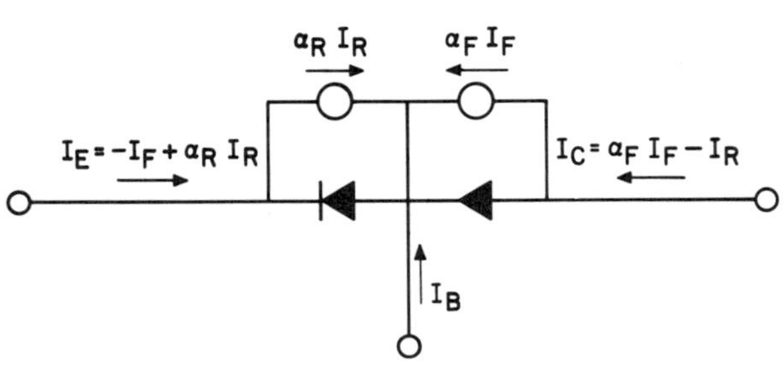

Fig. 10. Ebers–Moll equivalent current.

6.2.5.2 *Charge-Control Transistor Model*[3]

The base current continuity equation in the *p*-base is

$$\frac{dJ_B(t)}{dx} = q\frac{\partial n'(x)}{\partial t} + \frac{n'(x)}{\tau_{eff}}, \tag{62}$$

where $J_B(t)$ is the base current density, τ_{eff} is the effective base region lifetime, and $n'(x, t)$ is the excess electron density in the base. Integrating Eq. (62) between the limits of $x = 0$ and $x = W_B$, we obtain

$$I_B(t) = \frac{dq_B}{dt} + \frac{q_B}{\tau_{eff}}, \tag{63}$$

where q_B is the total base excess charge. Equation (63) holds for an arbitrarily doped base, and its advantage over (62) is that the position dependent excess electron density $n'(x)$ is replaced by the total excess charge density q_B. Moreover, the nonlinear differential equation is replaced by an approximate linear equation.

Some charges are also stored in the emitter region and in the emitter and collector transition capacitances that communicate with the base. Taking all these charges into account, Eq. (63) is modified to

$$I_B(t) = \frac{d(q_B + q_E)}{dt} + \frac{d(q_{TE} + q_{TC})}{dt} + \frac{q_B + q_E}{\tau_{eff}}, \tag{64}$$

where q_{TE} and q_{TC} are the electric charges of the emitter and collector transition regions, respectively (Fig. 11).

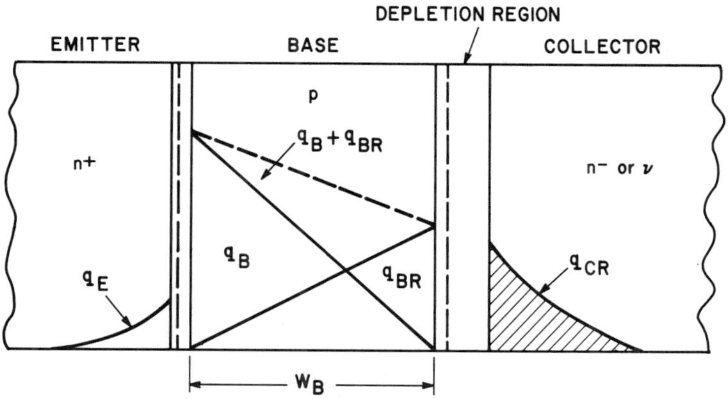

Fig. 11. Charge distribution of an n^+–p–ν–n^+ transistor in saturation.

[3] See [14–16].

The charge control model assumes that not only the base current but also the emitter and collector currents are linearly related to the base charge and other charges communicating with the base. The collector current density is related to the base charge by

$$I_C(t) = \frac{(q_B + q_E)}{\tau_F} - \frac{dq_{TC}}{dt}. \tag{65}$$

The proportionality constant τ_F is equal to the minority carrier's base transit time. The emitter current is obtained by adding (64) and (65) as

$$I_E = -\frac{q_B + q_E}{\tau_E} - \frac{d(q_B + q_E)}{dt} - \frac{dq_{TE}}{dt} \tag{66a}$$

with

$$\tau_E = \left(\frac{1}{\tau_F} + \frac{1}{\tau_{eff}}\right)^{-1}. \tag{66b}$$

An analogous set of equations can be obtained for the reverse active mode of the transistor.

The total control charge for a transistor operating in the saturation mode is the sum of the charges due to the *forward* and *reverse* modes of operation. In the simpler case, when the transition capacitance charges may be neglected,

$$I_B(t) = \frac{d(q_B + q_E)}{dt} + \frac{(q_B + q_E)}{\tau_{eff}} + \frac{d(q_{BR} + q_{CR})}{dt} + \frac{(q_{BR} + q_{CR})}{\tau_{Reff}}, \tag{67}$$

where q_{BR} and q_{CR} are the base and collector additional charges for the forward biased collector and τ_{Reff} is the effective lifetime in the base for the collector injected electrons. The location of the various charges of Eq. (67) for a HV transistor with a lightly doped collector and uniformly doped base is shown in Fig. 11.

The charge-control equations are particularly useful in the determination of the dynamic properties of lower power devices where the assumption of τ_F and τ_{eff} constancy is valid and where the two-dimensional effects may be disregarded.

6.2.5.3 Incremental Transistor Model

Power transistors are rarely operated at small signal levels; nevertheless, their characterization includes the incremental parameters that lead to linear equivalent circuit representation.

In the active mode the collector current is given as

$$I_C = I_S \exp(qV_{BE}/nkT) \tag{68}$$

with n usually close to unity for moderate current densities. The transistor transconductance is equal to

$$g_m = \left.\frac{\partial I_C}{\partial V_{BE}}\right|_{V_{BC} = \text{const}} \tag{69}$$

From (69) and (68), we obtain with $n = 1$

$$g_m = qI_C/kT. \tag{70}$$

Another small signal parameter is defined by

$$g'_{be} = \partial I_B/\partial V_{BE} \tag{71a}$$

or

$$g'_{be} = g_m/h_{FE} \tag{71b}$$

since $I_B = I_C/h_{FE}$. The diffusion capacitance is defined by

$$C_D = dq_B/dV_{BE}. \tag{72}$$

q_B, the total base excess charge, is given by

$$q_B = I_C\tau_F; \tag{73}$$

therefore,

$$C_D = g_m\tau_F. \tag{74}$$

The small signal equivalent circuit for a transistor in the active mode of operation and with the base spreading resistance r'_b is depicted by Fig. 12. This circuit is due to Giacoletto [17] and is known as the hybrid-π equivalent circuit.

The gain–bandwidth product f_T, i.e., the frequency at which h_{FE} becomes unity at low currents, is given by

$$f_T = g_m/2\pi(C_D + C_{TE} + C_{TC}). \tag{75}$$

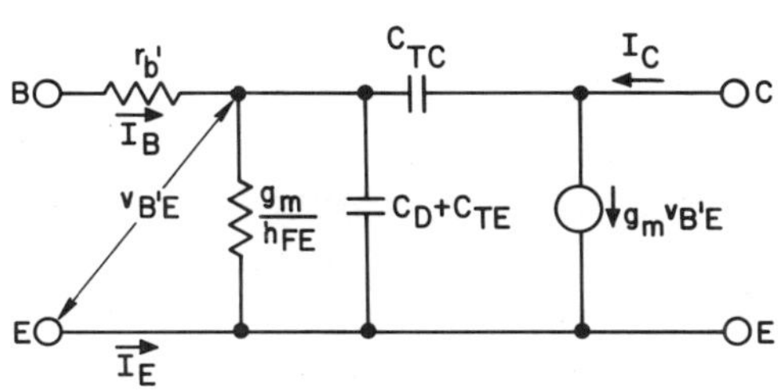

Fig. 12. Hybrid-π incremental equivalent circuit of a transistor. v'_{BE} is the small signal base-emitted voltage appearing at the intrinsic transistor input.

C_{TE} and C_{TC} are the emitter and collector junction capacitances, respectively. For negligible $(C_{TE} + C_{TC})$

$$f_T = 1/2\pi\tau_F, \tag{76}$$

where τ_F is the base transit time.

6.2.5.3.1 BASE TRANSIT TIME. The transit time of the minority carriers from the emitter to the transistor collector is related to the base excess charge q_B by

$$\tau_F = q_B/I_C, \tag{77}$$

where I_C is the collector current density. For low injection levels and for a uniformly doped base,

$$\tau_F = W_B^2/2\bar{D}_n, \tag{78}$$

where $\bar{D}_n$ is the averaged electron diffusion constant in the p-base. A similar relationship holds for holes in an n-type base of a p–n–p transistor. From the Moll–Ross equation for the collector current (11) and the transit time definition (78), we obtain for an arbitrarily doped base

$$\tau_F = q_B \int_0^{W_B} p\,dx \bigg/ q\bar{D}_n n_i^2 \exp\left(\frac{qV_{BE}}{kT}\right) \tag{79}$$

with

$$q_B = qA \int_0^{W_B} n'(x)\,dx \qquad \text{and} \qquad V_{BE} \gg \frac{kT}{q}, \tag{80}$$

where $n'(x)$ is the excess charge density in the base and A is the emitter area.

For a uniformly doped short base the excess charge may be assumed to have a linear distribution so that

$$\int_0^{W_B} p\,dx = \frac{1}{2}[n'(0) + N_A], \tag{81}$$

and Eq. (79) reduces at high injection levels for which $n'(0) \gg N_A$ to

$$\tau_F = W_B^2/4\bar{D}_n. \tag{82}$$

This result indicates that the transit time under very high-level conditions is reduced by 50%. This takes place because of the base conductivity modulation and the appearance of the transport aiding field, resulting in doubling of the diffusivity $\bar{D}_n$, known as the Webster effect [18]. If there is a built-in field in the base due, for example, to the exponential impurity profile, then the transit time τ_F may be considerably reduced under the low injection level conditions.

6.2.6 Limitations of Transistor Models

All of the above described models share one serious drawback especially troublesome in the case of high power transistors: they all *neglect two- and three-dimensional effects* such as emitter crowding, localized base widening, and current modulation of the spreading base resistance. Most importantly, they ignore the presence of the quasi-saturation region in the $I-V$ characteristics of the $n^+-p-\nu-n^+$ high voltage transistors. Nevertheless, they serve a useful purpose for those instances where accuracy is less important than the feel for the direction in which device design should proceed. Second- and third-order models [8] consider many parameters, taking into account the effects of ohmic resistances, temperature changes, nonconstant cross sectional area, etc., and are useful in transistor computer simulation.

The extension of the Moll–Ross and Gummel-Poon models to the transistor quasi-saturation region is treated in Section 8.2.1 on the power transistor $V-I$ characteristics.

REFERENCES

1. A. Blicher and B. R. Czorny, U.S. Patent 3427515 (1969).
2. R. Denning and D. A. Moe, *IEEE Trans. Electron Devices* **ED-17,** 711 (1970).
3. C. F. Wheatley, RCA Solid State Technical Center, Somerville, New Jersey. Private communication.
4. J. L. Moll and I. M. Ross, *Proc. IEEE* **44,** 72 (1956).
5. H. K. Gummel, *Proc. IRE* **49,** 837 (1961).
6. H. K. Gummel and H. C. Poon, *Bell System Tech. J.* **49,** 115 (1970).
7. H. K. Gummel and H. C. Poon, *Bell System Tech. J.* **49,** 827 (1970).
8. I. Getreu, "Modeling the Bipolar Transistor." Tectronix, Beaverton, Oregon, 1976.
9. J. M. Early, *Proc. IRE* **40,** 1401 (1952).
10. F. A. Lindholm and D. J. Hamilton, *Proc. IEEE* **59,** 1377 (1971).
11. O. Manck, H. H. Heimeier, and W. L. Engl, *IEEE Trans. Electron Devices* **ED-21,** 403 (1974).
12. R. N. Hall, *Proc. IRE* **40,** 1512 (1952).
13. J. J. Ebers and J. L. Moll, *Proc. IRE* **42,** 1761 (1954).
14. R. Beaufoy and J. J. Sparks, *Automat. Telephone Electron. J.* **13,** 310 (1957).
15. J. J. Sparks and R. Beaufoy, *Proc. IRE* **45,** 1740 (1957).
16. J. J. Sparks, *Proc. IRE* **45,** 1740 (1957).
17. L. J. Giacoletto, *RCA Rev.* **15,** 506 (1954).
18. W. M. Webster, *Proc. IRE* **42,** 914 (1954).

CHAPTER 7

CURRENT GAIN AT HIGH CARRIER CONCENTRATIONS

Power transistor current gain depends appreciably on the emitter and base doping levels, as well as on the emitter injected current density. It is strongly affected by the emitter geometry because of the current crowding effect and by the transistor vertical structure because of the base widening effects.

7.1 Emitter Efficiency versus Emitter Concentration

Heavy emitter doping results in the reduction of the silicon energy gap, increased effective intrinsic concentration, and lower lifetime (Chapter 2). All of these effects contribute, to some degree, to the lowering of the emitter injection efficiency.

7.1.1 Energy Gap Narrowing and Emitter Efficiency

The minority current density J_p flowing into the n-emitter of an n–p–n transistor is due to the hole diffusion and drift from the base and is given by

$$J_p = q\mu_p pE - qD_p \frac{dp}{dx} \tag{1}$$

with the electric field

$$E = -\frac{kT}{q}\frac{1}{N(x)}\frac{dN(x)}{dx}, \tag{2}$$

where μ_p and D_p are the hole mobility and diffusivity, respectively, and p is the excess hole concentration. $N(x)$ is the position dependent emitter impurity concentration. Combining (1) and (2) and using Einstein's relationship

$$D_p = (kT/q)\mu, \tag{3}$$

we obtain

$$J_p = qD_p\left[-\frac{dp}{dx} + \frac{p}{N(x)}\frac{dN(x)}{dx}\right]. \tag{4}$$

At heavy emitter doping, the effective intrinsic concentration n_{ie} is related to the intrinsic concentration n_i by (Chapter 2)

$$n_{ie}^2 = n_i^2 \exp(\Delta E_g/kT). \tag{5}$$

This change of the intrinsic concentration is due to the band gap narrowing ΔE_g and results in the creation of an additional electric field component in the diffused emitter. This component is given by Mertens *et al.* [1] as

$$\Delta E = \frac{kT}{q}\frac{1}{n_{ie}^2}\frac{dn_{ie}^2}{dx}, \tag{6}$$

and it acts in the opposite direction of the field introduced by the impurity gradient so that the total emitter field is given by

$$E = \frac{kT}{q}\left(-\frac{1}{N}\frac{dN}{dx} + \frac{1}{n_{ie}^2}\frac{dn_{ie}^2}{dx}\right). \tag{7}$$

The current J_p injected into the emitter is therefore changed from (4) to [1]

$$J_p = qD_p\left[-\frac{p}{N(x)}\frac{dN}{dx} + \frac{p}{n_{ie}^2}\frac{dn_{ie}^2}{dx} - \frac{dp}{dx}\right]. \tag{8}$$

Thus the level of the emitter doping strongly influences the electric field distribution throughout the emitter quasi-neutral region and helps the minority carriers to reach the emitter contact.

In transistors with reasonably high current gains ($h_{FE} > 10$), the transport factor α_T can be assumed to be equal to unity, and in the absence of avalanche multiplication,

$$h_{FE} \simeq \gamma_E. \tag{9}$$

The electron current density injected into the base for the base–emitter voltage V_{BE} much greater than kT/q can be obtained from

$$J_n(0) = q\bar{D}_n n_i^2 \exp\left(\frac{qV_{BE}}{kT}\right) \bigg/ \int_0^{W_B} N_A\, dx, \tag{10}$$

where $\bar{D}_n$ is the averaged electron diffusivity in the base, N_A is the base acceptor concentration, and W_B the neutral base width. An analogous equation can be written for the hole current $J_p(0)$ injected by the base into the emitter so that the common emitter current gain can be expressed as

$$h_{FE} = \frac{J_n(0)}{J_p(0)} = \left(\frac{\bar{D}_n}{\bar{D}_p}\right)\left(\int_0^{W_E} N_D(x)\, dx \bigg/ \int_0^{W_B} N_A(x)\, dx\right) \tag{11}$$

with $N_D(x)$ being the emitter concentration and $\bar{D}_p$ the averaged hole diffusivity in the emitter. W_E is the active emitter width. Integrations in (11) are carried over the emitter and base widths, respectively. By integrating Eq. (8) and assuming the validity of (9), it is found that for heavy doping

$$J_p(0) = q\bar{D}_p \exp\left(\frac{qV_{BE}}{kT}\right) \bigg/ \int_0^{W_E} \frac{N_D\, dx}{n_{ie}^2} \tag{12}$$

so that the emitter efficiency or the current gain can be expressed by [1] as

$$\begin{aligned} \gamma_E \simeq h_{FE} &= \frac{\bar{D}_n}{\bar{D}_p}\left(\int_0^{W_E} N_D \left(\frac{n_i}{n_{ie}}\right)^2 dx \bigg/ \int_0^{W_B} N_A\, dx\right) \\ &= \frac{\bar{D}_n}{\bar{D}_p}\left(\int_0^{W_E} N_D \exp\left(\frac{-\Delta E_g}{kT}\right) \bigg/ \int_0^{W_B} N_A\, dx\right) \end{aligned} \tag{13}$$

or

$$h_{FE} = (\bar{D}_n/\bar{D}_p)(Q_E/Q_B) \tag{14}$$

with

$$Q_E = \int_0^{W_E} N_D \left(\frac{n_i}{n_{ie}}\right)^2 dx, \tag{15}$$

which introduces the effective emitter doping profile

$$N'(x) = N_D[n_i/n_{ie}(x)]^2. \tag{16}$$

The effective emitter doping profile is compared in Fig. 1 [1] to the emitter net physical doping profile. We see that the effective emitter concentration becomes a few orders of magnitude smaller than the actual doping level. The results of computation and experimentation [1] show that, in the cases

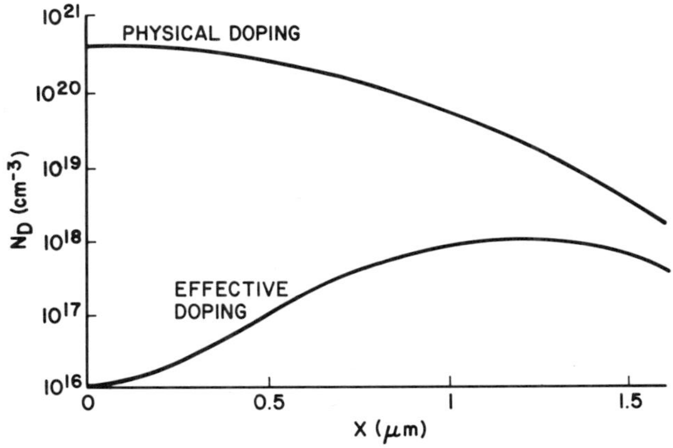

Fig. 1. Physical and effective emitter doping as functions of emitter depth x. (From Mertens *et al.* [1].)

studied, the effect of the band gap narrowing dominated over the effects of the minority carrier lifetime in the emitter.

7.1.2 Auger Recombination Effect on Emitter Efficiency

The emitter injection efficiency is affected by the minority carrier diffusion length in the emitter and, consequently, by the carrier lifetime, which in the low concentration part of the diffused emitter is governed by the Shockley–Read–Hall (SRH)-type recombination. In the heavily doped emitter region, the Auger-type recombination should also be taken into account according to Sheng [2]. Since the Auger recombination lifetime τ_A decreases with the square of the minority carrier concentration, it will start dominating over the SRH recombination lifetime at some crossover points x_0 at which

$$\tau_A = \tau_{SRH}. \tag{17}$$

The emitter region can therefore be divided into SRH and Auger dominated regions as shown in Fig. 2 [2]. For the approximation of the upper limit of the diffusion length L_p, it is possible to neglect the presence of the emitter electric field and write for the hole diffusion in the emitter

$$d^2p/dx^2 = p/D_p\tau_A. \tag{18}$$

The emitter Gaussian impurity distribution follows the expression

$$N(x) = N_O \exp(-x^2/L^2), \tag{19}$$

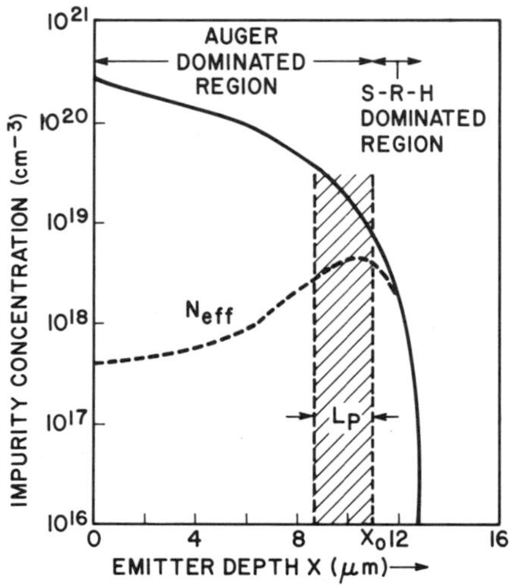

Fig. 2. The solid line represents the emitter physical profile; the broken line represents the effective profile calculated by [2]. The shaded area represents the active portion of the emitter. (From Sheng [2].)

where L is the impurity diffusion length and N_O is the surface concentration. Introducing this relationship into (18) gives

$$d^2p/dx^2 = (p/L_S^2)\exp(-2x^2/L^2), \tag{20}$$

where L_S is the minority carrier diffusion length at impurity concentration equal to that at the surface. Solving (20) to obtain p as a function of distance x from the emitter surface leads to an expression which can be solved to obtain the effective diffusion length L_p:

$$\operatorname{erf}\left(\frac{x_0 - L_p}{L}\right) = \operatorname{erf}\left(\frac{x_0}{L}\right) - \frac{2}{\pi^{1/2}}\frac{L_S}{L}. \tag{21}$$

It follows from (21) that the effective diffusion length L_p varies very slowly with small changes of x_0 (Fig. 2); consequently, the shaded area remains unchanged making it more or less independent of the exact SRH recombination lifetime value. The conclusion drawn by [2] is that as a result of the Auger recombination, only a part of the effective profile of Fig. (2) (shaded area) is active as an emitter. For devices with emitter depths between 10 μm and 14 μm considered by [2] with $x_0 = 10$ μm and $L_p = 2$–3 μm, the Auger recombination could not be neglected. It could be disregarded, however, when $L_p \gg x_0$.

On the other hand, the work of Adler *et al.* [4] shows that the Auger process is dominated by the band gap narrowing and that the Auger recombination alone cannot predict by itself actual measured injection efficiencies (see Section 7.14).

7.1.3 Emitter Efficiency of a Two-Dimensional Emitter

While the analysis of Mertens *et al.* [1] was performed for a one-dimensional transistor, the study of a two-dimensional case was undertaken by Mock [3], whose transistor gain model is represented by Fig. 3. The device base was diffused with a Gaussian profile into a homogeneous background. The emitter was diffused through a mask and had a complementary error function profile in its central part. The physical and effective doping profiles in the emitter and in the base differed strongly and are depicted by Figs. 4 and 5, which show the profiles along the device central part (line AB of Fig. 3) and near the emitter along the semiconductor surface (line CD of Fig. 3). Calculations of [3] show that the relative importance of the sidewall portion of the emitter on the current gain is increased by the heavy doping effects. The electrons injected into the heavily doped base along the emitter sidewall can reach the base contact because the built-in electric field, due to the base acceptor gradient, is overbalanced by the heavy doping effects, i.e., large n_{ie}/n_i ratio. The holes injected into the emitter region at any point of the emitter periphery can readily reach the emitter metallic contact. The SHR carrier recombination is proportional to n_{ie}^2, and leakage current becomes particularily large at the sidewalls where the impurity compensation is the greatest.

The meaning of the three current gain parameters of Fig. 3 is as follows: β_1 is the ratio of the total collector current to the hole current injected into the emitter in the central portion, β_p is the ratio of the collector current to

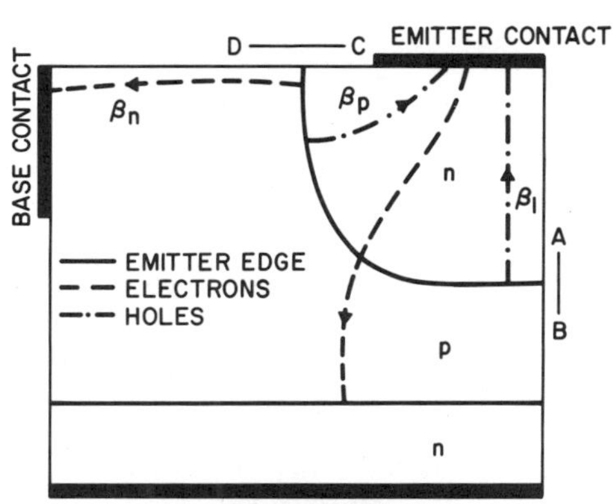

Fig. 3. Two-dimensional *n–p–n* transistor gain model. (From Mock [3].)

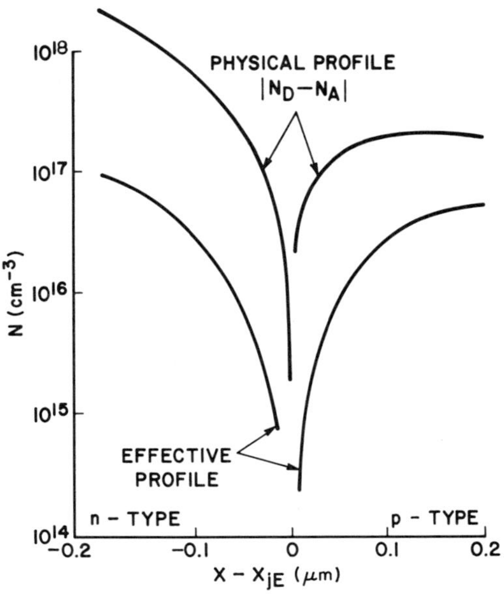

Fig. 4. Physical and effective emitter doping profiles along the *AB* line (see Fig. 3) in the device central part. x_{jE} is the emitter junction depth. (From Mock [3].)

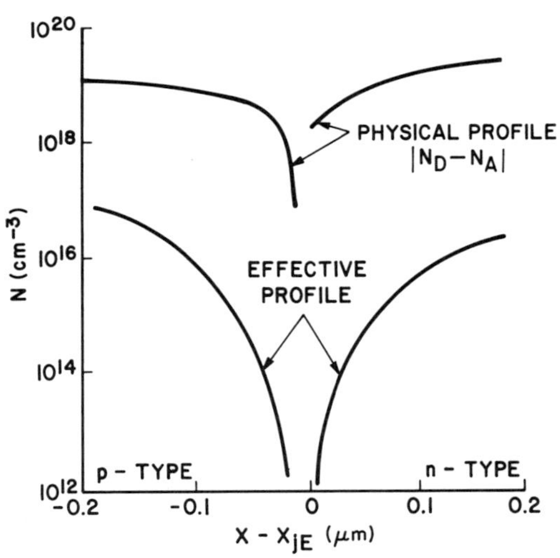

Fig. 5. Physical and effective emitter doping profiles along the *CD* line (see Fig. 3) at the emitter surfaces. x_{jE} is the emitter junction depth. (From Mock [3].)

the hole current injected along the emitter *sidewall*, and β_n is the ratio of the collector current to the electron current flowing to the base contact. The injection efficiency limited current gain h_{FE} is then given by

$$h_{FE}^{-1} = \beta_1^{-1} + \beta_p^{-1} + \beta_n^{-1}. \tag{22}$$

Mock's calculations show a rapid decrease of β_p (sidewall) with increasing injection level. For instance, at 1.2×10^2 A cm^{-2}, $\beta_p = 230$; but at 2.2×10^4 A cm^{-2}, $\beta_p = 20$. β_n, which arises from the electron current flow to the base contact, decreases from 4100 to 450 for the same change in the current density.

7.1.4 Relative Importance of Various Effects on Emitter Efficiency

There is no general agreement, at this time, about the relative importance of each effect described in Sections 7.1.1–7.1.3. In one study, made by Adler *et al.* [4], a one-dimensional emitter model that included the band gap narrowing, current dependent diffusion constants, and Auger and Shockley–Read–Hall recombinations was developed. The theoretical model was compared with measurements made on transistors fabricated with various emitter surface concentrations and depths of 1 μm to 8 μm. The studied devices had doping profiles of the type represented by Fig. 6.3.

An additional *two-dimensional* study of one of the transistors indicated that less than 1% of the emitter current was injected through the emitter sidewalls. However, the base surface concentration was low (about 2×10^{17} cm^{-3}), resulting in a very low base impurity concentration at the emitter periphery. Consequently, there was very little compensation between the base and emitter diffusions, and no appreciable band gap narrowing. As a result, the sidewall injection was very small. The one-dimensional study showed that electron current flow to the base contact decreases for the same current density from 4100 to 450. The heavy doping effects reduce h_{FE} at moderate or high injection levels.

The emitter efficiency is primarily affected by the effective intrinsic concentration n_{ie}. Although the accuracy of the quantitative results are doubtful because of the many approximations and assumptions (the validity of Boltzmann's relationship, of quasi-neutrality, and of quasi-equilibrium conditions), the *qualitative* results of Mock appear to be valid, based on previously reported experimental observations. Mock's calculations of the current gain of the bipolar transistors and its dependence on the injection level show that the most important single mechanism limiting the current gain is the injection of holes into the emitter region along the sidewall portion of the emitter; the effective sink for these holes is the emitter contact, not the recombination. This mechanism causes the current fall-off, particularly at high injection levels. At low injection levels, the recombination in the emitter near the semiconductor surface may be of great importance, even if moderately good lifetimes are assumed.

The Auger recombination had a noticeable effect on the current gain of only the very shallow emitters for which it added up to the band gap narrowing effect. This is in conflict with the results of Sheng [2]. Generally, the dominant mechanisms lowering the emitter injection efficiency, according to Adler *et al.* [4], are the SRH recombination and band gap narrowing. Moreover, for emitter depths of 1 μm or less the band gap narrowing dominates all other effects. For transistors with emitters 4 μm deep or more, the predominant mechanism is the SRH recombination, which is supported by the experimental observation of Adler *et al.* [4] that for these emitter depths the current gain increases with injection level. Among all of the effects included in the model, only the effective SRH lifetime increases with current.

Figure 6 [4] is a plot of the current gain versus the emitter surface concentration, which shows a distinct maximum value for the 1 μm deep emitter at 5×10^{19} cm^{-3}. Deeper emitters do not exhibit any maximum.

It should be pointed out that the results of the Adler *et al.* study depend greatly on the values of physical data used in computation. The use of data originating with sources other than those used in [4] may considerably shift the dividing line between the emitter with the dominant energy gap effect and the dominant recombination effect.

McGrath and Navon [5] studied transistors with a structure similar to

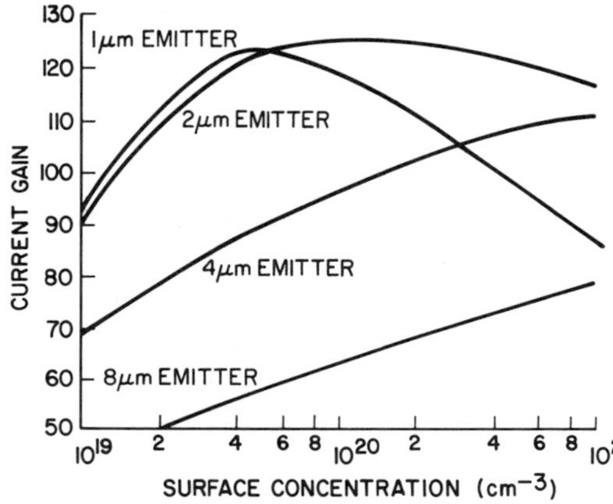

Fig. 6. Current gain as a function of the emitter surface concentration with 1, 2, 4, and 8 μm emitter depths. (From Adler [4].)

that used in the work of Adler *et al.* [4] with the emitter depths of 4, 7.5, and 14 μm. The study uses two-dimensional analysis; it is based on Mock's [3] *band gap narrowing* formulation and represents an extension of the one-dimensional Adler *et al.* model [4]. According to [5], the sidewall emitter region affects the emitter efficiency only when the base regions are highly doped so that there is a heavy diffusion compensation.

At low current levels, SHR recombination dominates over Auger recombination for all emitter depths studied. For moderate and high injection levels, on the other hand, Auger recombination dominates, particularly when the band gap narrowing is taken into account. Because of the band gap narrowing the emitter efficiency decreases, the base current increases, and the emitter current tends to crowd at the edges. Due to the high current density in the crowded condition, the Auger recombination rate is significantly increased. Auger recombination in the Adler *et al.* study [4], on the other hand, was shown to have much less effect on the emitter efficiency.

7.2 Base Widening at High Currents

The phenomenon of the transistor base widening at high current densities was first described by Kirk and is known as the Kirk effect [6]. Base widening is particularily troublesome in high voltage devices of the type, shown by Fig. 7, provided with a n^- (or ν) low concentration region necessary for the support of the requisite high collector voltage. A similar effect exists in the n^+–p–n^+ transistor structure in which the collector breakdown is mainly supported by the collector depletion region extending into the p-base. The increase in the base width profoundly affects device properties such as the emitter efficiency, current transport factor, and transistor voltage drop in the "on" condition.

In the case of the n^+–p–ν–n^+ transistor, the collector voltage considered

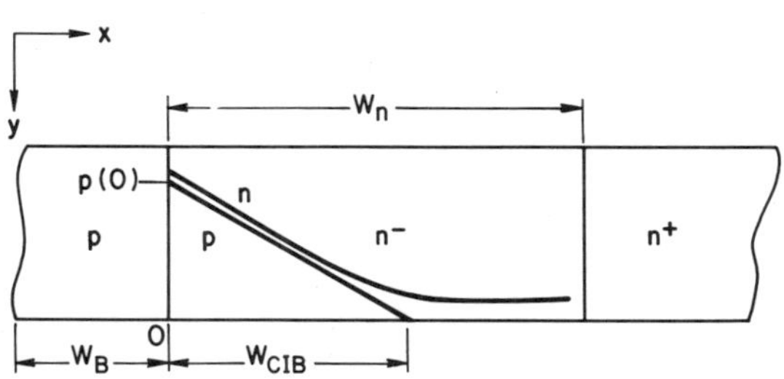

Fig. 7. Base and collector regions of the n^+–p–ν–n^+ transistor. Curves marked p and n represent hole and electron profiles in the n^--region, respectively, for the low field case.

in the base widening analysis may be either high or low, resulting in the high or low electric field in the ν high resistivity collector region. For the high voltage case, almost all potential appears across the collector depletion region, and the voltage drop in the ν region is of no consequence. In the low field case, the voltage drop becomes of the same magnitude as the collector voltage, seriously affecting the phenomenon and its analysis.

7.2.1 Base Widening in n^+–p–ν–n^+ Structure (High Field Case)

At the high current densities commonly encountered in power transistors, the mobile carrier density (electrons for the n–p–ν–n structure) in the collector space-charge region becomes comparable to or larger than the fixed impurity density (donors in this case). If we limit ourselves to the epitaxial *uniformly* doped ν-region, then the net collector depletion region charge under the current flow condition becomes

$$\rho = q[N_D - n(x)], \tag{23}$$

where N_D is the donor concentration of the ν-region and $n(x)$ is the position dependent electron concentration. Poisson's equation for the depletion region with collector current flowing can be written as

$$dE(x)/dx = q[N_D - n(x)]/\epsilon_s, \tag{24}$$

where ϵ_s is the silicon permittivity. At sufficiently high collector voltages, the electric field in the depletion region exceeds 10^4 V cm^{-1}, and the carriers move at the maximum saturated velocity v_s, which for electrons is equal to about 10^7 cm s^{-1}. The collector current density is therefore given by

$$J_C = qv_s n(x), \tag{25}$$

and

$$n(x) = J_C/qv_s. \tag{26}$$

Substituting (26) into (24) for $n(x)$, we obtain

$$\frac{dE(x)}{dx} = \frac{q}{\epsilon_s}\left(N_D - \frac{J_C}{qv_s}\right). \tag{27}$$

Integrating (27), we obtain [7]

$$E(x) = E(0) + \frac{q}{\epsilon_s}\left[N_D - \frac{J_C}{qv_s}\right]x, \tag{28}$$

where $E(0)$ is the field at $x = 0$. Thus the field distribution in the depleted ν-region is linearly dependent on position x with a slope dependent on the current density J_C; see Fig. 8 [7]. At a current density

$$J_C = J_0 = qv_s N_D, \tag{29}$$

the field in the depletion region becomes uniform (line c of Fig. 8 [7]) since the slope is zero. As the collector current I_C is increased beyond J_0, the slope changes sign, and the field eventually becomes equal to zero at $x = 0$ [$E(0) = 0$], but the field at the ν–n^+ junction is maximum. In this situation the current becomes entirely space-charge limited, and the base is fully widened. If the voltage drop in the depleted ν-collector region of width W_n (Fig. 8) is V_{BC}, then the electric field at the ν–n^+ interface is

$$E(W_n) = -2V_{BC}/W_n. \tag{30}$$

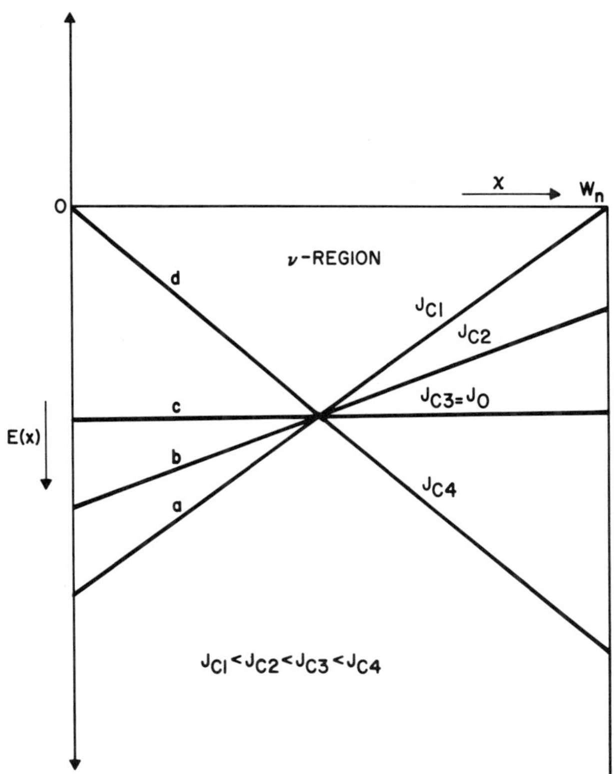

Fig. 8. Electric field in the ν-region at various current densities. (From Whittier and Tremere [7].)

The critical current at which the field reaches this value at $x = W_n$ is obtained [7, 8] from (28) for $E(0) = 0$ *as*

$$J'_0 = qv_s[N_D + 2\epsilon_s|V_{BC}|/qW_n^2]. \tag{31}$$

This current is plotted for an n^+–p–ν–n^+ (Fig. 9 [7]) transistor as a function of the collector voltage drop V_{BC} with the ν-region concentration as a parameter and $W_n = 10\ \mu$m. Formula (31) agrees with more accurate computer generated data only at higher field values in excess of 10^4 V cm^{-1}.

As the collector voltage is increased and the collector current starts exceeding J'_0, the number of electrons in the space-charge region increases, and the slope dE/dx increases (Fig. 10 [7]). The areas above the lines are equal to the constant voltage drop V_{BC} so that the distance over which the

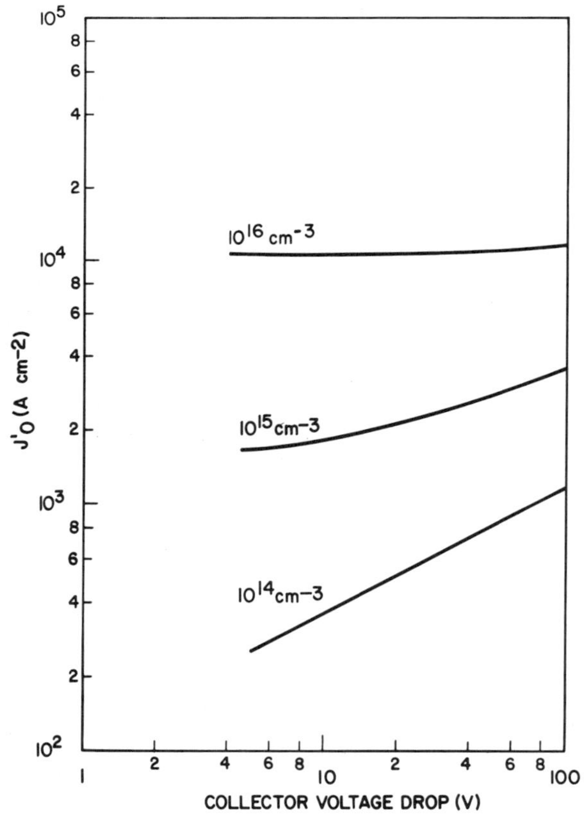

Fig. 9. Collector current density as a function of the collector voltage drop. (From Whittier and Tremere [7].)

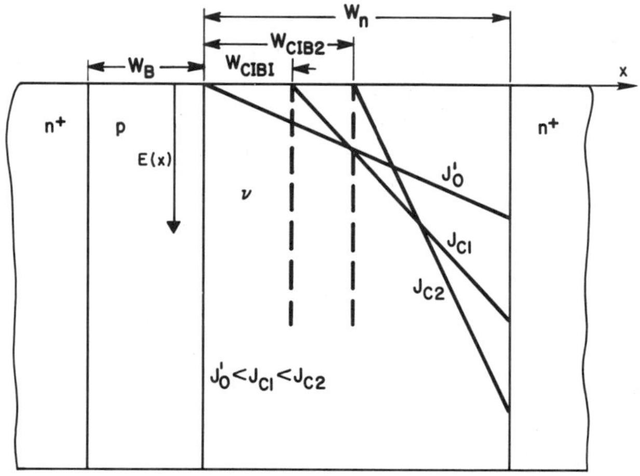

Fig. 10. Collector current-induced base W_{CIB} versus collector current density with fixed collector voltage. Lines marked J_0', J_{C1}, and J_{C2} represent the electric field $E(x)$ as a function of the distance x from the p-base. (From Whittier and Tremere [7].)

field extends decreases with the increasing current and is equal to $W_n - W_{CIB}$. Consequently

$$V_{BC} = \tfrac{1}{2}E_{max}(W_n - W_{CIB}). \tag{32}$$

W_{CIB} is the width of the neutral region representing the *collector-induced base*. The total base thickness now becomes $W_{Btot} = W_B + W_{CIB}$. The value of W_{CIB} can be obtained by solving Poisson's equation (27), utilizing expression (32) for V_{BC} with $E(0) = 0$ and utilizing $W_{CIB} = W_n$ at $J_C = J_0'$ as integration limits [7]:

$$W_{CIB} = W_n\{1 - (J_0' - qv_sN_D)^{1/2}/(J_C - qv_sN_D)^{1/2}\}. \tag{33}$$

Figure 10 depicts the current-induced base width versus the collector current density for a constant collector bias.

7.2.2 Base Widening in n^+–p–ν–n^+ Structure (Low Field Case)

The electric field in the ν-region becomes low when the collector voltage is low, as happens for the device operating in saturation, or when the high resistivity ν-collector region is very wide or both. Due to the ν-region high resistivity, the collector undepleted portion $W_n - W_{\mathrm{CIB}}$ (Fig. 7) gives rise to the ohmic voltage drop $V_{n\mathrm{C}}$. The electric field in the undepleted region is

$$E_{n\mathrm{C}} = \rho_\mathrm{C} I_\mathrm{C}/A_\mathrm{E} = \rho_\mathrm{C} J_\mathrm{C}, \tag{34}$$

where $\rho_\mathrm{C} = (q\mu_n N_\mathrm{D})^{-1}$ and A_E is the emitter effective area. The undepleted region voltage drop is

$$V_{n\mathrm{C}} = E_{n\mathrm{C}}(W_n - W_{\mathrm{CIB}}). \tag{35}$$

From (35) and (34),

$$J_\mathrm{C} = |V_{n\mathrm{C}}|/(W_n - W_{\mathrm{CIB}})\rho_\mathrm{C}. \tag{36}$$

As the collector current increases, the voltage drop $V_{n\mathrm{C}}$ eventually becomes equal to the collector supply voltage V_{BC}, neglecting the collector junction built-in potential. The voltage left to create the depletion region becomes equal to 0 so that the depletion region collapses [6]. At that point, the collector current is equal to

$$J_0'' = q\mu_n N_\mathrm{D}|V_{\mathrm{BC}}|/W_n. \tag{37}$$

For $J_\mathrm{C} > J_0''$, the collector–base junction becomes forward biased, and holes are injected from the p-base into the ν-region. A part or the entire ν-region becomes conductivity modulated, and an additional majority carrier charge is stored in the ν-region neutralizing the injected carriers. This neutral region becomes the collector-induced base W_{CIB}. Introducing (37) into (36) yields [6], for $V_{n\mathrm{C}} = V_{\mathrm{BC}}$,

$$W_{\mathrm{CIB}} = (1 - J_0''/J_\mathrm{C})W_n. \tag{38a}$$

Substituting expression (37) for J_0'', we obtain

$$W_{\mathrm{CIB}} = W_n - V_{\mathrm{BC}}/J_\mathrm{C}\rho_\mathrm{C}, \tag{38b}$$

where $\rho_\mathrm{C} = (q\mu N_\mathrm{D})^{-1}$. The base starts increasing as soon as the collector current J_C exceeds the critical current J_0''. The unmodulated region continues shrinking regardless of the increased number of electrons injected into the ν-region.

7.2.3 Base Widening in the n^+–p–n^+ Transistor

Devices with the n^+–p–n^+ structure (Fig. 6.1) are used for low voltage power transistors since a p-base would have to be made very long to support higher voltages.

In the presence of electrons flowing through the depletion region with the limiting velocity v_s at the electric field exceeding 10^4 cm s^{-1} m, the total net charge density in the depletion layer consists of fixed acceptors and electrons, i.e.,

$$N = N_A + n. \tag{39}$$

Since the total net charge density is increased, the collector depletion region width in the p-base shrinks, and the base width increases.

The total number of electrons flowing in the space-charge region with the limiting velocity is given by

$$n = J_C/qv_s. \tag{40}$$

The critical current density in the depletion region is given by

$$J_0 = qv_sN_A. \tag{41}$$

From (39)–(41), we obtain

$$N = N_A(1 + J_C/J_0). \tag{42}$$

The depletion region width W_d can be computed from the well-known equation

$$x_d = (2\epsilon_s V_{BC}/qN)^{1/2}. \tag{43}$$

The difference between the depletion region width with no current flow ($N = N_A$) and that with current flow can be easily computed as being the base widening increment

$$\Delta W_B = \left(\frac{2\epsilon_s V_{BC}}{qN_A}\right)^{1/2}\left(1 - \frac{1}{1 + J_C/J_0}\right)^{1/2}. \tag{44}$$

7.2.4 Two-Dimensional Base Widening Model

In all previous considerations regarding the base widening effects it was assumed that the current flowing from the emitter torward the collector does not diverge and that the one-dimensional analysis is therefore valid. In a real device there must be at least some diverging of the current and one-dimensional analysis cannot be entirely accurate. Van der Ziel and Agourdis [9] assume that the current from the emitter spreads laterally until the current density is lowered to the critical current density level J_0'. Below J_0', the current flow is unidimensional; above J_0', it is two dimensional. The current density in the ν-collector region, therefore, never exceeds the J_0' value. Consequently, we can write

$$A_C J_0' = A_E J_C, \tag{45}$$

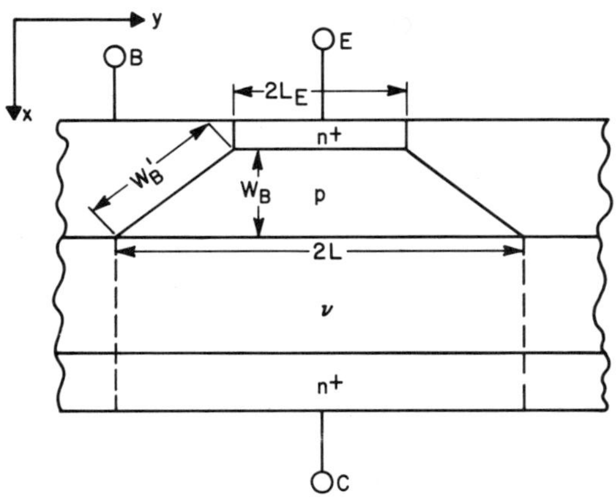

Fig. 11. Model for the two-dimensional current spreading. (From Van der Ziel and Agourdis [9].)

assuming that the current gain $\alpha = 1$. A_E is the emitter area equal to $2L_EZ$ (Fig. 11 [9]), where $2L_E$ is the emitter length and Z the emitter depth in the z direction. The collector current area is given by $A_C = 2LZ$, where L is defined by Fig. 11. From (45), we conclude that

$$L = (J_C/J_0')L_E. \tag{46}$$

The new effective base width W_B can be calculated from the geometry of Fig. 11 as

$$W_B'^2 = W_B^2 + (L - L_E)^2 \tag{47}$$

or, using expression (45), as

$$W_B' = [W_B^2 + L_E^2(J_C/J_0') - 1)^2]^{1/2}. \tag{48}$$

The base starts to widen at $J_C \geqslant J_0'$. The Van der Ziel and Agourdis analysis disregards completely any base widening into the n^--region. A more realistic analysis of the base widening would have included, of course, both effects.

7.3 Emitter Current Crowding

The base current I_B flowing laterally beneath the forward biased emitter causes a voltage drop along the emitter junction (Fig. 12). Consequently, the base–emitter voltage $V_{BE}(y)$ varies along the emitter (y direction) and is the smallest at the emitter center so that the emitter current tends to crowd

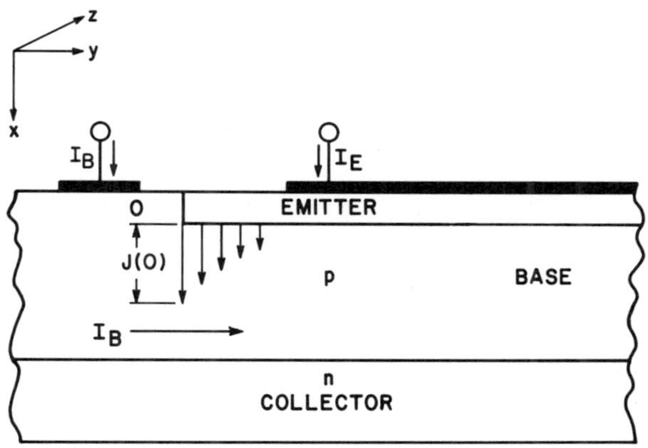

Fig. 12. Current crowding at the semi-infinite emitter edge.

at the emitter edges. This effect is very pronounced at high current levels and for transistors with low dc or rf current gains and is responsible for localized base widening and nonuniform current gain along the emitter junction. The current crowding condition is therefore undesirable and should be avoided, especially since it may lead to the forward second breakdown.

The emitter current crowding effect under forward bias was studied by Fletcher [10] for a semi-infinitely wide emitter and by Hauser [11] for an emitter with a final width. More recently, Hower and Eindhoven [12] applied the integral charge control model to a HV transistor with n^+–p–ν–n^+ structure exhibiting base widening.

7.3.1 Distributed Analysis

The analysis of the current distribution in a single emitter stripe with one base contact (Fig. 12) can be greatly simplified if it is assumed, following Fletcher [10], that the emitter is semi-infinitely long in the y direction, that no base conductivity modulation takes place, and that the current gain is the same all along the emitter length. In addition, the emitter efficiency is supposed to be equal to unity, the lifetime is current independent, and no base widening takes place. The injected current density follows the expression

$$J_E = J_s \exp[qV(y)/kT] \tag{49}$$

for $qV(y)/kT \gg 1$. In (49), J_s is the saturation current and $V(y)$ the emitter–base voltage at y. For the common situation of a highly conductive and therefore equipotential emitter, voltage $V(y)$ can be found from the following equation [10]:

$$\frac{d^2V(y)}{dy^2} = \frac{\rho_B}{W_B}(1 - \alpha)J_s \exp\left(\frac{qV(y)}{kT}\right), \tag{50}$$

where W_B is the device metallurgical base width, ρ_B the average resistivity of the base, and α is the low frequency common base current gain. The solution of (50) is

$$V(y) \simeq -\frac{2kT}{q}\ln\left\{\left[\exp -\left(\frac{qV_0}{2kT}\right)\right] - \frac{q}{2kT}\left[\frac{2\rho_B}{W_B}(1 - \alpha)J_s\frac{kT}{q}\right]^{1/2} x\right\}, \tag{51}$$

where $V_0 = V$ at $y = 0$. Substituting expression (51) for $V(y)$ into (49), we obtain Fletcher's expression for lateral current distribution in the emitter stripe in the proximity of the base contact as

$$J_E(y) \simeq J_E(0)(1 + y/y_0)^{-2} \tag{52}$$

with

$$y_0 = \left[\frac{2kT}{q}\frac{W_B}{(1 - \alpha)}\frac{1}{\rho_B J_E(0)}\right]^{1/2}, \tag{53}$$

where $J_E(0)$ is the current density at the emitter edge (Fig. 12).

The characteristic length y_0 defines the relative utilization of the emitter stripe. For instance, at a distance $y = y_0$ from the emitter edge the current $J_E(y_0) = J_E(0)/4$.

In Fletcher's derivation, the emitter and base temperatures were assumed to be uniform. This may lead to serious discrepancies in the case of high dissipations and wide emitter stripes for which the thermal resistance at their centers is higher than that at their edges. The resulting temperature gradient with higher temperature at the center may either cause more uniform current distribution than expected or even reverse the current crowding, making its density higher toward the emitter center [13]. In the case of a significant base conductivity modulation, it was shown by Ghandi [14] that the current along the emitter decays exponentially with a characteristic length

$$y'_0 = \left[\frac{2kT}{q}\frac{W_B}{(1 - \alpha)}\frac{1}{\rho_B J_{\text{mod}}^{(0)}}\right]^{1/2}, \tag{54}$$

where $J_{\text{mod}}^{(0)}$ is the current at which the high level injection begins and which can be determined from

$$J_{\text{mod}}^{(0)} \simeq 2q\text{D}_n(N_A/W_B), \tag{55}$$

where N_A is the base acceptor concentration. Any increase in the current density above $J_{\text{mod}}^{(0)}$ will lead to the heavy base conductivity modulation with

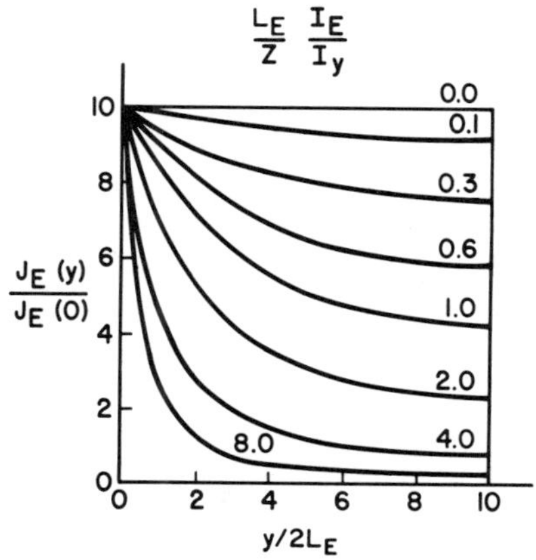

Fig. 13. Emitter current density $J_E(y)$ normalized to the current density at the emitter edge $J_E(0)$ as a function of the normalized distance $y/2L_E$ with $(L_E/Z)(I_E/I_y)$ as a parameter. $2L_E$ is the emitter length, Z the emitter width in the z direction. (From Hauser [11].)

the result that the lateral base voltage drop will become practically current independent.

Substitution of (55) into (54) gives

$$y'_0 = KW_B(h_{FE})^{1/2} \tag{56}$$

with $K = 1.4$. Olmstead *et al.* [15] have obtained a similar expression, but with the constant $K = 0.54$, by assuming that in a well-designed emitter the voltage drop in the crowded region at the emitter edge should not exceed $\frac{1}{2}(kT/q)$.

Hauser [11] analyzed the emitter current distribution with assumptions similar to those of Fletcher, but for a *finite* length emitter in the direction and two-stripe geometry. The results of his analysis are shown in Fig. 13 [11], where the current density $J_E(y)$ normalized to the current density $J_E(0)$ at the emitter edge is plotted as a function of distance y normalized to the emitter half-width L_E with $(L_E/z)(I_E/I_y)$ as a parameter. Z is the emitter width in the z direction, perpendicular to the page surface, and

$$I_y = (2kT/q)(W_B h_{FE}/\rho_B).$$

W_B is the base width.

7.3.2 Current Crowding Integral Charge-Control Analysis

Hower and Einthoven [12] have studied current crowding of a high voltage power transistor with n^+–p–ν–n^+ structure (Fig. 14 [12]) applying

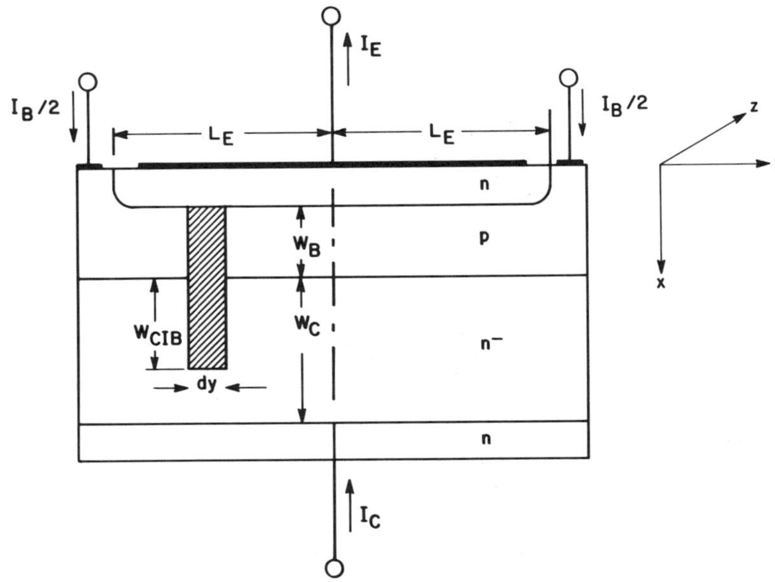

Fig. 14. Cross section of an n–p–n^-–n transistor showing various dimensions. (From Hower and Einthoven [12].)

the modified charge-control model of [16]. The device is assumed to be operating in that region of the current gain versus current plot where the slope of the curve becomes -2 (Fig. 15), i.e., when the current crowding causes base widening (see Section 7.4). In such a structure the injecting emitter area may become considerably reduced. Under the crowded conditions, the base is fully widened, and the effective base width is

$$W_{\mathrm{Beff}} = W_{\mathrm{B}} + W_{\mathrm{CIB}}. \tag{57}$$

The base charge per unit area at any point y (Fig. 14) is therefore given by

$$\frac{Q(y)}{\bar{D}_n} = \frac{J_{\mathrm{E}}(y) W_{\mathrm{Beff}}^2}{4qD_{\mathrm{c}}^2}, \tag{58}$$

where $\bar{D}_n$ is the average electron diffusion constant, D_{c} is the high injection value of the electron diffusion constant D_n; and $J_{\mathrm{E}}(y)$ is the emitter current density at a distance y from the emitter edge. The base current is mainly carried by the majority and minority carriers' drift, and the diffusion effects are neglected without seriously affecting the analysis results [12]. The base current for an emitter stripe $Z/2$ long is therefore given by

$$J_{\mathrm{B}}(y) = -\left[\frac{q}{kT}\,\frac{b+1}{b}\,qZD_{\mathrm{c}}Q(y)\right]\frac{dV_{\mathrm{BE}}(y)}{dy}. \tag{59}$$

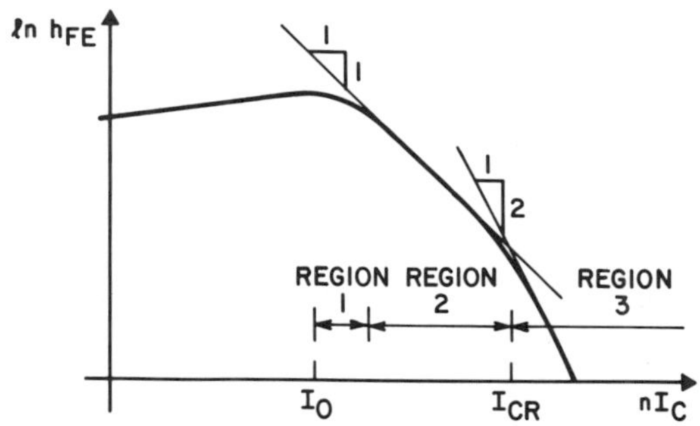

Fig. 15. Current gain plotted versus collector current with constant collector voltage. (From Hower and Einthoven [12].)

In (59), b is the μ_n/μ_p ratio, $Q(y)$ is the total majority carrier concentration per unit area, and $V_{BE}(y)$ is the base–emitter voltage at y. The emitter current density from the charge control analysis is given by

$$J_E = [qn_i^2 D_c/Q(y)] \exp[qV_{BE}(y)/kT]. \tag{60}$$

The base current increment is related to J_E and h_{FE} and is given by

$$dJ_B(y) = (ZD_p/D_c Q_E) J_E(y) Q(y)\, dy, \tag{61}$$

where D_p is the low level hole diffusion constant in the n-type emitter. The solution of the three equations (59), (60), and (61) yields an expression for the emitter current density as a function of position y

$$J_E(y) = J_1\{1 + 1.7 \tan^2[\phi(y)/2]\}, \tag{62}$$

where J_1 is the emitter current density at L_E, i.e., at the emitter center, and ϕ is the amplitude of the incomplete elliptic integral of the first kind. ϕ is a function of the y/L_E ratio. The current density J_E normalized to J_I versus position y normalized to L_E is shown by Fig. 16 with the edge-to-center current density ratio $\zeta = J_E(0)/J_1$ as a parameter. Figure 17 shows h_{FE} normalized to crossover gain h_{CR} as a function of the collector current normalized to crossover current I_{CR}. The crossover gain is defined as

$$h_{CR} = \tfrac{1}{9}[b/(1 + b)](L_E/W_C), \tag{63}$$

and the crossover current as

$$I_{CR} = 36q \frac{Z}{L_E}\left(\frac{1 + b}{b}\right)\frac{Q_E}{D_p} D_c^2. \tag{64}$$

Q_E is the emitter effective concentration per unit area. Referring to Fig. 17, h_{FE1} and h_{FE2} are the low current ($h_{FE1} \propto I^{-1}$) and high current ($h_{FE2} \propto I^{-2}$) asymptotic gain values, respectively. It can be seen from Fig. 17

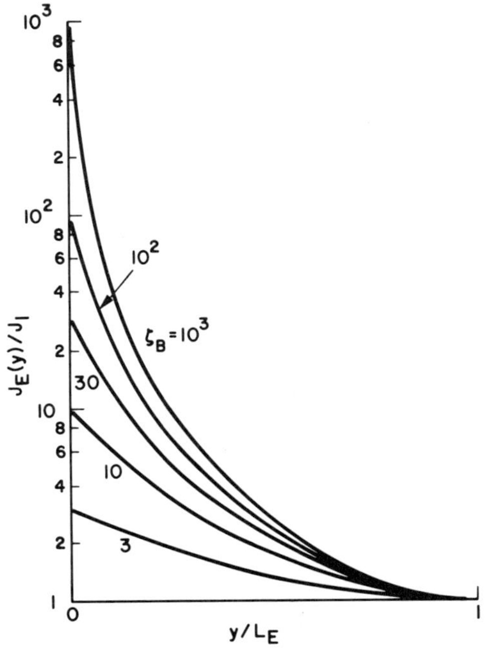

Fig. 16. Normalized emitter current density plotted versus normalized distance from the stripe edge, with ζ_B as a parameter. (From Hower and Einthoven [12].)

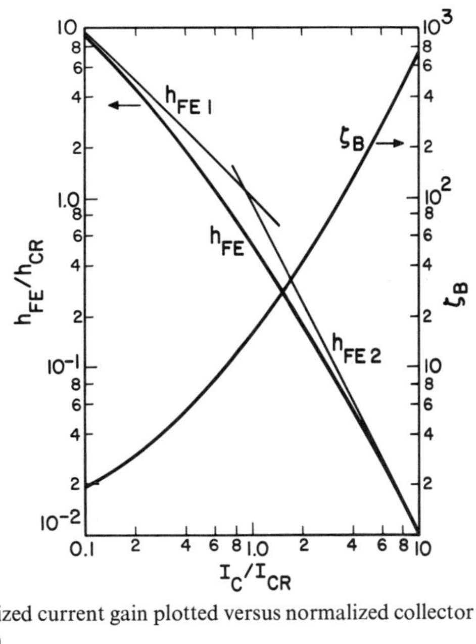

Fig. 17. Normalized current gain plotted versus normalized collector current. (From Hower and Einthoven [12].)

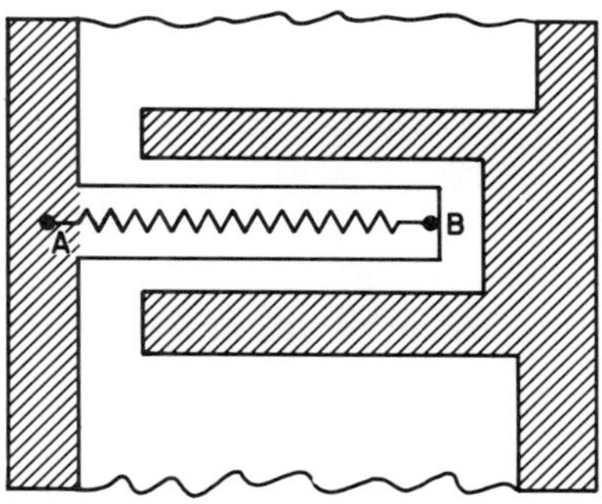

Fig. 18. Ohmic path along an emitter stripe.

that $h_{FE1} = h_{FE2} = h_{CR}$ at the collector current $I_C = I_{CR}$, hence the name "crossover." The crossover current I_{CR} may serve as a figure of merit for any particular emitter geometry. The current gain h_{FE} degradation with current can be avoided by making I_{CR} high enough by designing the emitter stripe width $2L_E$ small and its length Z large. A high emitter efficiency and a high transport factor are, of course, necessary to minimize the base current and the voltage drop along the emitter junction.

It should be noted that the effective emitter area under crowded conditions, i.e., for $I_C \gg I_{CR}$, is given by

$$A_{Eeff} = A_E(I_{CR}/I_C), \tag{65}$$

where A_E is the geometrical emitter area.

The current crowding analysis of [12] is useful in the design of power transistor switches facilitating the determination of the emitter stripe width as a function of the metallurgical emitter area.

7.3.3 Longitudinal Emitter Debiasing

In the interdigitated transistor with long emitter stripes there exists a voltage drop along the emitter and base fingers (Fig. 18) which may lead to the emitter debiasing. The current density becomes highest in the emitter stripe region close to the emitter contact and lowest at the far end of it. Adequately thick, highly conductive metallization and a greater number of contacting leads alleviate this problem.

7.4 Current Gain Fall-Off at High Currents

The overall h_{FE} common-emitter current gain is a sensitive function of the collector current I_C (Fig. 15). At low current levels, the gain is low because of the space-charge generation–recombination current; and as the current increases, h_{FE} eventually reaches the peak value h_{FE0} due to the diffusion component that prevails over the generation–recombination current. After this flat peak region is reached, however, further collector current increase causes the gain to fall, an effect first studied by Webster [17] and since then further studied by several authors [7, 15, 16, 18] trying to explain the behavior of the current gain versus current curve in the high current level region.

The current gain fall-off phenomena can be analyzed by using the Gummel–Poon (Srction 6.2) charge control model as was done by Olmstead *et al.* [15]. The current gain in the common-emitter configuration is given by

$$h_{FE} = \alpha_T \gamma/(1 - \alpha_T \gamma), \tag{66}$$

where the emitter efficiency in the grounded base configuration is given by

$$\gamma = \gamma_E/(1 + \gamma_E) \tag{67a}$$

or

$$\gamma \simeq [1 + (1/\gamma_E)]^{-1}. \tag{67b}$$

$1/\gamma_E$ is known as the emitter injection efficiency defect factor. The base transport factor can be expressed at high injection levels by

$$\alpha_T = 1 - (W_B^2/4\bar{D}_n\tau_{eff}) = 1 - \delta, \tag{68}$$

where $\bar{D}_n$ is the averaged electron diffusivity in the p-base of an n–p–n transistor and τ_{eff} is the electron lifetime in the base. The second term on the right-hand side of the Eq. (68) is called the transport defect factor δ. Since in a good transistor $\delta \ll 1$ and $\gamma_E \gg 1$, it is easy to show that

$$h_{FE} \simeq [\delta + (1/\gamma_E)]^{-1}. \tag{69}$$

From the charge control analysis, the emitter efficiency was shown to be (Section 6.2)

$$\gamma_E = \bar{D}_n Q_{E0}/\bar{D}_p Q_{B0}, \tag{70}$$

where Q_{B0} and Q_{E0} are the equilibrium majority electrical charges per unit area due to the ionized impurities in the base and the emitter, respectively. $\bar{D}_n$ and $\bar{D}_p$ are properly averaged electron and hole diffusion constants, respectively.

As the result of the minority carrier injection by the emitter, the base majority charge increases in order to preserve the base neutrality. Minority carriers are also injected from the base into the emitter, but since the emitter is normally highly doped, the majority charge increase may be neglected.

If we call the additional base charge ΔQ_B, then the emitter efficiency at high levels becomes

$$\gamma_E = D_n Q_{E0}/D_p(Q_{B0} + \Delta Q_B). \tag{71}$$

But the base charge increment is related to the collector current by

$$\Delta Q_B = \tau_F J_C, \tag{72}$$

where τ_F is the base transit time given by

$$\tau_F = W_B^2/4\bar{D}_n. \tag{73}$$

Substitution of Eqs. (68), (71), and (72) into Eq. (69) provides the following current gain expression:

$$h_{FE} = \left[\frac{W_B^2}{4D\tau_{eff}} + \left(\frac{\bar{D}_p}{\bar{D}_n}\right)\left(\frac{Q_{B0}}{Q_{E0}}\right) + \left(\frac{D_p}{D_n}\right)\left(\frac{W_B^2 J_C}{4D_n Q_{E0}}\right)\right]^{-1}, \tag{74}$$

essentially similar to Webster's theory results. The last term of this expression is much larger at high current levels than all other terms, which can be then neglected, so Eq. (74) reduces to

$$h_{FE} \simeq 4D_n^2 Q_{E0}/\bar{D}_p W_B^2 J_C \tag{75}$$

or

$$h_{FE} W_B^2 I_C \simeq 4(D_n)^2 Q_{E0}/\bar{D}_p. \tag{76}$$

From (76),

$$h_{FE} = k_E/W_B^2 J_C, \tag{77}$$

where J_C is the average current density and k_E is a constant equal to the right-hand term of (76). This current gain dependence is shown in Fig. 15 as that part of the curve which has a slope of -1. A very similar equation was obtained by Clark [18], who gives for the gain the following expression:

$$h_{FE} \simeq 4D^2 q^2 n_i^2/J_S J_C W_B^2, \tag{78}$$

where D is the minority carrier diffusion constant in the base and J_S the emitter saturation current density. J_S current density values reported by [17] were in the range of $5\text{–}20 \times 10^{-13}$ A cm^{-2} at 27°C.

Under base widening conditions, when the collector current exceeds the critical current, the base width at low collector voltages (Section 7.2.2) is augmented by the collector induced base

$$W_{CIB} = (1 - J_0''/J_C)W_n. \tag{79}$$

Substituting this value into the expression for h_{FE} (77), we obtain

$$h_{FE} = \frac{k_E}{(W_B + W_{CIB})^2 J_C} \tag{80}$$

or

$$h_{FE} = \frac{k_E}{[W + (1 - J_0''/J_C)W_n]^2 J_C}. \tag{81}$$

For $J_C \gg J_0''$ the base is fully widened to

$$W_B' = W_B + W_n \tag{82}$$

and

$$h_{FE} = k_E/W_B'^2 J_C. \tag{83}$$

Again, the current gain under the base widened condition is inversely proportional to the collector current density J_C. In deriving the h_{FE} expressions (81) and (83) it was assumed that the lifetimes and diffusivities in the metallurgical and the collector induced base W_{CIB} are the same. A more accurate analysis would differentiate between these values.

Under current crowding conditions, the current flows only at the emitter edges, whereas there is practically no flow at the emitter central portion. If the *effective* emitter half-width is y_0 (we consider here a stripe geometry) and the emitter perimeter is P_E, then we can write for the average emitter current density

$$J_C = I_C/A_E = I_C/P_E y_0. \tag{84}$$

The emitter effective half-width y_0 is given by

$$y_0 = \text{const} \cdot W_B' h_{FE}^{1/2}. \tag{85}$$

Substituting (84) into (83) yields

$$h_{FE} = \text{const} \cdot (W_B'^2 I_C^2)^{-1}, \tag{86}$$

which implies that the current gain for the severely crowded conditions tends to fall off with the increasing current with a slope of -2.

Experimental results show a very good agreement with the derived -1 slope value in the noncrowded case. In the crowded case at very high current densities, the slope may exceed the -2 value.

7.5 Current Gain Temperature Dependence

Transistor emitter efficiency in the common-emitter configuration is expressed as (Section 7.1)

$$\gamma_E = \frac{\bar{D}_n}{\bar{D}_p}\left(\int_0^{W_E} N_D \exp\left(\frac{-\Delta E_g}{kT}\right) \Big/ \int_0^{W_B} N_A\, dx\right), \tag{87}$$

and the transport factor in the common-base configuration is

$$\alpha_T = [\cosh(W_B/L_n)]^{-1}. \tag{88}$$

Generally, h_{FE} increases appreciably (Fig. 19) with increasing temperature, mainly because of the improved emitter efficiency. Expression (87) predicts an exponential temperature dependence of γ_E on account of the presence of the $\exp(-\Delta E_g/kT)$ term, which arises due to the energy gap reduction of a highly doped emitter [19, 20]. For moderately or lightly doped emitters (concentrations below $\sim 5 \times 10^{19}$ cm^{-3}), ΔE_g is very small, and the temperature effect on the current gain is negligible.

For dopings above 10^{19} cm^{-3}, the hole mobility in the emitter is relatively constant with respect to temperature. It follows, then, from Einstein's relationship that the diffusion constant must increase with temperature. The hole lifetime in the emitter also increases with temperature; consequently, the emitter diffusion length $L_p = (D_p\tau_p)^{1/2}$ goes up with temperature and adds up to the ΔE_g effect by increasing the overall transistor gain.

Transistor bases are always doped far below the 10^{19} cm^{-3} concentration for which the electron diffusion constant is inversely proportional to the temperature. But since the electron lifetime goes up with temperature and dominates over D_n, the diffusion length L_n in the base is increased, and the transport factor α_T increases with temperature. All of the described effects are very much dominated by the energy gap narrowing ΔE_g [20].

Figure 19 shows that at some crossover point the high temperature common-emitter current gain starts falling below the low temperature gain. This may indicate that the electron diffusion constant in the widened base decreases rather rapidly with temperature because of the presence of the carrier–carrier scattering effect.

The temperature sensitivity of the current gain is a contributing factor to

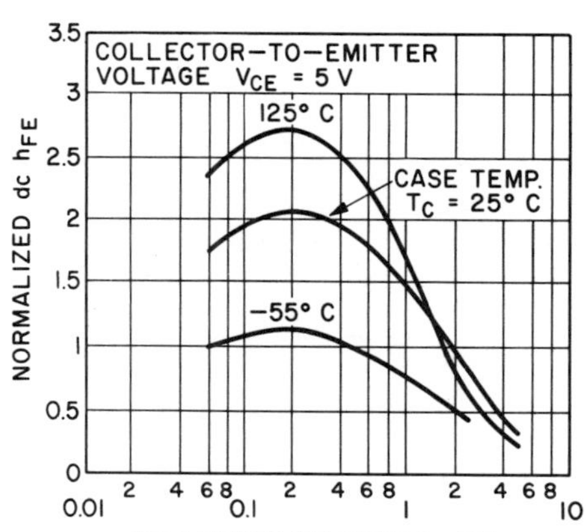

Fig. 19. h_{FE} versus temperature. (Courtesy RCA Solid State Division, Somerville, New Jersey.)

the hot spot formation [20] and affects the second breakdown energy limit (Section 10.5). It is therefore advantageous to reduce the current-gain temperature dependence, e.g., by lightly doping the transistor base and limiting the phosphorus-doped emitter surface concentration to about $7 \times 10^{19}\,\mathrm{cm}^{-3}$.

7.6 Methods of Gain Improvement

7.6.1 Emitter Geometric Design

The effective half-width of the emitter stripe is proportional to the effective base width and the square root of the common-emitter current gain [expression (56)]. Making the emitter stripe width appreciably more than twice the effective half-width does not improve the device performance and, as a matter of fact, makes it worse in some respects by increasing unnecessarily the emitter transition capacitance and the path for carrier withdrawal. Moreover, it increases the device cost because of poor "real estate," i.e., silicon surface utilization. Good silicon chip area utilization with low current densities and good current gains can be achieved making the emitter stripes narrow and very long. This is accomplished, for instance, by the so-called comb structure (Fig. 20), where all emitter contacts, on one hand, and all the base contacts, on the other, are connected together. There are many variations of the interdigited geometries which present some

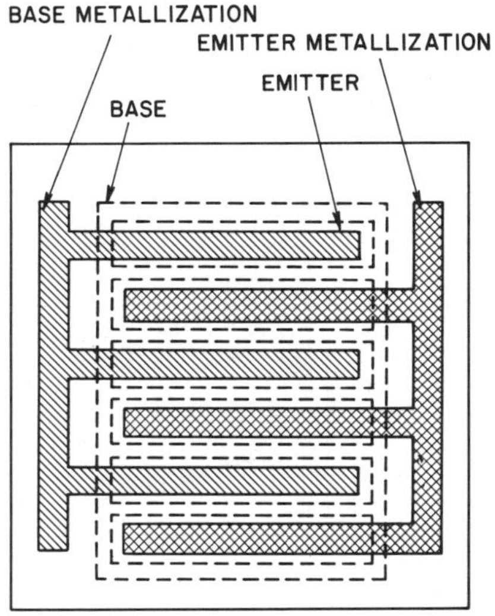

Fig. 20. Emitter comb structure.

special advantages such as the easiness of inclusion of emitter ballasting resistors. In all these designs the purpose should be to *maximize* the emitter perimeter to the emitter area ratio and simultaneously the emitter perimeter to the base area ratio. In rf transistors the high frequency gains are more difficult to obtain than in the low frequency devices, and a very high interdigitation is essential. These devices are usually designed with the so-called overlay geometries consisting of multiple cells consisting of small emitters surrounded with base contacts and all connected by a metallization overlying the silicon oxide (see Section 9.3).

7.6.2 Emitters with Low Impurity Concentrations

7.6.2.1 LEC Emitter

It was suggested by Yagi *et al.* [21] that the emitter efficiency may be improved by using an emitter structure (Fig. 21) consisting of a highly doped n^+-region and a very low doped epitaxial n^--layer. This kind of an emitter is known by the acronym LEC (light emitter concentration). The base doping level of the transistor may be an order of magnitude higher than that of the n^--region. Despite the low emitter concentration, it is possible to achieve high injection efficiency because the minority carriers injected from the base into the n^--emitter cannot reach the emitter contact as they are reflected from the energy barrier created by the n^+–n high–low junction. Transistors fabricated with the structure of Fig. 21 have indeed shown high current gains [21]. It was pointed out by DeGraaf and Slotboom [22] that the base current in a LEC transistor, similar to that of a double diffused transistor, is mainly due to the recombination in the n^+-region at the medium and high current levels and that the main advantage of the LEC structure may just

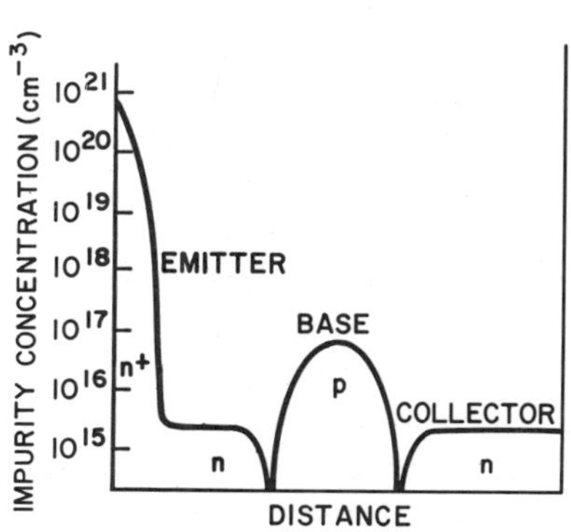

Fig. 21. LEC emitter. (From Yagi *et al.* [21].)

consist of low charge recombination rate in the emitter–base depletion region.

The presence of the charge storage in the lightly doped emitter has a disadvantage of lowering the transistor gain–bandwidth [22].

7.6.2.2 LCE—Low Concentration Phosphorus Emitter

Working on the assumption that highly doped emitters may result in low emitter efficiency, Martinelli and Jetter [23] studied an *n–p–n* power transistor whose diffused emitter surface concentration was in the range of 5 to 7 $\times$ 10^{19} cm^{-3}. Compared with an emitter with 5 $\times$ 10^{20} cm^{-3} surface concentration, the low concentration emitter has higher minority carrier lifetime and a negligible band gap narrowing. The equilibrium majority charge per unit area of the high concentration emitter was about 3 $\times$ 10^{17} cm^{-2} and that of the LCE (low concentration emitter) was 3 $\times$ 10^{16} cm^{-2}.

The comparison of transistors with high and low emitter surface concentration over a temperature range from 25 to 150°C has shown a significant superiority of the low doped emitter over a broad range of collector currents. For instance, the current gain at $I_C = 10$ A was about 1.5 greater for the low concentration emitter than for the high concentration emitter.

7.6.3 Arsenic-Doped Emitter

Arsenic can be used in place of more commonly used phosphorus for the emitter doping. It gives the possibility of obtaining steeper gradients at the emitter–base junction and eliminates the so-called emitter "push-out" effect characteristic of phosphorus diffusions, particularly troubling in the very high frequency transistors with very narrow bases. High emitter efficiencies that increase with surface concentrations of up to 1.5 $\times$ 10^{20} cm^{-3} can be obtained with arsenic [24, 25]. The experimental work of [26] demonstrated that the transistor's gain with an arsenic-doped emitter was limited not by the injection efficiency but by the base transport factor. Similar results to those obtained with As can be obtained by using a phosphorus and arsenic mixed diffusion source with three parts of As to one part of phosphorus. The emitter efficiency increases as the concentration of arsenic in the mixture increases above this ratio. The arsenic-doped emitter efficiency is very much dependent on the way a transistor is processed. Fair [25] has pointed out that if the arsenic-doped emitter and the boron-doped base are sequentially diffused from chemical sources, then there exist several factors that retard boron diffusion. These factors include the presence of internal fields, ion pairing, and vacancy undersaturation. These effects are unimportant, however, in shallow arsenic–boron–doped n–p junctions processed by ion implantation predeposition followed by diffusion. In this case there is no interaction between the two diffusion processes. The calculated values of the emitter efficiency γ_E versus the emitter surface concentration C_{SE} for the noninteracting and interacting diffusions are plotted in Fig. 22a,b [24], respectively. Figure 22b demonstrates the very high sensitivity of the emitter efficiency to the emitter surface concentration for the range of C_{SE} values between 1×10^{20} and 2.2×10^{20} cm^{-3}. As C_{SE} is increased, the total base width and charge is decreased due to the enhanced boron diffusion retarding effects.

Experimental data of Fair [25] agree well with his calculations based on the difference in the effective emitter and base energy band gaps. This difference is negligibly small until the effective emitter concentration reaches its maximum value, which occurs for the emitter surface concentration $C_{SE} = 1.5 \times 10^{20}$ cm^{-3} (Fig. 23). Above the maximum effective emitter doping the energy gap differential becomes significant, and the emitter effective concentration starts decreasing with increasing concentration.

For an implanted–diffused emitter in the absence of impurity interaction, the rate of γ_E increase with C_{SE} is much slower, about one-fifth as great as that for the double diffused case (Fig. 22a). Ion implantation, therefore, provides a superior method of current-gain control in device production. Since there is no energy gap narrowing for arsenic emitters doped with surface concentrations $C_{SE} \lesssim 1.5 \times 10^{20}$ cm^{-3}, one can expect that

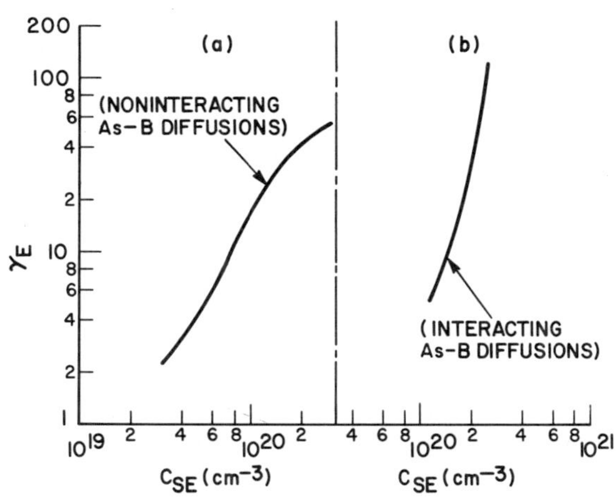

Fig. 22. Emitter efficiency plotted versus arsenic emitter surface concentration (a) with noninteracting sequential As and B diffusions and (b) with interacting diffusions. (From Fair [25].)

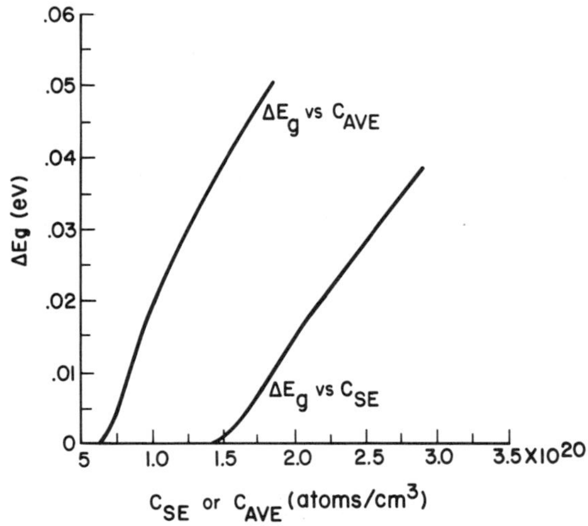

Fig. 23. Change in the effective energy band gap in silicon as a function of the average and surface arsenic doping. (From Fair [25].)

transistors with such an emitter will exhibit current gains with negligibly small temperature sensitivity. The experimental results of [25] confirm that that is actually the case for $C_{SE} < 1.4 \times 10^{20}$ cm^{-3}.

7.6.4 Heterojunction-Type Emitter

Two semiconductors with different energy gaps can be combined to form a heterojunction. On the other hand, a homojunction consists of one type of material.

The germanium–gallium arsenide heterojunction probably has been the most studied, since the lattice constants (atomic spacing) of these two semiconductors are not very much different, 0.5657 and 0.5653 nm for germanium and gallium–arsenide, respectively. A good lattice match prevents the creation of a large number of interface states. On the other hand, the energy gaps of the two semiconductors are much different: the energy gap of Ge is about 0.8 eV and that of GaAs is 1.43 eV at 300°K.

Two different semiconductors can be combined to form four types of a heterojunction, e.g., *n*Ge–*n*GaAs, *n*Ge–*p*GaAs, *p*Ge–*p*GaAs, and *p*Ge–*n*GaAs. The energy diagram for *p*Ge–*n*GeAs is shown in Fig. 24 [27]. Similar diagrams are valid, of course, for heterojunctions consisting of other materials and/or other polarities.

The energy band bending at the two semiconductor interfaces occurs

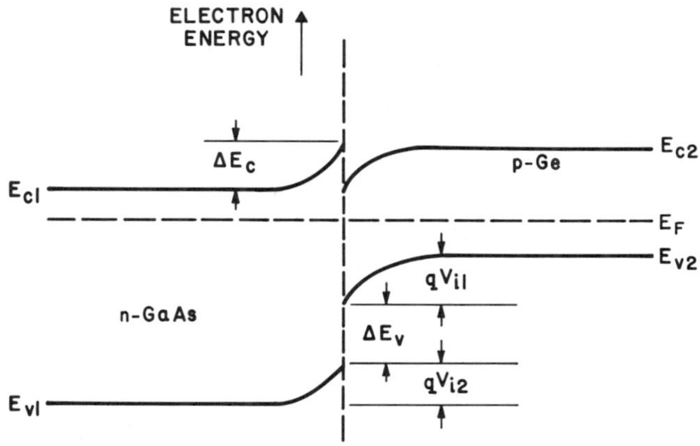

Fig. 24. Band diagram of a heterojunction. (From Anderson [27].)

because of the depletion of charges taking place to align the Fermi levels of two different materials. The total built-in voltage of a heterojunction is

$$V_i = V_{i1} + V_{i2}, \tag{89}$$

where V_{i1} and V_{i2} are the built-in voltages on either side of the junction (Fig. 24). It is obvious from the diagram that the barrier to the electrons flowing from the wide-band to the narrow-band semiconductor is much lower than the barrier to the holes flowing in the opposite direction; the barrier height for electrons is ΔE_c, but for holes it is equal to $\Delta E_v + qV_{i1} + qV_{i2}$. The current flow, therefore, will be entirely dominated by electrons injected from GaAs into Ge. The injection of holes from the low energy gap semiconductor is negligible so that the injection efficiency approaches unity and stays very high even for a highly doped Ge *p*-region. When this structure is used as an emitter, it is possible to have a highly doped base without impairing the injection efficiency and to reduce considerably the base current crowding. This improves second breakdown characteristics and increases transistor maximum oscillating frequency. Until recently, wide-band heterojunction-type emitters were very difficult to achieve mainly because of high interface recombination velocity. Recently, however, a breakthrough was made by Oh-uchi *et al.* [28], who used an *n*-emitter consisting of the semi-insulating polycrystalline silicon (Sipos) highly doped with phosphorus. The Sipos layer, (10 nm to 1 μm thick), was covered by a polysilicon film of 0.3–0.5 μm metallized with an aluminum contact. According to [28], the oxygen-doped polycrystalline silicon (Sipos) has an energy gap of 1.5 eV, much higher than that of monocrystalline silicon used for the *p*-type base. Figure 25 [28] compares

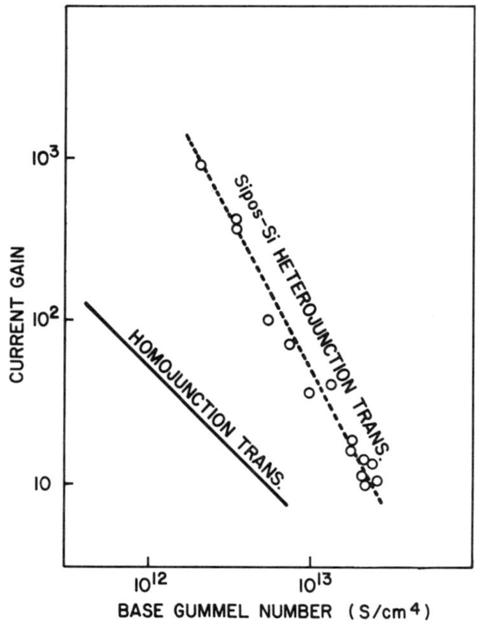

Fig. 25. Sipos and homojunction current gains as functions of the base Gummel number. Circles represent experimentally obtained data. (From Oh-ouchi *et al.* [28].)

maximum current gain of an *n–p–n* Sipos–silicon transistor for various base Gummel numbers. The superiority of the wide-band emitter is obvious even at the highest base doping levels. Figure 26 compares the Sipos–silicon transistor current gain with that of a homojunction transistor with the same base width of 20 μm at various collector currents. The superiority of the heterojunction emitter extends over a wide range of the collector currents.

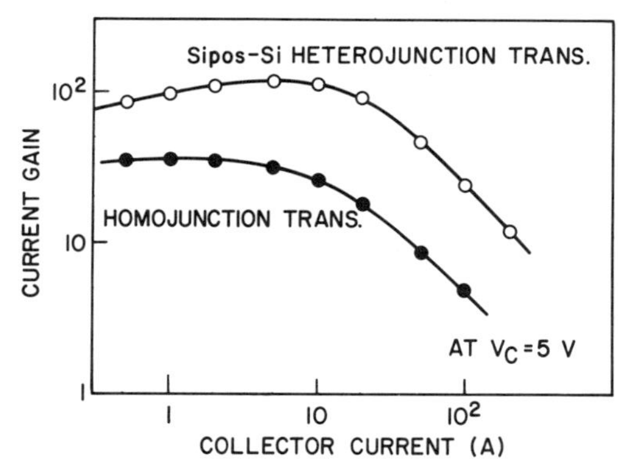

Fig. 26. Common emitter gains of the Sipos and homojunctions as functions of the collector current. Circles represent experimentally obtained data. (From Oh-ouchi *et al.* [28].)

At this time the Sipos–silicon transistors are at an early developmental stage, and further work is needed to uncover their possible limitations.

7.6.5 n^+–p–π–ν–n^+ Transistor Structure[1]

As we have seen in Chapter 6 in the n^+–p–π–ν–n^+ structure, the collector depletion region extends into the ν- and π-regions. For the same collector voltage V_{BC}, the electric field at the π–ν interface is lower than that at the p–ν interface of the n^+–p–ν–n^+ structure. This gives the device designer the possibility of some trade-off between the breakdown voltage and current gain fall-off by increasing the doping level of the collector ν-region. Higher concentration increases the value of the critical current $J''(0)$ [Eq. (37)] at which the base starts widening. Moreover, the degree of base widening [Eq. (38)] affecting the current gain is decreased for the lower resistivity ν-region.

7.6.6 Monolithic Darlington Power Transistors

High voltage power transistors exhibit rather low common-emitter current gains (h_{FE} below 10) at high current levels caused by the base widening effects. A Darlington pair (Fig. 27 [31]) remedies this situation by enormously increasing the gain, but not without some compromises, however, such as increased saturation voltage and decreased switching speed. The device of Fig. 27 is integrated on one silicon chip to reduce the fabrication cost over the cost of two individual transistors. It is possible, of course, to use more

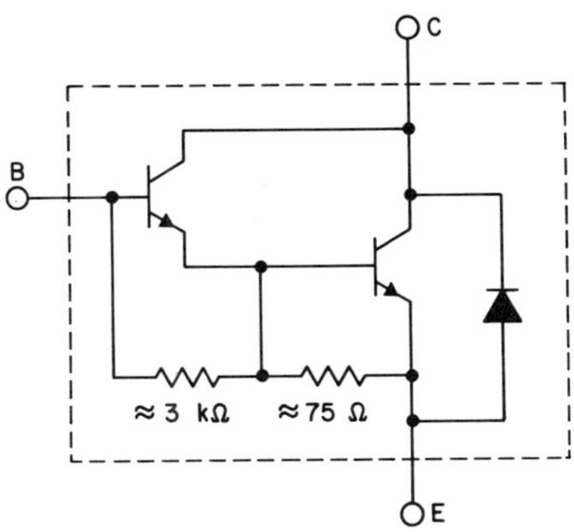

Fig. 27. Monolithic Darlington couple. (Courtesy RCA Solid State Division [30].)

[1] See Refs. [29, 30].

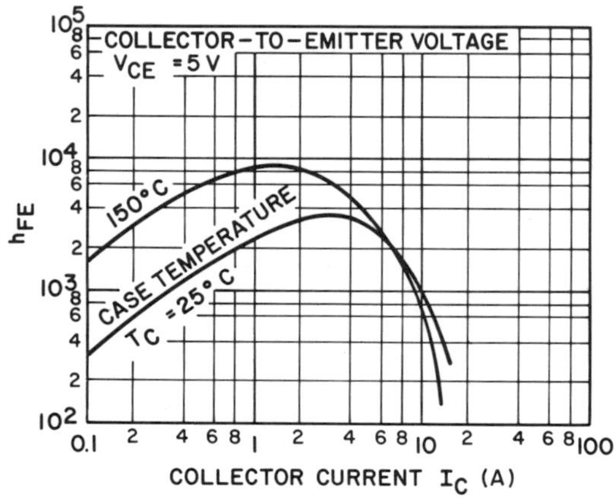

Fig. 28. Current gain of a Darlington pair. (Courtesy RCA Solid State Division [30].)

than two transistors in the Darlington-type connection [32]; the limitation, however, lies in the overall saturation voltage $V_{CE(sat)}$.

Figure 28 [31] shows a typical dc current gain characteristic plotted versus the collector current of a monolithic Darlington pair at two different temperatures. Because of the very high gains (1000 or more) at high currents, these devices can be operated directly from the I_C circuits without the use of additional drivers. Darlington circuits are useful as switches and audio amplifiers.

If the driving and output transistors of the Darlington pair have current gains equal to h_{FE1} and h_{FE2}, respectively, the overall gain is given by

$$\mathrm{h}_{FE} = h_{FE1} + h_{FE2}(h_{FE1} + 1). \tag{90}$$

The total collector current is given by the sum of two collector currents

$$I_C = I_{C1} + I_{C2}, \tag{91}$$

and the overall collector-to-emitter voltage is

$$V_{CE2} = V_{CE1} + V_{BE2}. \tag{92}$$

REFERENCES

1. R. P. Mertens, H. J. DeMan, and R. J. Van Overstraeten, *IEEE Trans. Electron Devices* **ED-20,** 772 (1973); R. J. Van Overstraeten, H. J. DeMan, and R. P. Mertens, *IEEE Trans. Electron Devices* **ED-20,** 290 (1973).
2. W. W. Sheng, *IEEE Trans. Electron Devices* **ED-22,** 25 (1975).
3. M. S. Mock, *Solid-State Electron.* **16,** 1251 (1973), and **19,** 819 (1974).
4. M. S. Adler, B. A. Beatty, S. Krishna, V. A. K. Temple, and M. L. Torreno, Jr., *IEEE Trans. Electron Devices* **ED-23,** 858 (1976).

5. E. J. McGrath and D. H. Navon, *IEEE Trans. Electron Devices* **ED-24,** 1255 (1977).
6. C. T. Kirk, *IEEE Trans. Electron Devices* **ED-9,** 164 (1962).
7. R. J. Whittier and D. A. Tremere, *IEEE Trans. Electron Devices* **ED-16,** 39 (1969).
8. J. L. Moll, "Physics of Semiconductors." McGraw-Hill, New York, 1964.
9. A. Van der Ziel and D. Agourdis, *Proc. IEEE* **54,** 411 (1966).
10. N. H. Fletcher, *Proc. IRE* **43,** 551 (1955).
11. R. J. Hauser, *IEEE Trans. Electron Devices* **ED-11,** 238 (1964).
12. P. L. Hower and W. G. Einthoven, *IEEE Trans. Electron Devices* **ED-25,** 465 (1978).
13. D. Navon and R. E. Lee, *Solid-State Electron* **13,** 981 (1970).
14. S. K. Ghandi, "Semiconductor Power Devices." Wiley, New York, 1977.
15. J. Olmstead, W. Einthoven, S. Ponczak, and P. J. Kannam, *RCA Rev.* **32,** 221 (1971).
16. P. L. Hower, *IEEE Trans. Electron Devices* **ED-23,** 863 (1976).
17. W. M. Webster, *Proc. IRE* **42,** 914 (1954).
18. L. E. Clark, *IEEE Trans. Electron Devices* **ED-17,** 661 (1970).
19. N. L. Kauffman and A. A. Bergh, *IEEE Trans. Electron Devices* **ED-15,** 732 (1968).
20. D. Buhanan, *IEEE Trans. Electron Devices* **ED-16,** 117 (1969).
21. H. Yagi, T. Tsuyuki, K. Koma, and T. Miyazawa, *IEDM Tech. Digest*, p. 282 (1974).
22. H. C. DeGraaf and J. W. Slotboom, *Solid-State Electron.* **18,** 809 (1976).
23. R. V. Martinelli and E. Jetter, *RCA Rev.* **38,** 60 (1977).
24. H. N. Ghosh, A. S. Oberai, J. J. Chang, and M. B. Vau, *IBM J. Res. Develop.* **15,** 472 (1971).
25. R. B. Fair, *IEEE Trans. Electron Devices* **ED-20,** 642 (1973).
26. R. D. Lillard, S. L. Saltich, and C. E. Volk, *J. Electrochem. Soc.* **123,** 1116 (1973).
27. R. L. Anderson, *Solid-State Electron.* **5,** 341 (1962).
28. N. Oh-uchi, H. Hayashi, H. Yamoto, and T. Matsushita, *IEDM Tech. Digest*, p. 522 (1979).
29. A. Blicher and B. Czorny, U.S. Patent 3427515 (1969).
30. R. Denning and D. A. Moe, *IEEE Trans. Electron Devices* **ED-17,** 711 (1970).
31. "RCA Power Transistors, Data Book SSD-220B." RCA Solid State Division, Somerville, New Jersey, 1978.
32. C. F. Wheatley, Jr. and W. G. Einthoven, *IEEE Trans. Electron Devices* **ED-23,** 870 (1976).

CHAPTER 8

CURRENT–VOLTAGE CHARACTERISTICS OF POWER TRANSISTORS

The current–voltage characteristics of a bipolar transistor (Figs. 1 and 2), besides the active region, consist essentially of the saturation, quasi-saturation, and avalanche breakdown regions, each of which will be discussed separately.

8.1 Saturation Region

In the low voltage power transistor with structures that do not include high resistivity n^- (or ν) collector regions (see Fig. 6.1), the total voltage drop $V_{CE(sat)}$ taking place when the device is operated in the saturation region can be determined using the first-order Ebers–Moll transistor model. If the current crowding and base widening may be disregarded, the voltage drop is expressed by [1] as

$$V_{CE(sat)} = \pm \frac{kT}{q} \ln\left\{\alpha_R\left(1 - \frac{I_C}{h_{FE}I_B}\right)\Bigg/\left[1 + \frac{I_C}{I_B}(1 - \alpha_R)\right]\right\} + R_{tot}I_C, \quad (1)$$

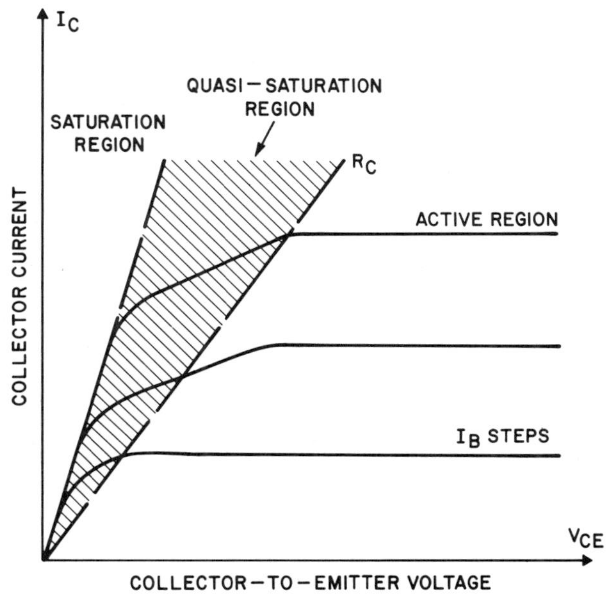

Fig. 1. The three regions of the bipolar transistor collector *I–V* characteristics.

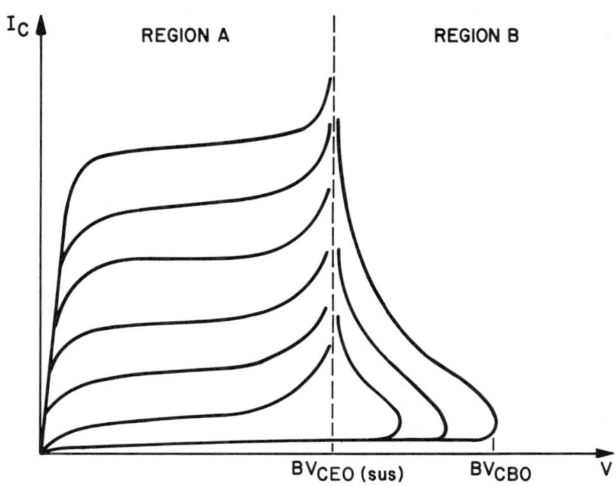

Fig. 2. Bipolar transistor breakdown characteristics.

where α_R is the inverted transistor current gain in the common-base configuration, I_B is the base turn-on current, and R_{tot} is the total series resistance that includes the emitter and collector resistances. The positive sign applies to the *n–p–n* and the negative to the *p–n–p* transistors. Relationship (1) is only valid for $I_C/I_B > h_{FE}$, i.e., for hard driving conditions.

8.2 Quasi-Saturation Region

High-voltage power transistors with the $p^+-p-\nu-n^+$ structure exhibit collector current versus collector voltage characteristics (Fig. 1) with the low current gain region extending from the saturation to active region. This so-called quasi-saturation region arises due to the base widening effects discussed in Chapter 7. As the base widens into the ν-region (Fig. 7.7), the injection efficiency and transport factor deteriorate and decrease the current gain.

An increase of the collector current beyond the critical current J_0'' (low field case) causes the current induced base W_{CIB} (Fig. 7 of Chapter 7) to grow rapidly so that the remaining portion of the ν-region shrinks. A large portion of the applied collector-to-base potential is dropped in the ohmic region of $W_n - W_{CIB}$. There is no appreciable field, though in the extended base W_{CIB}, where local charge quasi neutrality exists. It is maintained by holes moving from the p-base into the W_{CIB} region (Fig. 7 of Chapter 7). The hole injection is made possible by the metallurgical base–collector junction becoming forward biased. The current widened base becomes conductivity modulated even if the metallurgical ***p-base*** is not.

In the $p^+-p-\nu-n^+$ structures, base widening results from the ohmic voltage drop, normally so high that it alone limits the maximum collector current density to such values at which other voltage drops can be neglected. The only voltage drop, therefore, will be as shown by **Olmstead** *et al.* [2]

$$V_{CE1} \simeq J_C\rho_C(W_n - W_{CIB}), \tag{2}$$

where ρ is the ν-region resistivity.

Since the current-induced base width is given by

$$W_{CIB} = W_n - V_{BC}/J_C\rho_C, \tag{3}$$

the onset of base widening takes place when $W_{CIB} = 0$, i.e., when

$$V'_{BC} = \rho_C W_n J_C. \tag{4}$$

Equation (4) indicates that the locus of the critical voltage V'_{BC} as a function of collector current is simply determined by the ν-region resistance R_C (Fig. 1).

A more complete model of the quasi-saturation region was developed by Chudobiak [3], who gives the following expression for the total collector-to-emitter voltage drop in the quasi-saturation region:

$$V_{CE} = V_{BE} - \frac{kT}{q}\ln\frac{N_D P(0)}{n_i^2} - \frac{kT}{q}\ln\left[\frac{J_C W_{CIB}}{2qD_n p'(0)} + 1\right] - J_C\rho(W_n - W_{CIB}), \quad (5)$$

where $p'(0)$ is the injected hole density and $P(0)$ is the net hole density at the edge of the p–ν junction on the collector side. V_{BE} is the voltage drop across the base–emitter junction; the second term on the right-hand side is the voltage drop across the p–ν junction; the third term is the voltage drop in the modulated part of the ν-region (W_{CIB}); and the last dominant term is due to the voltage drop in the unmodulated portion of the ν-region. The third term, arising from the presence of the extended base, is usually on the order of 100 mV and can be neglected in some cases.

The transition from the saturated to the quasi-saturated region [4] takes place when $W_{CIB} = W_n$ at the collector current which can be evaluated on the basis of the ν-region linear hole distribution as (Fig. 7 of Chapter 7)

$$J_T = 2qD_n p'(0)/W_n. \quad (6)$$

Chudobiak's [4] analysis is based on the assumption that the electron density injected by the n^+-emitter into the region between the widened base and n^+-collector is very small, i.e., $n' < N_D$. This limits the model validity to current densities on the order of $qv_s N_D$. Since v_s is the limiting electron velocity, an increase in the collector current density can take place only by the increase of the carrier density n', and the conduction becomes space-charge limited, invalidating the analysis.

8.2.1 Integral-Charge Control Model of Quasi-Saturation Region

From the transistor design point of view, it is desirable to have the saturation region model based on quantities that can be easily determined from simple measurements, as was done by Hower [5, 6], who derived an expression for the collector-to-emitter saturation voltage of an n^+–p–ν–n^+ transistor on the basis of the Moll–Ross and Gummel–Poon models (Chapter 6).

The total number of charges in the widened base per unit area is

$$Q_{\mathrm{Beff}} = Q'_{\mathrm{BO}} + \frac{J_n}{q}\frac{W^2_{\mathrm{CIB}}}{4D_{\mathrm{c}}}, \tag{7}$$

where J_n is the electron current and D_{c} is the effective diffusion constant in the collector region. The second term on the right-hand side of Eq. (7) is the product of the transit time through W_{CIB} and the electron current. This product is equal to the electric charge in W_{CIB}. The collector-induced base width is

$$W_{\mathrm{CIB}} = W_n(1 - I''_0/I_{\mathrm{C}}), \tag{8}$$

as was computed for the low-field base widening case in Section 7.2.2.

The collector current in the high resistivity ν-region is assumed to consist primarily of electrons.

It is also assumed that the transistor transport factor α_{T} is equal to unity, that there is no significant emitter current crowding and that the base is more heavily doped than the collector ν-region. The collector current of an n–p–n triple-diffused transistor is given by

$$I_{\mathrm{C}} = \frac{qn_{\mathrm{i}}^2 A}{Q'_{\mathrm{BO}}/D_n} \exp \frac{qV_{\mathrm{BE}}}{kT}. \tag{9}$$

Under conditions of base widening in an n^+–p–ν–n^+ transistor, Q'_{BO} in Eq. (9) should be replaced by Q_{Beff} (Eq. 7). On the other hand, Q_{EO}, which enters the emitter efficiency expression [Eq. (36) of Chapter 6], is almost unaffected by base widening.

Under these conditions and for a highly interdigitated emitter, the current gain in the common-emitter configuration is given by [6] as

$$h_{\mathrm{FE}} = h_{\mathrm{FEO}} \Big/ 1 + \frac{Q_0 D_n}{Q'_{\mathrm{BO}} D^*_{\mathrm{c}}} \left(\frac{I_{\mathrm{C}}}{I''_0} + \frac{I''_0}{I_{\mathrm{C}}} - 2 \right) \tag{10}$$

for $I_{\mathrm{C}} > I''_0$. In expression (10) h_{FEO} is the current gain in the active region, and Q'_{BO} is the metallurgical base impurity concentration per unit area. The quantities I''_0 and Q_0 are related to the collector–emitter voltage V_{CE}, reflecting the fact that the base widening occurs when the collector–base junction is forward biased:

$$I''_0 = Aq\mu_n N_{\mathrm{D}} V_{\mathrm{C}} / W_n, \tag{11}$$

$$Q_0 = N_{\mathrm{D}} W_n (qV_{\mathrm{CE}}/4kT). \tag{12}$$

Q_0 is a number of charges per unit area. In the expressions (9)–(11) A is the emitter effective area, N_{D} the ν-layer impurity concentration, W_n the ν-layer thickness, and μ_n the electron mobility at low injection levels. D_c^* is related to the electron effective diffusivity D_c in the ν-region by

$$D_c^* = D_c^2/D_n. \tag{13}$$

Because of the carrier–carrier scattering mechanism, D_c is smaller than D_n at high currents. Equation (10) can be solved for V_{CE} utilizing relationships (11)–(13) with the result [6]

$$V_{CE} = I_C R_C \left\{ 1 - \left[\frac{4qAD_c^2}{W_n^2 I_C} \frac{Q'_{B0}}{D_n} \left(\frac{h_{FE0} I_B}{I_C} - 1 \right) \right]^{1/2} \right\}, \tag{14}$$

where R_C is the series resistance of the collector ν-region in the absence of conductivity modulation and is given by

$$R_C = W_n / q\mu_n N_D A. \tag{15}$$

Equation (14) is valid in the quasi-saturation region for $I_C < h_{FE0} I_B$ and $V_{CE} < 0.1$ V. The total saturation voltage is obtained by adding to (14) the small voltage drops in the emitter and collector n^+-regions. All the parameters in (14) can be obtained from measurements, e.g., W_B, W_n, and N_D can be determined from the spreading resistance data, the Gummel number Q'_{B0}/D_n from the V_{BE} versus I_C data in the active region using Eq. (9), and the Gummel number Q_E/D_p from the h_{FE0} measurements in the active region. Equation (14) confirms the rather obvious conclusion that the emitter-to-collector voltage drop can be minimized by making W_n as small as is compatible with the required voltage $V_{CEO(sus)}$. Large emitter areas (high degree of interdigitation) and large h_{FE0} values accomplish the same goal, but increase the device cost. Large Gummel numbers Q'_{B0}/D_n for the base are not advisable as they affect adversely the emitter efficiency. High base doping levels are acceptable, however, in the wide-band emitter devices.

8.3 Collector-to-Emitter Breakdown Voltage

The empirical relationship for the p–n junction carrier multiplication factor (Chapter 3),

$$M = \frac{1}{1 - (V_R/BV)^n} \tag{16}$$

with V_R being the applied reverse potential, is satisfactory for germanium, but is inaccurate for silicon because of the considerable difference existing between the silicon ionization coefficients α_n and α_p for electron and holes.

Moll *et al.* [7] have proposed another, more accurate empirical expression for the multiplication factor M. It introduces an additional coefficient f_n for electrons and f_p for holes so that the avalanche multiplication factor M given by expression (16) is modified to

$$M = \frac{1}{1 - (V_R/fBV)^n} \tag{17}$$

where $f = f_p$ or $f = f_n$ according to carrier type.

The collector-to-emitter breakdown voltage with open base BV_{CEO} is related to the open-emitter breakdown BV_{CBO} for germanium by

$$BV_{CEO}/BV_{CBO} = (1 - \alpha)^{1/n}, \tag{18}$$

where α is the transistor common-base current gain. In view of (17), this relationship for silicon should be modified to

$$BV_{CEO}/BV_{CBO} = f(1 - \alpha)^{1/n}, \tag{19a}$$

where $f = f_p$ or $f = f_n$. Coefficients n and f are dependent on the carrier type and the breakdown voltage BV range. Figure 3 [7] shows the dependence of the coefficients f_p and f_n on junction breakdown voltage for various silicon junction types, such as step, linearly graded (LG), and p–i–n. The coefficient f_n has a range of 0.5 to 0.8, whereas f_p is almost equal to unity. The values of the exponent $n = n_p$ for holes and $n = n_n$ for electrons

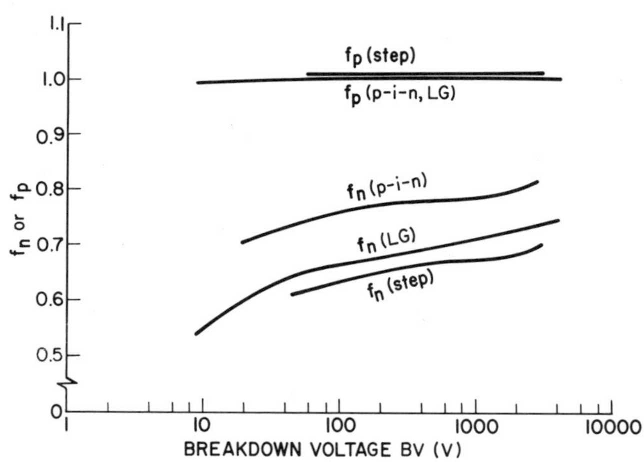

Fig. 3. Dependence of coefficients f_n and f_p on junction breakdown voltage. (From Moll *et al.* [7].)

are given by Fig. 4 [7]. The values of n range between 4 and 20 with the larger values occurring at higher voltages.

Equation (19a) relating the two breakdowns can also be expressed approximately as

$$BV_{\text{CEO}}/BV_{\text{CBO}} = f(1 + h_{\text{FE0}})^{-1/n}, \tag{19b}$$

where h_{FE0} is the common-emitter peak current gain. The BV_{CEO} value, as given by (19a,b), is the *minimum* collector-to-emitter breakdown voltage, i.e., the sustaining collector breakdown voltage in the common-emitter configuration $BV_{\text{CEO(sus)}}$. For reasonably high h_{FE0}, $BV_{\text{CEO(sus)}}$ will be considerably lower than the collector-to-base breakdown voltage (Fig. 2).

The effect of the avalanche multiplication on the common-base current gain alpha is given by $M\alpha$, consequently, the common-emitter current gain becomes

$$h_{\text{FE}} = \text{M}\alpha/(1 - M\alpha). \tag{20}$$

As $M\alpha$ tends to unity, h_{FE} tends to infinity, and the transistor goes into breakdown. When the emitter current is very small, both α and h_{FE} are very small due to low emitter efficiency, and as a consequence, BV_{CEO} approaches the BV_{CBO} value. As the collector voltage is increased slightly above the BV_{CBO} value, the current rises rapidly due to the avalanche multiplication; the current gain increases; and the breakdown voltage decreases to the minimum sustaining value. Transistor operation beyond this point may lead to destructive second breakdown effects.

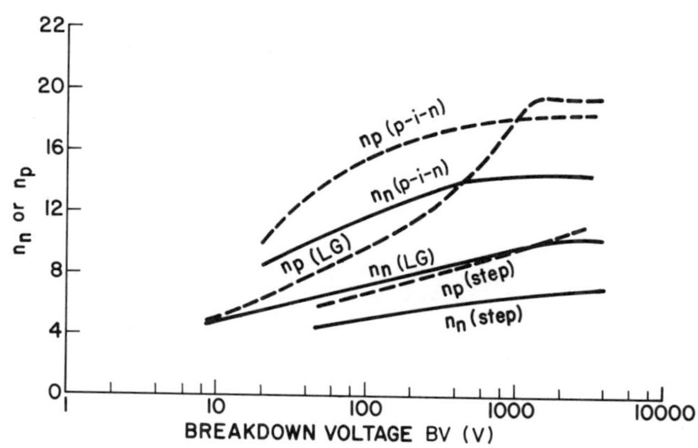

Fig. 4. Dependence of exponents n_p and n_n on junction breakdown voltage. (From Moll *et al.* [7].)

REFERENCES

1. J. J. Ebers and J. L. Moll, *Proc. IRE* **42,** 1716 (1954).
2. J. Olmstead, W. Einthoven, S. Ponczak, and P. J. Kannam, *RCA Rev.* **32,** 221 (1971).
3. W. J. Chudobiak, *IEEE Trans. Electron Devices* **ED-17,** 843 (1970).
4. L. E. Clark, *IEEE Trans. Electron Devices* **ED-16,** 113 (1969).
5. P. L. Hower, *IEEE Trans. Electron Devices* **ED-20,** 426 (1973).
6. P. L. Hower, *IEEE Trans. Electron Devices* **ED-23,** 863 (1976).
7. J. L. Moll, J. L. Su, and A. C. M. Wang, *IEEE Trans. Electron Devices* **ED-17,** 420 (1970).

CHAPTER 9

FREQUENCY RESPONSE. SWITCHING TRANSIENT. MICROWAVE TRANSISTORS

9.1 Bipolar Power Transistors Frequency Response

The gain–bandwidth product f_T largely determines the transistor frequency response. For instance, the knowledge of f_T permits the calculation of the transistor maximum oscillation frequency as

$$f_{max} \simeq \left(\frac{f_T}{8\pi r'_b C_{TC}}\right)^{1/2}, \tag{1}$$

where r'_b is the input resistance of the hybrid π equivalent circuit and C_{TC} is the collector junction capacitance. The maximum available power gain (MAPG) is also related to f_T by

$$\text{MAPG} = f_T/8\pi f^2 r'_b C_{TC} = f_{max}^2/f^2. \tag{2}$$

9.1.1 Gain–Bandwidth Product f_T of n^+–p–ν–n^+ Transistor

The angular gain–bandwidth product at low current levels was given in Chapter 6 by an expression repeated here for convenience

$$\omega_T = g_m/(C_D + C_{TE} + C_{TE}), \tag{3}$$

where g_m is the bipolar transistor transconductance, C_D its diffusion capacitance, and C_{TE} and C_{TC} are the emitter and collector junctions averaged capacitances. The angular bandwidth product is related to the total emitter-to-collector delay time τ_{EC} by

$$\omega_T = \tau_{EC}^{-1}. \tag{4}$$

τ_{EC} is comprised of four terms—the emitter junction capacitance charging time τ_{TE}, the base transit time τ'_F, the collector depletion region transit time τ_x, and the collector junction capacitance charging time τ_{TC}—as shown in the following equation:

$$\tau_{EC} = \tau_{TE} + \tau'_F + \tau_x + \tau_{TC}. \tag{5}$$

The time necessary to charge the emitter transistor capacitance from zero or reverse bias to the forward bias is equal to

$$\tau_{TE} = n(kT/qI_E)C_{TE}(0), \tag{6}$$

where I_E is the emitter current and $C_{TE}(0)$ is the total emitter junction capacitance at 0 V. The n-factor in (6) is introduced to take account of the fact that C_{TE} varies with the emitter voltage during the charging time. The factor $n \simeq 4$ for very approximate calculations.

The base transit time at high injection levels when the collector current exceeds the critical current for base widening is given by

$$\tau'_F = (W_B^2/4D_n) + (W_{CIB}^2/4D_c), \tag{7}$$

where W_B and W_{CIB} are the metallurgical and collector induced base widths, respectively. D_n is the electron diffusion constant and D_c is the effective electron diffusivity in the collector-induced base W_{CIB}.

The transit time τ_x through the space-charge region is equal to

$$\tau_x = (W_n - W_{CIB})/2v, \tag{8}$$

where W_n is the ν-region thickness (Fig. 7 of Chapter 7) and v the carrier velocity in the collector depletion region. The velocity v is related to the base-to-collector voltage V_{BC} by

$$v = \mu_n E = \mu_n(V_{BC} + V_i)/(W_n - W_{CIB}), \tag{9}$$

where E is the electric field in the depletion region and V_i is the built-in voltage. Expression (9) is valid for low collector fields. In high collector fields the mobile carrier velocity becomes equal to the limiting velocity v_s.

The collector junction capacitance charging time τ_{TC} can be calculated from

$$\tau_{TC} = R_C C_{TC}, \tag{10}$$

where R_C is the collector series resistance.

The total base transit time τ'_F is strongly related [Eq. (7)] to the collector-induced base width W_{CIB}, which increases with the increasing current density until it reaches its maximum value W_n equal to the ν-region width. As a result, the transit time increases, and f_T decreases with increasing current density. This phenomenon is known as Kirk's effect [1]. In the case of the two-dimensional widening model [2], all the terms of Eq. (5) remain essentially unchanged with the exception of the effective base transit time, which now becomes

$$\tau''_F = \frac{W_B^2}{4D_n} + \left(\frac{I_C}{I_0} - 1\right)^2 L_E^2 \Big/ 4D_c, \tag{11}$$

where W_B and L_E are defined by Fig. 11 of Chapter 7.

The transit time τ_x through the collector depletion region should always be based on the limiting velocity v_s since the lateral spreading, according to [43], can take place only when the collector voltage is high enough to make the space-charge region electric field greater than 10^4 V cm^{-1}. For fields lower than 10^4 V cm^{-1}, Kirk's effect dominates over the Van der Ziel–Agourdis effect.

The first-order gain–bandwidth f_T fall-off theory shows a considerable discrepancy with experimental results. Kumar and Hunter [4] attribute this discrepancy to the presence of an extra charge unaccounted for in Kirk's theory. When an n^+–p–ν–n^+ transistor operates with currents greater than the critical current J''_0 for base widening, the collector-to-base junction is forward biased. Consequently, this junction is responsible for electron injection into the p-base and hole injection into the ν-collector region. The presence of the charges stored in these regions increases the collector capacitance by more than one order of magnitude. This results in a considerable increase of the collector charging time, which accounts largely for the above noted discrepancy.

9.2 Bipolar Power Transistors Switching Transient

With a purely resistive load, the switching transient can be described by Fig. 1. Upon the application of the on-pulse I_{B1}, the current starts charging the emitter junction transition capacitance. This charging time and the time necessary for the carriers to reach the collector contribute to the delay time t_d, which, from the practical point of view, is definec as the time necessary for the collector current to reach 10% of its peak value.

The time needed to reach 90% of the peak value is known as the rise time t_r. When the base current pulse polarity is reversed (I_{B2}), the current continues

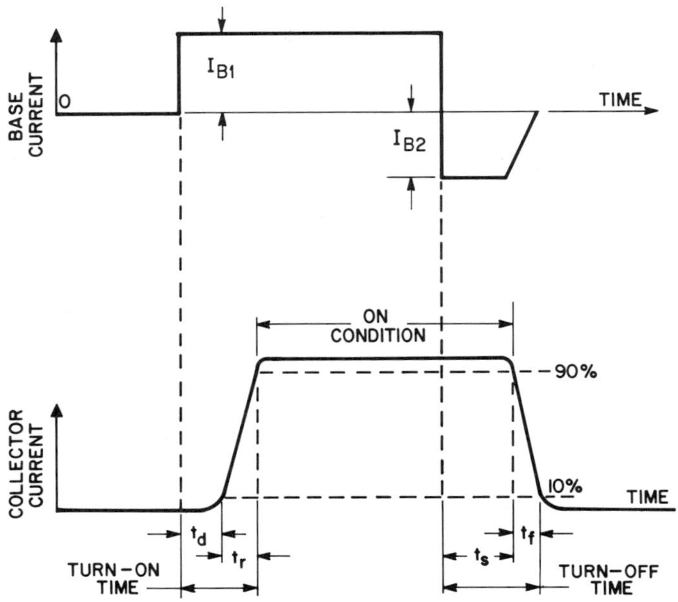

Fig. 1. Idealized bipolar transistor switching waveforms.

to flow for a time t_s known as the saturation time, due to device-stored charges. After the stored charges are removed, the current starts falling off, reaching 10% of its maximum value after the time interval known as fall time t_f.

Large signal transient response for transistors operating at low current densities without base widening and for one-dimensional structures was treated by Moll [5] and by Ebers and Moll [6], who derived comparatively simple expressions for all transistor time delays.

The transient response of high voltage one-dimensional transistors with n^+–p–ν–n^+ structures was studied by Hower [7], who applied the basic charge control concepts to predict the HV power transistor turn-on characteristics. As we have seen in Chapter 6, the charge-control analysis leads to the expression of the time dependent base current as

$$I_B(t) = \frac{dq_B}{dt} + \frac{q_B}{\tau_{\text{eff}}}, \tag{12}$$

where q_B is the total excess charge in the base and τ_{eff} is the minority carrier lifetime in the base. Equation (12) can be rewritten as

$$I_B(t) = \frac{dq_B}{dt} + \frac{q_B}{h_{FE}\tau_F'}, \tag{13}$$

since $h_{FE} \simeq \tau_{eff}/\tau'_F$, where transit time τ'_F is given by Eq. (7), which shows its dependence on the collector current through that term which includes the collector-induced base width W_{CIB}. The time dependent collector current can be determined from the charge-control analysis by

$$J_C(t) = q_B/\tau'_F. \tag{14}$$

Since the base transit time τ'_F and minority carrier lifetimes are collector current functions, the base and collector current expressions (13) and (14) are nonlinear so that the simplicity of the charge-control analysis for the HV transistor case is lost. To recover the charge-control simplicity, Hower [7] uses a piecewise approach using separate sets of *linear* relationships for the active region and the saturation region with widened base. The same equations, (13) and (14), describe the base and the collector currents in both regions, but with different constants. In the active region, τ'_F is replaced by τ_F, the transit time in the metallurgical base; and h_{FE0}, the current-gain peak value, is substituted for h_{FE}. In the saturation region, the transit time in the effective base is assumed to be current independent and equal to

$$\tau'_{Feff} = m(W^2_{CIB}/4D_c). \tag{15}$$

The constant m is positive and less than unity and accounts for the fact that

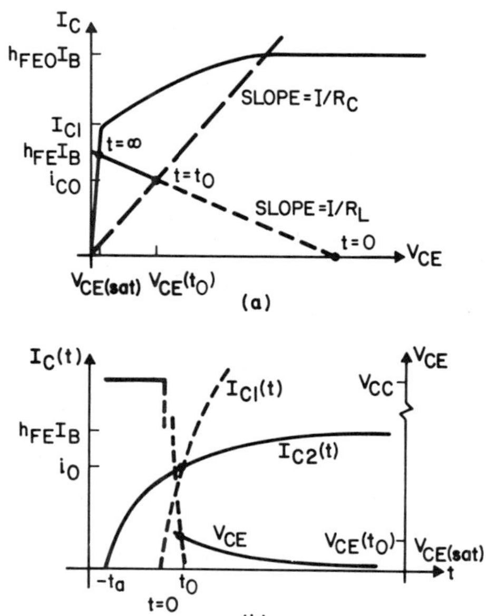

Fig. 2. (a) Static and (b) dynamic current–voltage characteristics of the n^+–p–ν–n^+ transistor. (From Hower [7].)

during the turn-on the induced base width varies from zero to W_{CIB}. If W_{CIB} reaches its maximum value W_n, then $m \simeq 0.5$. In (13) the current forced gain I_C/I_B replaces h_{FE}.

Figure 2 [7] illustrates the static and dynamic characteristics of the $n^+-p-\nu-n^+$ transistor. Upon the application of the turn-on pulse, both the collector current $I_C(t)$ and the charge q_B increase with a time constant $h_{FE0}\tau_F$ up to the time $t = t_0$. At t_0, the situation changes since the collector current reaches the critical current I_0 at which the base starts widening. This condition is given by

$$I_{C1}(t_0) = I_0 = V_{CC}/(R_2 + R_C), \tag{16}$$

where R_2 and R_C are the load and collector resistances, respectively, and V_{CC} is the collector supply voltage. For $t > t_0$, the transit time through the effective base is equal to $\tau_F' > \tau_F$ so that the collector current $I_C(t)$ continues increasing with a slower rate pictured by the $I_{C2}(t)$ curve of Fig. 2b.

Prior to the base widening for $t < t_0$, the solution of Eqs. (12) and (13) gives the collector current as a function of time as

$$I_{C1}(t) = h_{FE0}I_B[1 - \exp(-t/\tau_1)] \tag{17}$$

with the time constant $\tau_1 = h_{FE0}\tau_F$. I_B is the turn-on base current pulse.

The solution of the charge control equations (12) and (13) in the second region for $t > t_0$ give for the collector current

$$I_{C2}(t) = h_{FE}I_B[1 - \exp - (t + t_a)/\tau_2], \tag{18}$$

where $\tau_2 = h_{FE}\tau'_{F(\mathrm{eff})}$ and t_a (Fig. 2b) ensures the collector current continuity at $t = t_0$. This constant is given by

$$t_a = -[t_0 + \tau'_{\mathrm{Feff}} \ln(1 - I_0/h_{FE}I_B)]. \tag{19}$$

In the derivation of Eq. (18), the effective transit time in the widened base τ'_{Feff} was taken into account.

The instantaneous collector-to-emitter voltage $V_{CE}(t)$ is also plotted in Fig. 2b. Its value can be determined from

$$V_{CE}(t) = V_{CC} - I_C(t)R_L, \tag{20}$$

where V_{CC} is the collector supply voltage and R_L is the collector load resistance.

The results of the charge-control transient analysis of a one-dimensional transistor without base widening show that the collector rises with a single time constant. On the other hand, Hower's two-region model predicts the existence of two time constants. The time constant arising from the existence of base widening dominates for high current densities for which W_{CIB} is nearly equal to the collector ν-region width W_n.

To turn off the $n^+ - p - \nu - n^+$ transistor, it is necessary to remove all the charges stored in the p-base and ν-collector regions and to discharge the emitter and collector capacitances. The charge removal from a realistic device is a two-dimensional process. Upon the application of the reverse base–emitter bias, the base current flowing laterally under the emitter toward the base contact causes a lateral voltage drop, which tends to concentrate the emitter current at the emitter stripe center. The removal of the stored charge in the base under the wide emitter stripe would therefore be much longer than the one expected from one-dimensional analysis.

Figure 3 [8] represents the rise, storage, and fall times of the 2N6693 high voltage (650 V) transistor as a function of the collector current at the junction temperature $T_j = 25°C$.

Figure 4 [8] shows the effect of the junction temperature on various switching times. The increase of all of them with temperature is due to the increase with temperature of the minority carrier lifetime and transistor gain.

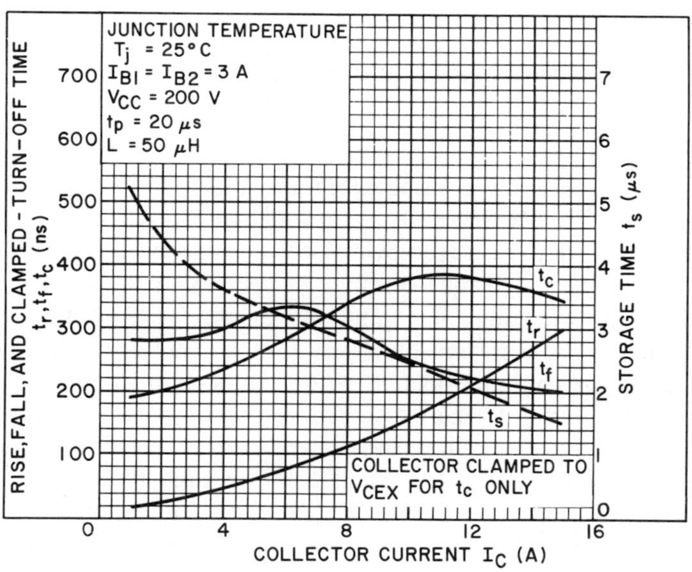

Fig. 3. Rise (t_r), storage (t_s), fall (t_f), and clamped-turn-off (t_c) times of the 2N6693 transistor. (Courtesy of RCA Solid State Division [8].)

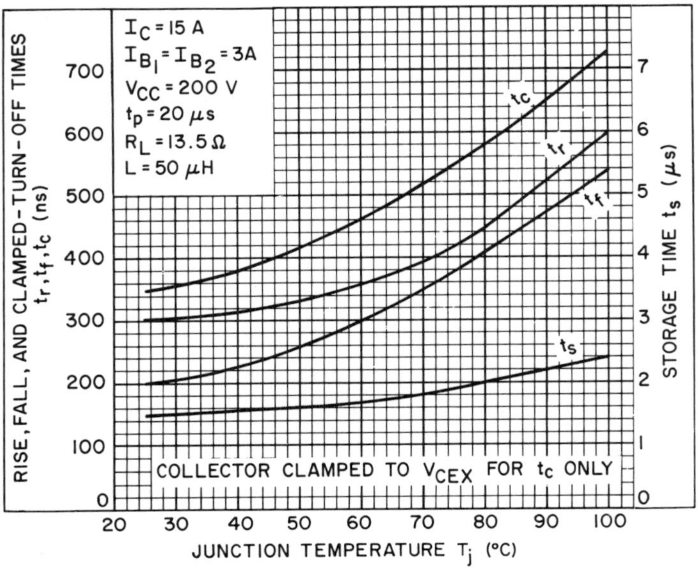

Fig. 4. Effect of the temperature on the various switching times. (Courtesy of RCA Solid State Division [8].)

9.3 Bipolar Microwave Transistors

At microwave frequencies (f > 1 GHz), most of the current is injected near the emitter edges as a result of comparatively low current gains and resulting transverse debiasing effects. In order to increase the current capability, microwave devices are therefore constructed with very narrow emitters with long peripheries for the purpose of decreasing the current density, minimizing base widening effects, and maximizing f_T. The gain–bandwidth product will also be increased by minimizing the emitter transition capacitance C_{TE}, whose charging time, as we saw in Section 9.1, affects the total delay time. The same is true for the collector transition capacitance C_{TC}. The small emitter area requirements lead to a very high ratio of the emitter periphery to emitter area. The small collector area requirement, on the other hand, necessitates a very large ratio of the emitter periphery to the base area [9]. Among the major design geometries meeting these requirements are the overlay with the base-grid structure and the mesh with the emitter-grid structure. These designs are adaptable to emitter ballasting, to good rf current distribution, and to the reduction of capacitances. The

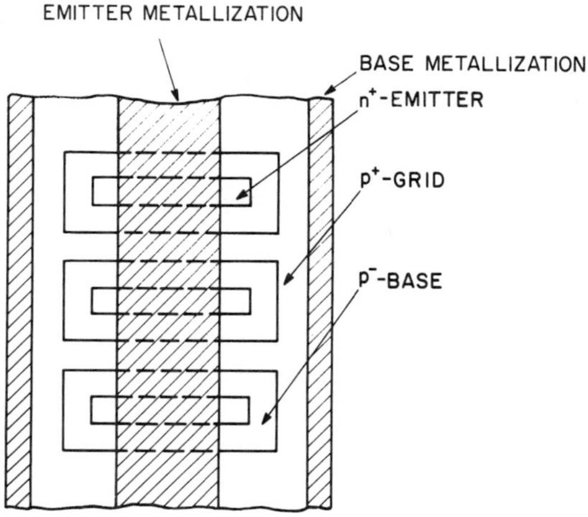

Fig. 5. Method of connecting multiple emitter sites in an overlay-type transistor.

overlay structure [10] is schematically represented by Figs. 5 and 6. It uses a large number of discrete emitter sites connected in parallel by a thin overlaying metallization as shown in both figures. The base surface consists of a p^+-diffused grid. The base current is collected from the p^+-grid into the base metallization. The emitter ballasting consists of a resistive polycrystalline layer introduced between the n^+-emitters and the emitter metallization in the fashion shown in Fig. 6. Each emitter is ballasted individually to ensure uniform rf current distribution, thus avoiding hot spot formation.

The mesh-type geometry (Fig. 7) is essentially an inverse of the overlay geometry: the n^+-emitter consists of a grid, but the p^+-base contact areas consist of multiple, discrete sites. Each of these geometries has its own ad-

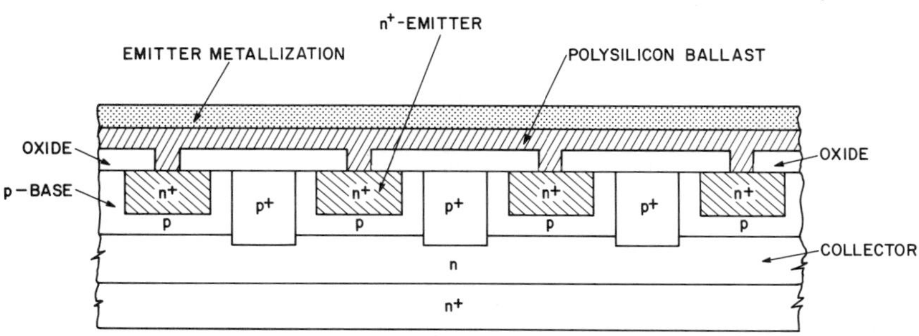

Fig. 6. A schematic cross section of an overlay-type microwave transistor, showing polysilicon ballasting emitters.

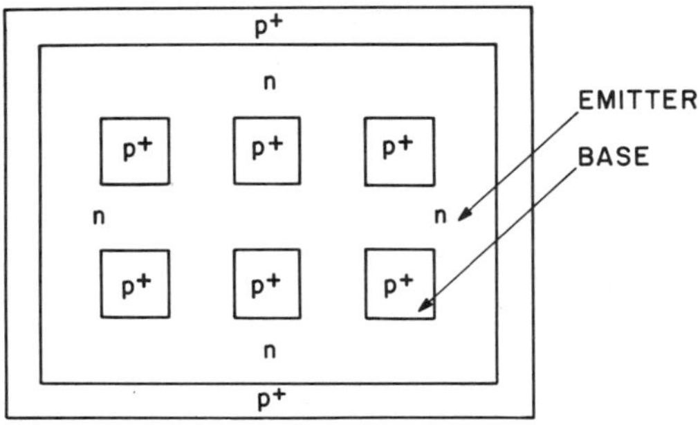

Fig. 7. Mesh-type geometry with p^+ multiple base contact areas.

vantages and drawbacks. The mesh-type geometry, for example, requires fewer processing steps than the overlay. The current handling capability of both designs is on the order of 0.5 A cm^{-1} of the emitter periphery. At present, silicon microwave transistors are mainly used for frequencies below 4 GHz. GaAs MESFETs (Chapter 11) are generally preferred for frequencies above 4 GHz.

REFERENCES

1. C. T. Kirk, Jr., *IEEE Trans. Electron Devices* **ED-9,** 164 (1962).
2. A. van der Ziel and D. Agourdis, *Proc. IEEE* **54,** 411 (1966).
3. D. L. Bowler and F. A. Lindholm, *IEEE Trans. Electron Devices* **ED-20,** 257 (1973).
4. R. Kumar and L. P. Hunter, *IEEE Trans. Electron Devices* **ED-22,** 51, 1031 (1975).
5. J. L. Moll, *Proc. IRE* **42,** (1954).
6. J. J. Ebers and J. L. Moll, *Proc. IRE* **42,** (1954).
7. P. L. Hower, *IEEE Trans. Electron Devices* **ED-23,** 863 (1976).
8. File Number 1165, RCA Solid State Division, Somerville, New Jersey (1979).
9. D. S. Jacobson, *Microwaves* **11** (July), 2 (1972).
10. D. R. Carley, P. L. McGeough, and J. F. O'Brien, *Electronics* **70,** 38 (August 23, 1965).

CHAPTER 10

TRANSISTOR THERMAL PROPERTIES. INSTABILITIES

Power transistor circuit performance is critically limited by the device's thermal behavior under steady state and transient conditions. Filamentary current conduction and hot spot formation may result in electrical and thermal instabilities, leading to forward and reverse second breakdowns, which also restrict the device's operation.

10.1 Junction Temperature

Power dissipation in a transistor takes place in any operating conditions; it is the highest, however, when the transistor remains either in the on condition or is being switched off from the highly conducting into the non-conducting state. At high frequency repetitive switching, the turn-on and turn-off losses may even exceed the on-state losses.

Under steady state conditions, the maximum temperature reached by the device is determined by the dissipated power and the overall thermal resistance of the device and its cooling system. Lower power devices are cooled by the heat convection and radiation of their package to the ambient air, whereas the higher power devices are mounted on readily conducting heat sinks.

The product of the dissipated power and junction-to-heat-sink thermal resistance determines the temperature rise above the heat sink temperature, i.e.,

$$\Delta T = P_T R_{th}(t), \tag{1}$$

where P_T is the dissipated power and $R_{th}(t)$ is the time dependent thermal resistance. $R_{th}(t)$ is somewhat dependent on the temperature, but this is disregarded in this discussion.

For silicon, the maximum allowable junction temperature is about +200°C above which the leakage currents become excessive and the high resistivity silicon crystal becomes intrinsic.

For a transistor subjected to a single, nonrepetitive, short-duration power pulse, the maximum allowable power dissipation during the transient period is substantially greater than the transistor steady state dissipation capability. This comes about because the device has a certain thermal capacitancc C_{th} and its temperature may never reach its allowed peak value if the devices thermal time constant

$$\tau_{th} = R_{th(max)} C_{th} \tag{2}$$

is larger than the pulse duration. $R_{th(max)}$ in (2) is the steady state thermal resistance.

The time dependent thermal resistance, known also as thermal impedance, is a more general concept than the steady state thermal resistance. Figure 1 [1] represents transient thermal resistance normalized to its steady state value as a function of time for a transistor mounted on an infinite heat sink. The transient thermal resistance is always smaller than the steady state value. In practice, the heat sink is not infinite, and the R_{th} versus the time curve deviates from the plot of Fig. 1 [2, 3] and depends on the actual heat sink dimensions. The peak thermal resistance can be computed in degrees Kelvin per watt from the relation

$$R_{th(max)} = L/kA, \tag{3}$$

where k is the material thermal conductivity in watts per centimeter degree Kelvin and A is the area normal to the thermal path L in square centimeters. The silicon conductivity k is equal to about 1.41 W cm^{-1} °K^{-1} at room temperature. The thermal capacitance C_{th} is the product of the material specific heat C and mass m:

$$C_{th} = Cm. \tag{4}$$

C_{th} is expressed in watt seconds per degree Kelvin if C is given by watt

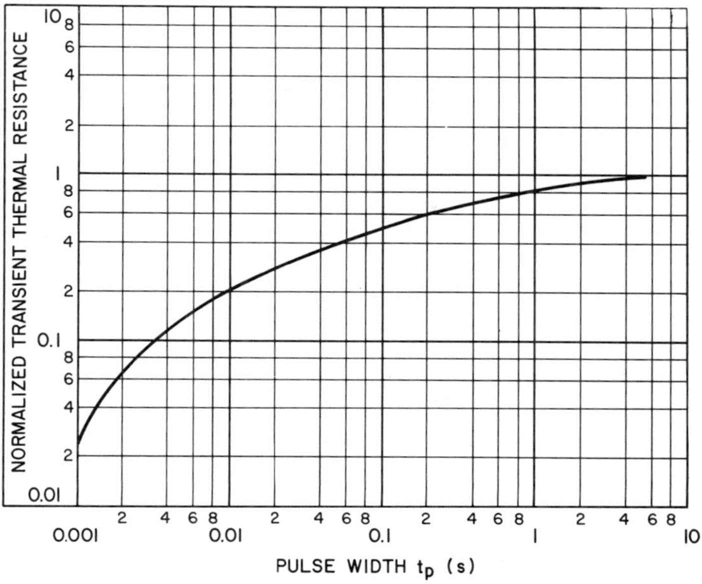

Fig. 1. Normalized transient thermal resistance as a function of pulse width. Transistor mounted on an infinite heat sink. (Courtesy of RCA Solid State Division [1].)

seconds per gram degree Kelvin and m in grams. The product of the thermal capacitance and the temperature increment ∇T above the heat sink temperature is equal to the amount of the heat absorbed by the body

$$Q_{\text{th}} = C_{\text{th}}\,\Delta T \quad \text{W s.} \tag{5}$$

10.2 Transistor Thermal Equivalent Circuit

The heat flow from the heat source to the heat sink is described by

$$\partial T/dt = D_{\text{th}}\,\nabla^2 T + q_{\text{th}}/Cd, \tag{6}$$

where T is the temperature, t the time, d the material density, and q_{th} the heat generated per unit volume. The thermal diffusivity D_{th} is given by

$$D_{\text{th}} = k/Cd. \tag{7}$$

Equation (6) is analogous to the semiconductor current continuity equation, but in the steady state reduces to

$$\nabla^2 T = -q_{\text{th}}/k, \tag{8}$$

analogous to Poisson's equation. The heat flow equations may be solved

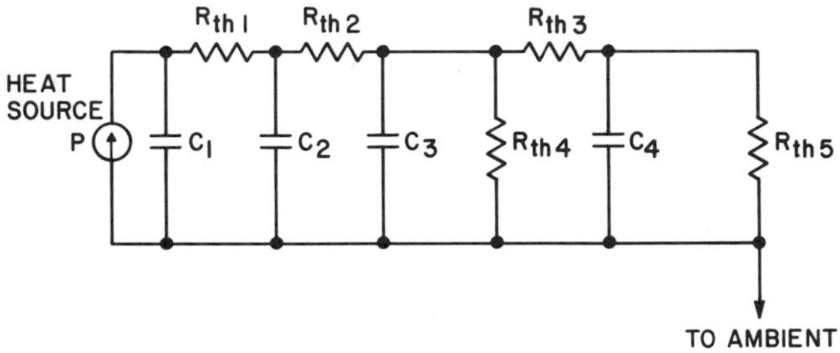

Fig. 2. Transistor's thermal equivalent circuit.

numerically on a computer. In most cases, very satisfactory solutions may be obtained by representing the transistor heal flow analogously to current flow in an electrical transmission line analog (Fig. 2) consisting of the heat source P and lumped thermal capacitances and resistances. C_1 is the silicon crystal thermal capacitance, and C_2, C_3, and C_4 are the solder, the case, and the heat sink thermal capacitances, respectively. $R_{th\,1}$ and $R_{th\,2}$ are the silicon crystal and solder thermal resistances, respectively. R_{th3}, the thermal resistance from the device case to the heat sink, may become appreciable if the case is insulated from the heat sink by the interposition of such insulating materials as mica or beryllia (BeO). The thermal conductivity of beryllia is good compared to other dielectrics. Its value is about 1.7 W cm^{-1} $°K^{-1}$. Mica's thermal conductivity, on the other hand, is rather low about 4×10^{13} W cm^{-1} $°K^{-1}$. R_{th4} is the case thermal resistance to the ambient air, and R_{th5} represents the heat sink thermal resistance to the ambient air. R_{th4} and C_3 are negligible when a good heat sink is used.

10.3 Thermal Runaway

The cooling rate of a power transistor should always be greater than the heating rate arising from the transistor dissipation. If the cooling system is inadequate and the junction temperature increases faster than the cooling rate allows, then a thermal runaway occurs because of thermal regeneration, i.e., positive thermal feedback. Let us assume that the transistor is provided with a heat dissipating element whose surface temperature is T_S and that the thermal resistance to the ultimate heat sink (ambient air) is equal to R_{sth}. If the ambient temperature is T_A, the dissipative element surface temperature, assumed to be uniform, is

$$T_S = T_A + R_{sth}(P_1 + P_2) \tag{9a}$$

with

$$P_1 + P_2 = P_T. \tag{9b}$$

P_T is total power dissipated, P_1 is its component that is temperature independent, and P_2 is the temperature dependent component. Dissipation P_2 that varies with T_S causes further increase of P_2, which in turn increases T_S and so on, leading to a thermal regeneration analogous to a positive electrical feedback. A transistor may exhibit thermal regeneration which depends largely on its bias arrangements. The most stable condition takes place for a constant emitter–current bias; a less stable situation exists for constant base–current biasing, and the least stability is obtained when the base–emitter voltage is provided by a constant voltage source. Differentiation of Eq. (9) with respect to T_S results in [4]

$$\frac{dT_A}{dT_S} + R_{sth}\frac{dP_T}{dT_S} = 1 \tag{10}$$

if the thermal resistance R_{sth} is treated as a temperature independent quantity. Equation (10) can be easily transformed to

$$dT_S/dT_A = (1 - R_{sth}dP_T/dT_S)^{-1}. \tag{11}$$

The thermal runaway situation exists when $dT_S/dT_A \to \infty$, i.e., when

$$dP_{2(max)}/dT_S = 1/R_{sth}. \tag{12}$$

The temperature sensitive power dissipation component of a transistor usually is of the form

$$P_2 = P_0 \exp[(T_S - T_0)/\theta], \tag{13}$$

where T_0 is the temperature at which $P_2 = P_0$ and θ is the thermal time constant. Combining Eqs. (12) and (13) gives for the temperature sensitive power dissipation

$$P_{2(max)} = \theta/R_{sth}. \tag{14}$$

As the device thermal resistance increases, the likelihood of a runaway is greater. A small thermal time constant θ that corresponds to a very fast rate of change of the temperature sensitive component P_2 may cause thermal runaway, even if its initial value is very small. On the other hand, a reasonably high fixed dissipation P_1 has no detrimental effect.

In Eq. (14) the P_2 value is referenced to the surface temperature T_S. However, the P_2 value can also be referenced to the temperature T'_S that is due only to P_1 and T_A. In this case [4],

$$P'_{2(max)} = \theta/eR_{sth}, \tag{15}$$

and thermal runaway will take place at the operating temperature T'_S if the temperature-sensitive dissipation P'_2 satisfies Eq. (15).

10.4 Principle of Least Entropy Generation. Current Filaments Formation. Hot Spots

There are two types of differential negative resistances: voltage controlled and current controlled. In the first one (Fig. 3a), the electric field E may have a few different values for the same device current. In the current-controlled differential resistance, on the other hand, the device current may be multivalued for the same electric field E (Fig. 3b [5]). The appearance of a voltage-controlled differential negative resistance leads to the electrical instabilities that may result in the formation of high and low field domains. In the case of the current-controlled differential negative resistance, the material is not split in domains, but in high current filaments is oriented in the electric field

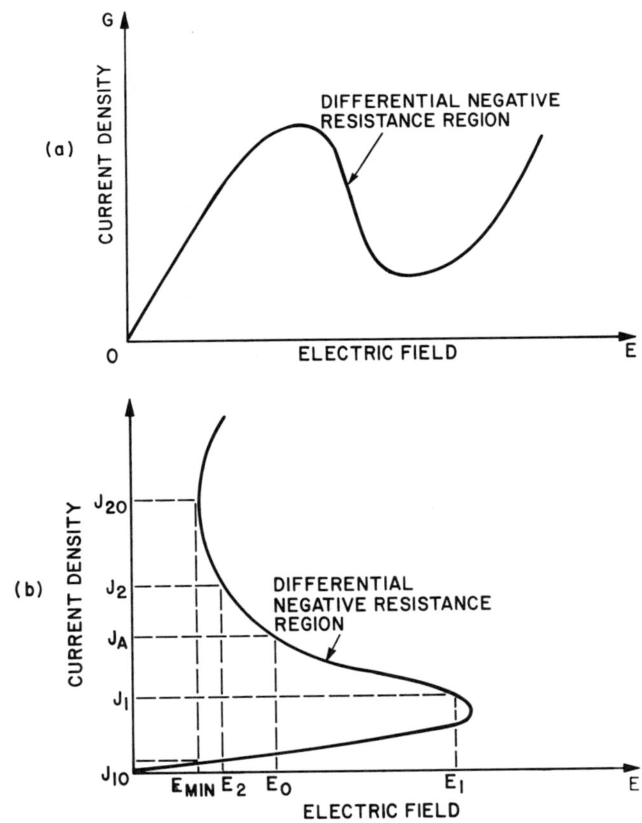

Fig. 3. (a) Voltage-controlled differential negative resistance (b) current-controlled differential negative resistance. (From Ridley [5], by permission of The Institute of Physics.)

direction (Fig. 4). The appearance of these domains and filaments was predicted by Ridley [5] on the basis of the principle of least entropy generation. This principle and the device dimensions determine the filament or domain sizes. The entropy S can be expressed as

$$S = C \ln W, \tag{16}$$

where C is a constant and W is a measure of disorder, i.e., the number of ways a given system can be rearranged. In a slab of electrically isotropic material of unit cross-sectional area and of unit length, the entropy flow J_s in the presence of an electric and thermal gradient is given by

$$-\operatorname{div} J_s = (dS/dt) - S_G, \tag{17}$$

where S is the entropy per unit volume and S_G is the entropy generation rate per unit volume. This equation is analogous to the electric current continuity equation with S_G being analogous to the carrier generation rate. In the steady state $dS/dt = 0$, whereas at equilibrium both the entropy flow J_s and the entropy generation rate S_G are zero. It is then expected that in order to be as close to the equilibrium state as possible, e.g., in the steady state, the entropy generation rate should be as small as possible. This argument is the basis of the least entropy generation principle, which we shall now apply to the case of current-controlled negative resistance often appearing in transistors.

If the semiconductor internal temperature gradients are small, then at times that are long compared to the material dielectric relaxation time, i.e., *at steady state, the entropy generation rate is due solely to Joule heating* and is given by

$$S_G = EJ/T, \tag{18}$$

where E is the applied electric field, J the current density, and T the absolute temperature. S_G is the sum of the entropy generation rate in the low and high current regions, i.e., in the outside and inside of the filament. If A is the

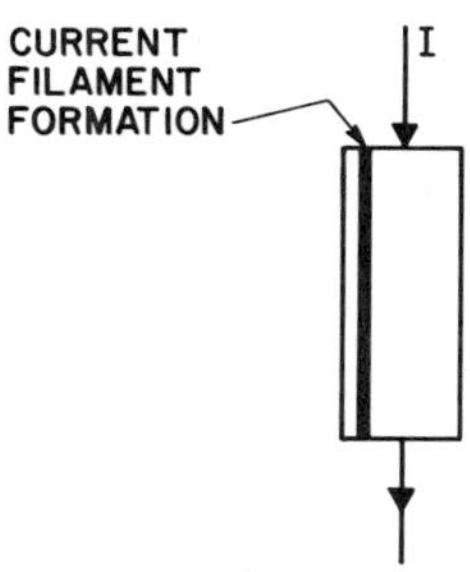

Fig. 4. Current filament formation.

total cross-sectional area of the semiconductor slab and a is the filament cross sectional area, then under steady state conditions (Fig. 3b)

$$TS_G = E_1 J_1 (A - a) + E_2 J_2 a \tag{19}$$

with

$$J_1 (A - a) + J_2 a = J_A, \tag{20}$$

where J_A is the applied current before the filament formation. J_2 and E_2 are the current density and electric field, respectively, in the filament of area a, and J_1 and E_1 are the current density and electric field, respectively, in $(A - a)$. Under steady state conditions, the field should acquire the minimum value $E_{\min}$. Solving Eq. (19) for a and substituting for $J_1 = J_{10}$ and for $J_2 = J_{20}$ corresponding to the minimum field, we obtain for the filament cross section the following expression:

$$a = A[(J_A - J_{10})/(J_{20} - J_{10})]. \tag{21}$$

10.4.1 Electrical Stability[1]

The filamentary current flow arises for two reasons: for the need of minimum entropy generation and for the sake of electrical stability. Before the current fluctuation filament occurs, the electric field is equal to E_0 (Fig. 3b). The appearance of the filament with its current density J_2 changes the field in the filament to a different value equal to E_2 (Fig. 3b). The total applied current J_A remains constant so that a complementary current fluctuation takes place in the remaining cross-sectional area $(A - a)$. Again, this current fluctuation causes a change in the electric field over the $(A - a)$ area. The two current density fluctuations are equal to

$$\Delta J = J_2 - J_A. \tag{22}$$

The new electric fields appearing in the filament and the rest of the material can be modified only if the space-charge adjustment takes place at the contacts. Poisson's equation in the space-charge region is

$$\nabla E = \rho/\epsilon_s \tag{23}$$

with ρ being the charge density and ϵ_s the material permittivity. The electric charge density is related to the current by

$$\nabla J = \partial\rho/\partial t. \tag{24}$$

[1] See Ridley [5].

Therefore

$$-\epsilon_s \nabla(\partial E/\partial t) = \nabla J. \quad (25)$$

Integration of Eq. (25) over the space-charge region gives

$$d/dt(E_2 - E_0) = (J_2 - J_A)/\epsilon_s, \quad (26)$$

where E_0 is the electric field in the absence of the current filament. From the geometrical considerations (Fig. 3b),

$$E_2 - E_0 = \rho_{diff} \Delta J_2, \quad (27)$$

where $\rho_{diff} = dE/dJ$ is the differential resistivity and $\Delta J_2 = J_2 - J_A$. Solving (27) for the current fluctuation in the filament gives [5]

$$\Delta J_2 = \delta J_A \exp(-t/\epsilon_s \rho_s), \quad (28)$$

where δJ_A is the initial current fluctuation and $\epsilon_s \rho_s$ is the dielectric relaxation time of the material. The initial current fluctuation, as can be seen from (28), grows exponentially with time for negative differential resistivities. For simplicity, we have assumed that the negative resistivity ρ_s is constant. In reality, this is not the case; it changes with the varying current, and as the current grows, it reaches its minimum absolute value and becomes equal to zero at point (E_{min}, J_{20}) (Fig. 3b). This point is stable, and there is no further current growth with time; an electrical steady state is reached at which the electric field again becomes uniform.

10.4.2 MICROPLASMAS

Due to the presence of several types of pointlike crystal defects, there are always some crystal regions where the current density becomes higher than in the rest of the crystal. These nonuniformities lead to current filament formations known as microplasmas. The microplasmas conduct electric currents both by electron and holes. Their diameter is typically on the order of a few micrometers. A mechanism of microplasma formation applicable to the *p–i–n* structures, which differs only in detail from that of Ridley [5], has been proposed by Muller [6]. As the diode reverse bias is increased beyond the avalanche breakdown voltage, the current density starts growing. The avalanche multiplication probability rises faster than the increasing electric field. This gives rise to current negative incremental resistance because the steady state voltage required to maintain the current becomes a decreasing function of the current density. As a consequence, the electric field resulting from the presence of excess voltage is largest at crystal defect points where the current density is highest, and as a result, a current filament known as microplasma is created. A steady state, i.e., a stable condition, will exist within the microplasma region if the region loss of carriers due to their out-diffusion is balanced exactly by avalanche multiplication.

10.4.3 Hot Spots (Mesoplasmas)

Stable microplasmas may initiate unstable phenomena, known as hot spots or mesoplasmas, when their temperature exceeds the intrinsic temperature characteristic of the given crystal doping density. The intrinsic carrier concentration in silicon is a strong function of temperature and is given by expression (29) of Chapter 2. At 300°K, $n_i \simeq 1.4 \times 10^{10}\ \text{cm}^{-3}$, and it increases by a factor of two for each 11°K increment. The temperature at which the intrinsic carrier concentration becomes equal to the crystal doping density is known as the intrinsic temperature. Its value may be obtained from Eq. (29) of Chapter 2. The intrinsic temperature for the $10^{14}\ \text{cm}^{-3}$ doping level is equal to about 230°C; for the $10^{16}\ \text{cm}^{-3}$ doping, it comes close to 450°C.

As the temperature of the microplasma exceeds the intrinsic temperature because of some localized heating, the current in the filament of the microplasma increases, establishing a positive thermal feedback that results in the formation of a large hot spot and possible device thermal runaway. This is one of the possible second breakdown mechanisms. Hot spots can also be found in a transistor which does not exhibit any microplasmas or crystal defects. The initiating mechanism may be just due to the current-controlled negative resistance as described by Ridley [5] (Section 10.4.1), which leads to current filament formation.

The hot spot temperature is high enough in most cases to destroy the device, either by melting metals used in the construction of the device or by melting the silicon itself. Very often, before reaching that point, the crystal cracks because of high temperature gradients [7].

The temperature at which mesoplasma is created depends on the crystal doping level and on the input pulse duration [8].

10.5 Second Breakdown

Transistor instabilities due to current constrictions are known as second breakdowns. Forward second breakdown (lateral thermal instability) takes place when the current is heavily crowded either at the emitter edges or at any other emitter–base regions while the transistor is forward biased. Another type of second breakdown takes place when the transistor is being turned off from the conducting state by the application of the reverse base–emitter bias. In this case, the hot spot may develop due to an electric instability. Both types of second breakdown lead to the device's degradation and, in most cases, to its destruction by melting the contacting metals and the silicon crystal.

10.5.1 Forward Second Breakdown ($I_{s/b}$)

With the *n–p–n* transistor emitter forward biased, the base electron current flows laterally from under the emitter toward the base contact, leading to the current crowding effect. The current may be focused into a comparatively small area with the current density high enough to initiate thermal instability and second breakdown. A similar effect may take place because of the existence of some inhomogeneity in the emitter–base junction region. The filamentary current flow may take place even in a perfect material under the right circumstances; it occurs, though, at lower current levels in the presence of pointlike crystal defects.

Power transistor emitters often consist of multiple emitter fingers (stripes) for the purpose of increasing the emitter periphery and decreasing the current density. It is, however, almost impossible to maintain uniform current distribution in such a structure without the use of special precautionary techniques such as emitter resistive ballasting (Fig. 5), providing each elementary emitter with a small series resistance. Such resistances are usually integrated together with the device. Without these precautionary steps, once one of the emitter fingers warms up more than the adjacent fingers, the current starts growing at the warmer emitter site at the expense of the colder neighboring

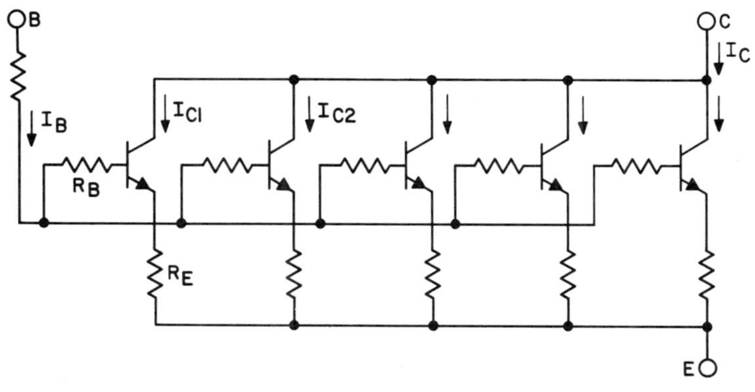

Fig. 5. Equivalent circuit of a transistor with multiple emitters. R_B and R_E are base and emitter series resistances.

fingers. The current will be robbed from other areas until it becomes concentrated in one hot spot.

10.5.1.1 Thermal Instability Model

The onset of the thermal instability in the one-dimensional transistor can be determined from the stability factor expression [9, 10]

$$S = R_{th} V_{CE}\, \partial I_C / \partial T, \tag{29}$$

where R_{th} is the transistor thermal resistance between its active region and the heat sink and T is the active region temperature. Expression (29) can be derived similarily to Eq. (12) with $dP_2 = V_{CE} dI_C$. It assumes that the dissipated power density is uniform and that the temperature T over the silicon die–heat sink interface is uniform. In reality, this is never the case, and both the power density and temperature very over the die surface. It appears, therefore, that the two-dimensional analysis would be more appropriate for the determination of the stability factor. Such a numerical analysis was performed by Hower and Govil [1], who solved the two-dimensional heat flow equations for a transistor mounted on a heat sink and have demonstrated both theoretically and experimentally the validity of the one-dimensional stability model by comparing it with a two-dimensional model.

The I_C versus V_{CE} instability locus can be determined from the relationship

$$I_C = R_{th}^{-1}\, \Delta T V_{CE}^{-1}, \tag{30}$$

where the transistor thermal resistance R_{th} is assumed to be temperature independent and

$$\Delta T = T - T_A \tag{31}$$

is the junction temperature increment above the constant ambient or heat sink temperature T_A. The absolute junction temperature T can be determined from the stability factor S expression by making it equal to unity. To determine S for a double diffused transistor, it is assumed that the current gain $h_{FE} \geqslant 5$, that the device is forward biased with $V_{BE} \gg kT/q$, and that the base is relatively heavily doped. Under this last condition, the base charge Q_B becomes more or less current independent, and there is no conductivity modulation. The device is supposed to operate without current crowding. Using the Moll–Ross model, the collector current can be expressed as

$$I_C = \frac{q A_E n_{i0}^2}{Q_{B0}/D_B} \exp\left(\frac{qV_{BE} - E_g}{mkT}\right). \tag{32}$$

A_E is the metallurgical emitter area, $n_{i0} = 2.46 \times 10^{19}\ \text{cm}^{-3}$ for silicon, D_B is the diffusion constant for the minority carriers in the base region. Designating the emitter and the base series resistances by R_E and R_B, respectively, the voltage drops in the emitter and the base add up to the applied emitter–base voltage V_{BB} as

$$V_{BB} \simeq V_{BE} + I_C[R_E + R_B/h_{FE}]. \tag{33}$$

The current gain is a function of temperature and can be determined from

$$h_{FE} = \text{const} \cdot \exp(-\Delta E/kT), \tag{34}$$

where ΔE is the activation energy. If the transistor operates at a point where the current gain varies only slowly with current, then

$$h_{FE} = h_{FE0} = (Q_{E0}/D_E)/(Q_{B0}/D_B) \tag{35}$$

where Q_{E0} and Q_{B0} are the emitter and base charges in equilibrium, respectively. By combining Eqs. (32)–(35), we obtain the stability factor expression [11]

$$S = \left(1 - \frac{T_A}{T}\right) \frac{m(kT/q)\ln(I_1/I_C) + (I_C R_B/h_{FE})(\Delta E/kT)}{m(kT/q) + I_C(R_E + R_B/h_{FE})}, \tag{36}$$

where $m = 1$ for the double diffused transistor and

$$I_1 = qA_E n_{i0}^2/(Q_{B0}/D_B). \tag{37}$$

Figure 6 [11] is the plot of the instability loci determined from (3) [11] with the base ballasting resistor R_B as a parameter and an emitter resistance

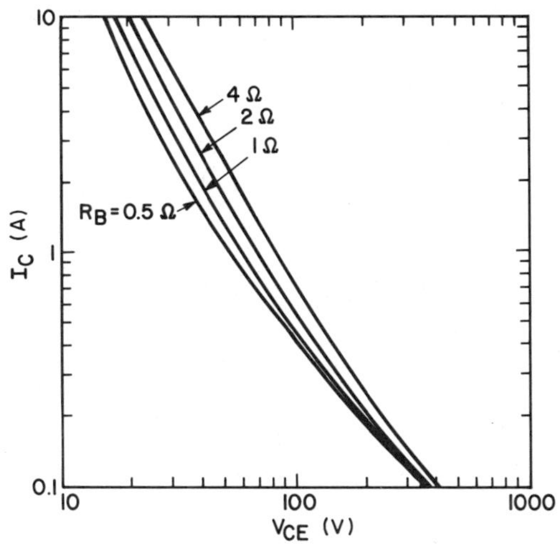

Fig. 6. Plot of the instability loci with base ballasting resistor R_B as a parameter. (From Hower and Govil [11].)

$R_E = 5\ \Omega$, $\Delta E = 0.1$ eV, $h_{FE0} = 50$, $A_E = 0.1\ \mathrm{cm}^2$, and $T_A = 298°\mathrm{K}$. The effect of the base ballasting resistance depicted by Fig. 6 is rather small. The base ballasting can appreciably increase the energy for the second breakdown only when the transistor is designed so that ΔE_g is negligibly small and the current gain is almost temperature independent. Figure 7 compares the theoretical and experimental values for two double diffused transistors designed with two different emitter ballasting resistances R_E. Both transistors, otherwise, had the same design parameters: $A_E = 0.18\ \mathrm{cm}^2$, $h_{FE0} = 20$, $R_B = 0.1\ \Omega$, $R_{th} = 0.62°\mathrm{C\ W}^{-1}$, and $\Delta E = 70$ meV. Increasing R_E from 10 to 20 mΩ increases I_C from about 1.5 to 4 A at $V_{CE} = 40$ V. The beneficial effect of the emitter ballasting is rather dramatic in the low voltage, high current region. At higher voltages, the locus approaches a hyperbola, $I_C V_{CE} =$ const, and can be modified mainly for a fixed maximum dissipation by minimizing the device thermal resistance.

To determine the instability locus for the single diffused transistor, Hower and Govil [11] assumed that the device base is lightly doped ($\sim 10^{14}\ \mathrm{cm}^{-3}$) and the metallurgical base width is in the range of 20 to 35 μm, which is reasonable for transistors with breakdown on the order of a few hundred volts. For this case, the base charge Q_B is proportional to the collector current I_C as the base is conductivity modulated. The collector current is given by the same expression (32) as for the double diffused case, but with $m = 2$. Taking into account base narrowing due to the Early effect,

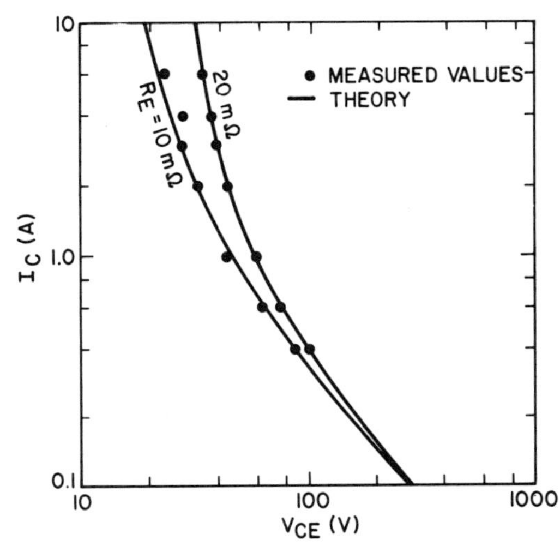

Fig. 7. Plot of the instability loci with emitter ballasting resistor R_E as a parameter. (From Hower and Govil [11].)

the stability factor according to [11] is the same as for the double diffused structure [Eq. (36)], but with $m = 2$ and

$$I_1 = I_2 = 2qA_E D_B n_{i0}/W_B. \tag{38}$$

The presence of factor $m = 2$ results in better single diffused transistor stability. Except at the highest V_{CE} values, the single diffused transistor is superior to the double diffused one from the thermal stability point of view.

10.5.2 Reverse Second Breakdown ($E_{s/b}$)

Forward-bias second breakdown is satisfactorily explained by the lateral thermal instability. In the case of the second breakdown occurring when the emitter–base junction is reverse biased, as during the device turn-off of the inductive loads, the initiating mechanism is attributed to electrical instability. The base current for the reverse biased emitter of an n–p–n transistor flows in the direction from the emitter toward the base contact (Fig. 8 [12]) so that the emitter–base junction becomes more forward biased at its center than at its periphery and the center injects more heavily. During the turn-off time, in the presence of the inductive load, the operating point may traverse the transistor breakdown region. Holes generated by the avalanche process are collected by the negative base contact, and the hole current adds up to the base current, reinforcing still further the current focusing effect. Figure 9 [13] depicts the I_C–V_C locus during an inductive switching. As the transistor is being turned off from the on-state, the current remains constant until the stored charges are withdrawn, and then rapidly falls off while the collector voltage rises to the breakdown value BV_{CEX}, which depends on the magnitude of the base series resistance R_B. BV_{CEX} is the collector-to-

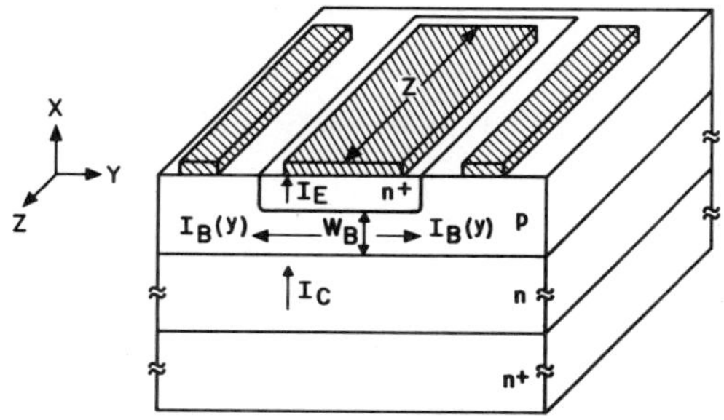

Fig. 8. Lateral current flow in transistor's base region. (From Beatty *et al.* [12].)

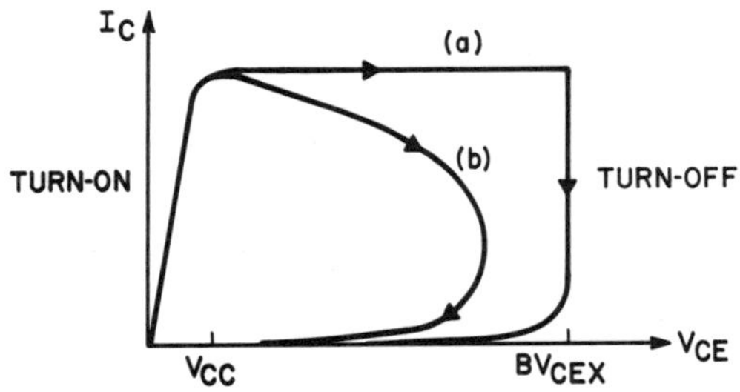

Fig. 9. I_C–V_C characteristic during inductive switching (a) for a short time constant T_C, (b) for a long time constant T_C. (From Krishna and Hower [13].)

emitter breakdown with specified circuit between the base and emitter. The I_C–V_C locus depends on the circuit time constant, which is equal to

$$T_C = L/R'_C, \tag{39}$$

where L is the collector circuit inductance and

$$R'_C = (BV_{CEX} - V_{CC})/I_C. \tag{40}$$

(V_{CC} is the collector supply voltage.) If the transistor fall time is small compared to T_C, then the voltage rise time is very rapid, and the current follows curve (a) of a Fig. 9. For fall times on the order of T_C or greater, the current follows the curve (b) of Fig. 9. In the case of the very rapid current variation, the high current and high voltage occur at the same time, leading to the high instantaneous dissipation and, more importantly, to the reverse second breakdown. As already mentioned, the current flows mainly from the emitter central portion because of the reversed base current. The mobile carrier density becomes higher than the doping concentration of the n–p–ν–n transistor ν-region so that the electric field and the avalanche breakdown are determined primarily by these mobile charges. The peak field shifts (Section 7.2.1) from the base–collector to the ν–n^+ junction, and BV_{CEO} decreases with the increasing current density. The transistor then exhibits the negative differential resistance: an increase in current lowers the voltage. Of course, the base control is lost under these circumstances. Once the collector-to-emitter voltage exceeds $V_{CEO(sus)}$, the current will be entirely focused at the emitter center, and the perimeter will be reverse biased.

The phenomena of avalanche breakdown due to the mobile carriers rather than to the collector fixed impurities is known as avalanche injection [12–14]. Avalanche injection is the cause of double injection so that the collector current is carried both by electrons and holes in a manner similar to that of the thyristor anode current [15]. It is important to know the voltage at which avalanche injection is initiated. This can be accomplished by solving Poisson's equation for the transistor collector region.

For the n–p–ν–n type power transistor structure, the electric field in the depleted ν-region in most cases is higher than 10^4 V cm^{-1} so that the electron drift velocity reaches its maximum value $v_s = 10^7$ cm s^{-1}. Due to the current constriction existing at the emitter stripe center, the collector current density will have some maximum value J_{max}, and the Poisson's equation that takes into account the presence of mobile carriers can be written as

$$dE/dx = (q/\epsilon_s)[N_D - J_{max}/qv_s], \tag{41}$$

where N_D is the ν-region doping concentration and y is the direction along the emitter (Fig. 8 [12]). In order to solve this equation for E, one must first find J_{max}.

The emitter current density may be expressed by the diffusion equation [12]:

$$J_E(y) = \frac{qD_n}{W_B(x, y)} n'(y)\left[1 + \frac{n'(y)}{N_{BO} + n(y)}\right], \tag{42}$$

where W_B is the metallurgical and induced base width, N_{BO} is the base impurity concentration under the emitter, and $n'(y)$ is the emitter injected electron concentration. The current at the emitter terminal is related to J_E by

$$I_E = 2Z \int_0^{\delta} J_E(y)\, dy, \tag{43}$$

with Z being the emitter length in the z direction. The integration is carried out from the emitter stripe center at $y = 0$ to $y = \delta$, where the lateral voltage drop per unit emitter length becomes $V(\delta) = 2kT/q$ (beyond this point, the injected current density becomes negligibly small).

The lateral voltage V per unit emitter length is given by

$$\frac{dV(y)}{dy} = \frac{1}{W_B(x, y)} \frac{1}{\sigma(y)} I_B(y), \tag{44}$$

where $I_B(y)$ is the lateral base current. The conductivity $\sigma(y)$ consists of the unmodulated component σ_0 and the modulated part σ_1, which was shown [12] to be

$$\sigma_1(y) = \tfrac{2}{3} q\mu_{eff} n(y) \tag{45}$$

with

$$\mu_{eff} = \tfrac{1}{2}\mu_n + \tfrac{3}{2}\mu_p. \tag{46}$$

μ_n and μ_p are electron and hole mobilities, respectively, that take into account the carrier–carrier scattering effects.

The injected electron density is given by

$$n'(y) = (n_i^2/N_{B0}) \exp[qV(y)/mkT], \tag{47}$$

where $m = 1$ for low injection levels and $m = 2$ for high injection levels. Equations (41)–(47) were computer solved with the boundary conditions at $y = 0$, $dv/dy = 0$, and $V = V(0)$, and for the n–p–v–n transistor with impurity concentration profile very similar to that of Fig. 3 of Chapter 6. The solution yielded emitter current density distribution J_E as a function of distance y from the emitter stripe center, as shown in Fig. 10 [12]. The maximum current density J_{Emax} occurs at the stripe center.

While computing J_E, it was important to also verify by computation whether or not the voltage developed across the emitter–base junction exceeded emitter junction avalanche breakdown for which the analysis is not valid. The computed value of J_{max} is introduced into Poisson's equation (41) to obtain the electric field E as a function of depth x as shown by Fig. 11 [12]. The E versus x straight lines have the slope given by the right-hand side of Eq. (41) with collector currents of 10 A cm^{-2}, 500 A cm^{-2}, and 1000 A cm^{-2} as parameters. The collector voltage was 500 V.

Knowing the maximum field, it is possible to compute the avalanche multiplication factor M and the voltage V_p that is necessary to initiate avalanche injection and second breakdown. The results of the numerical analysis compared very well with the experimental data [12] obtained by using an experimental test circuit which forced the transistor under test into second breakdown and sensed the exact time of its occurrence. To avoid device destruction and/or degradation, a special crowbar, bypassing inductive electrical energy, was automatically applied after the second break-

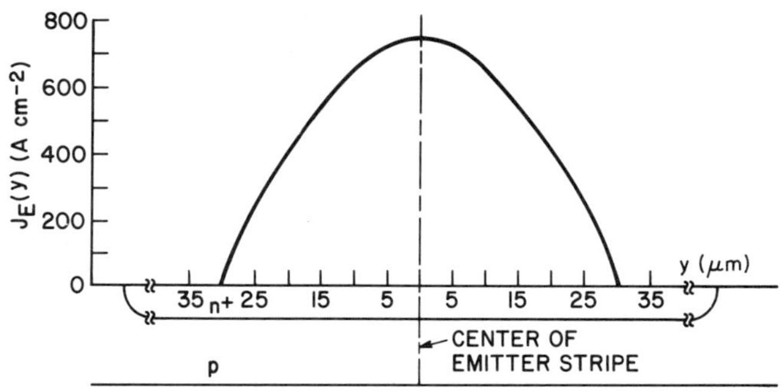

Fig. 10. Emitter current density $J_E(y)$ as a function of distance y from the emitter stripe center. (From Beatty *et al.* [12].)

down took place. To give an idea of the magnitudes involved, we quote here the physical parameters of one of these devices fabricated with diffusion profiles similar to that of Fig. 3 of Chapter 6: the emitter was diffused to a depth $x_{jE} \simeq 5.1\ \mu m$ and with a surface concentration $N_{E0} = 5 \times 10^{19}\ cm^{-3}$; the base surface concentration was $N_{B0} = 3 \times 10^{16}\ cm^{-3}$ and its metallurgical width $W_B = 7.38\ \mu m$; the ν-region width $W_n = 54\ \mu m$ and the collector voltage $V_{CB0} = 700$ V. The device was tested with collector current $I_C = 10$ A and turn-on and turn-off base currents $I_{BF} = I_{BR} = 1$ A. The computed J_{max} was equal to 755 A cm^{-2}, giving for the second breakdown initiating voltage $V_p = 715$ V. The measured V_p value was somewhat lower and equal to 680 V. With higher collector current $I_C = 12$ A, $J_{max} = 836$ A cm^{-2}, and $V_p = 635$ V, as compared to the experimentally obtained 600 V.

In an earlier work Krishna and Hower [13] gave an approximate expression for the critical current density that triggers avalanche injection as

$$J_{crit} = qv_s N_D + \epsilon_s v_s E_{crit}/W_n, \tag{48}$$

where $E_{crit} \simeq 1 \times 10^5\ V\ cm^{-1}$ is the critical field for avalanche carrier generation. Expression (48) is derived from Poisson's equation (41). For a transistor with the same values for N_D and W_n as above, the critical current is $J_{crit} = 537$ A, which differs significantly from the J_{max} value.

It was observed by Beatty *et al.* [12] that although the collector current and the test circuit load inductance were kept constant, the value of V_p varied

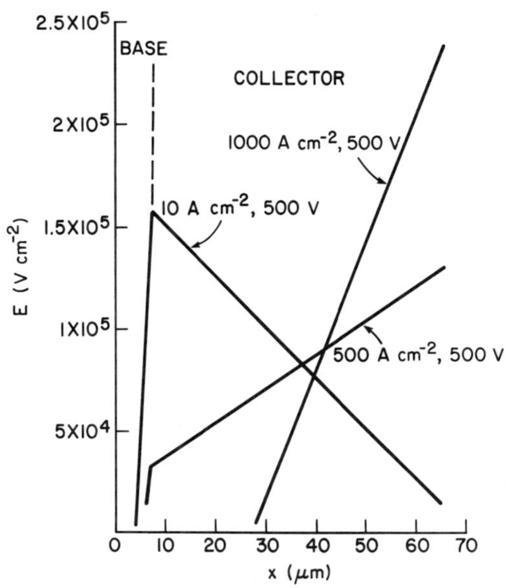

Fig. 11. Electric field as a function of depth x. (From Beatty *et al.* [12].)

from one device to the next, leading to the conclusion that the standard method of rating transistors for second breakdown with the reactive energy $1/2LI_C^2$ may be in error.

According to Beatty *et al.* [12], the main significant difference between the forward- and reverse-biased second breakdowns is the higher current density for the reverse case. It often is ten times higher than the current density for the forward-bias case and results in avalanche breakdown and injection in a localized region. It is also important to note that the threshold for forward-biased second breakdown is substantially reduced with increasing junction temperature, but the opposite is true for the reverse second breakdown: V_p increases with temperature, as expected for avalanche breakdown. This confirms that the initiating mechanism for the reverse second breakdown is due to electrical instability and not to thermal effects as in the case of the forward second breakdown.

10.5.2.1 Reverse Second Breakdown Improvement Techniques

Switching transistors may be protected from second breakdown by punch-through diodes (Section 10.5.2.2) or diode clamps connected in parallel with the collector. Additional components increase circuit cost so it is advantageous to design switching transistors with increased ability to withstand second breakdowns. One of the techniques used at present for the n^+–p–ν–n^+ transistors consists of a collector n-region sandwiched between the n^+- and ν-collector regions (Fig. 3 of Chapter 6). The doping level of the n-region and its thickness are adjusted to keep the electric field at the n–n^+ interface at a low value, avoiding premature breakdown when the mobile charge (electrons) density exceeds that of the ν-region concentration.

Owyang and Shafer [16] recently proposed an emitter structure that is useful in the reduction of the current focusing effect during the transistor turn-off. Figure 12 depicts this new structure in which the emitter central

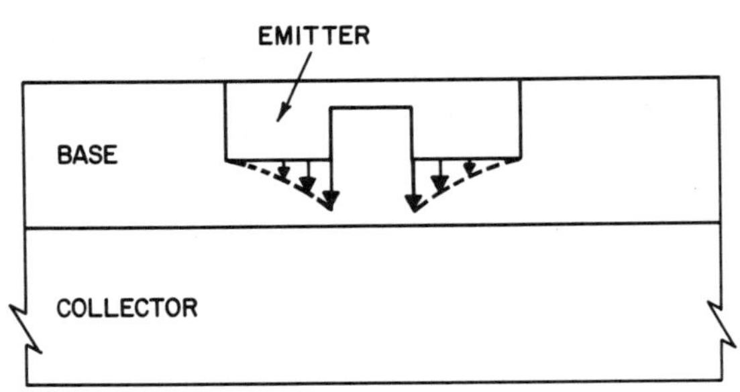

Fig. 12. Emitter with a narrow, lightly doped central portion. (From Owyang and Shafer [16].)

portion is narrower and much less doped than the edge regions. The doping level in the center is only slightly higher than that of the *p*-base, and the injection is negligible as compared to the rest of the emitter. With this emitter structure, current focusing at the device center is made impossible. Another advantage claimed for this emitter type is the reduction of the transistor fall time because the base charges are stored closer to the base contacts and can be removed faster and more efficiently than the charges at the emitter center. In the case of standard emitter stripe geometry, the base resistance is highly increased when the emission takes place in the emitter central portion because the regions between the base contacts and the emitter center cease to be conductivity modulated.

10.5.2.2 Punch-Through Diode as Collector Clamp

A punch-through limited diode may be used in place of a Zener diode to clamp the transistor collector-to-emitter voltage to a safe value protecting the device from second breakdown. Its advantage lies in its great power handling capability [17, 18] determined by the device thermal dissipation. In the punch-through diode, the whole device area conducts the current, whereas in the avalanche-type diodes, there is always a possibility of a localized current conduction and increased thermal resistance. The punch-through diode is designed so that the punch-through takes place much before the avalanche breakdown at the surface or bulk silicon.

The device structure (Fig. 13 [18]) consists of an n^-, high resistivity region sandwiched between two highly doped p^+-regions. The diode is mounted on a heat sink.

The punch-through voltage in the absence of the current flow can be

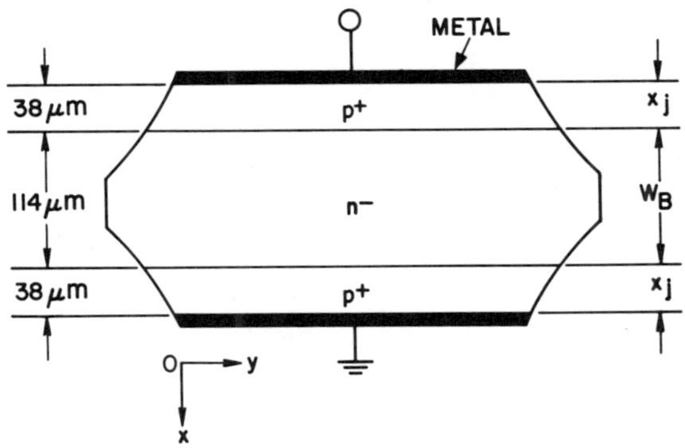

Fig. 13. Cross section of a punch-through diode. (From Kannam [18].)

determined from the well-known relation valid for abrupt junctions and uniformly doped n^--regions:

$$V_{pt} = (q/2\epsilon_s)N_B W_B^2, \tag{49}$$

where N_B is the n^--region impurity concentration and W_B its width.

In high current density J presence, the solution of Poisson's equation yields, for the punch-through voltage, the expression

$$V'_{pt} = (q/2\epsilon_s)(N_B + J/qv_s)W_B^2, \tag{50}$$

which assumes that the field in the high resistivity n^--region exceeds 10^4 V cm^{-1} and the carrier velocity reaches the limiting thermal velocity v_s.

The punch-through diode power handling capability is determined from

$$P_{max} = JV'_{pt}A, \tag{51}$$

where A is the device area for current conduction. Figure 14 [18] represents

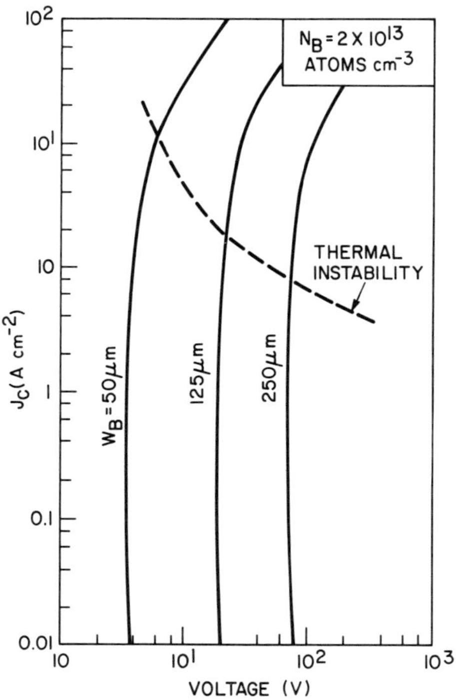

Fig. 14. V–I characteristics of a punch-through diode with base width W_B as a parameter. The interrupted line marks the thermal instability onset. (From Kannam [18].)

the punch-through diode V–I characteristics for $N_B = 2 \times 10^{13}$ cm^{-3} and various W_B values. The dotted line marked thermal instability was calculated by Kannam [18] on the basis of stability criteria of Navon and Miller [19].

10.6 Safe Operating Area (SOA)

It is necessary for the circuit designer to know what the safe limits of transistor operation are in the current–voltage (I_C–V_{CE}) plane. Beyond some limits, the device operation, either in the forward or reverse mode, may result in thermal and electrical instabilities leading to the device's degradation or destruction. The safe operating limits are provided by the forward or reverse SOA (an acronym for the safe operating area) published by device manufacturers.

10.6.1 Forward SOA

A typical SOA plot is shown in Fig. 15 [1] for an n–p–n power transistor. The upper line marked $I_{C\max}$-continuous indicates the limit to the collector current beyond which, regardless of the power dissipated in the transistor itself, the operation becomes unsafe because of the possible damage to the transistor connections and/or the package. The next limiting region belongs to the line marked dc operation. This region arises due to the device's thermal limitation imposed by the device's thermal resistance and maximum temperature of operation. The maximum permissible temperature rise above the heat sink reference temperature is given by

$$\Delta T_{\max} = (I_C V_{CE})_{\max} R_{th}. \quad (52)$$

If the thermal resistance of the transistor R_{th} is assumed to be temperature independent, then

$$(I_C V_{CE})_{\max} = \text{const} \quad (53)$$

or

$$I_C = \text{const} \cdot V_{CE}^{-1}, \quad (54)$$

which when plotted on the log–log scale results in a straight line with a -1 slope known as the SOA thermal limit. The assumption of a constant thermal resistance is usually reasonable enough in the high current low voltage operating region [20]. The forward second breakdown $I_{s/b}$ limit lies below the thermal limit region and refers to the maximum current allowed prior

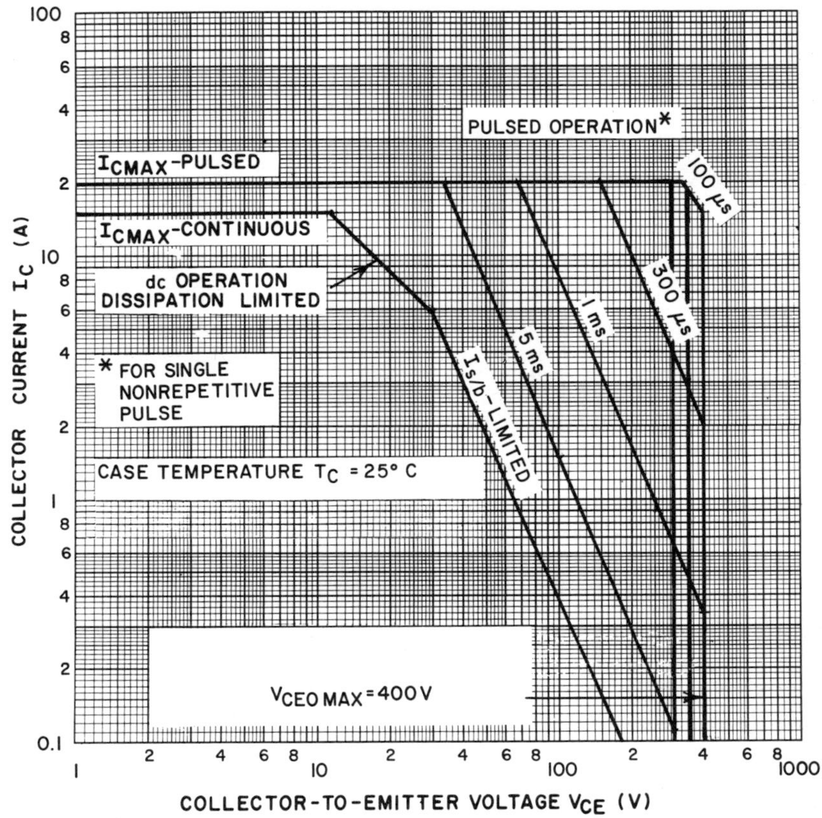

Fig. 15. Maximum safe operation areas (SOA) of a HV transistor. (Courtesy of RCA Solid State Division.)

to the occurrence of the forward second breakdown. Operation below the limiting curve assures freedom from the second breakdown; it is questionable, however, whether or not the transistor is adequately protected from the thermal overload, since in this region of high voltage and comparatively low current, the thermal resistance is usually strongly dependent on operating conditions [20].

The energy level at which the second breakdown occurs increases when current pulses rather than continuously flowing current are used. The shorter the nonrepetitive pulse duration is, the larger the maximum allowed energy as demonstrated by the SOA curves (Fig. 15). Pulse operation also increases the safe operating area for the regions limited by the thermal dissipation and the maximum current. The 25°C SOA graphs published by the manufacturers are accompanied by temperature derating curves (shown in Fig. 16 [1]), which indicate how much to derate the current at a constant

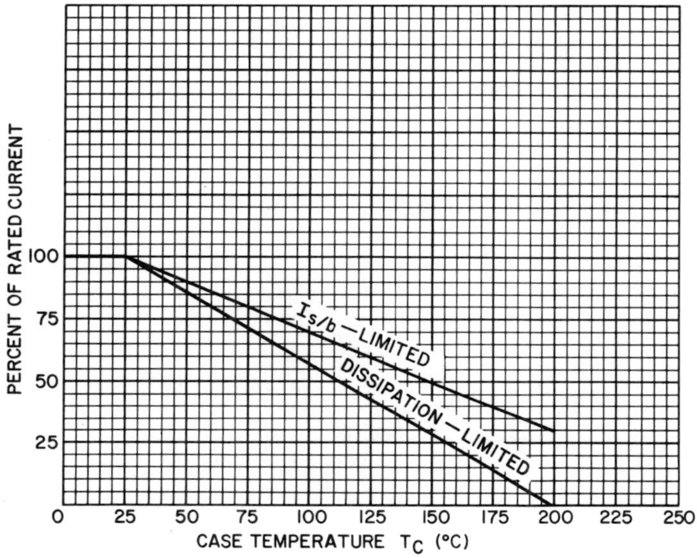

Fig. 16. Current derating curves in the dissipation-limited and $I_{s/b}$-limited portions of SOA curves. (Courtesy of RCA Solid State Division [1].)

voltage in the dissipation limited and forward second breakdown limited portions of the SOA curves. The curves do not derate the specified value of I_{Cmax}.

10.6.2 Reverse Bias Safe Operating Area (RSOA)

While switching inductive loads, it is possible to exceed the collector–emitter sustaining voltage $V_{CEO(sus)}$ at which avalanche generation becomes significant enough to increase the common-base current gain α above unity. The base current for $\alpha > 1$ is in the same direction as the reverse base current used to switch the transistor off. The total base current becomes large enough to cause current pinch-off at the emitter center, leading to the reverse second breakdown. If the power density is sufficiently high, the transient thermal gradient may exceed 10^5 °C cm^{-1}. It is quite conceivable that such temperature gradients may cause a mechanical stress leading to the silicon physical cracking [21]. The purpose of the RSOA is to give the circuit designer the information about the maximum permitted values of the emitter-to-collector voltage for a given base turn-on current I_{BF} with the base turn-off current I_{BR} as a parameter (Fig. 17 [22, 23]). The second breakdown plot is obtained using a circuit which permits transistors to operate for a short time in the reverse-bias second breakdown without device degradation.

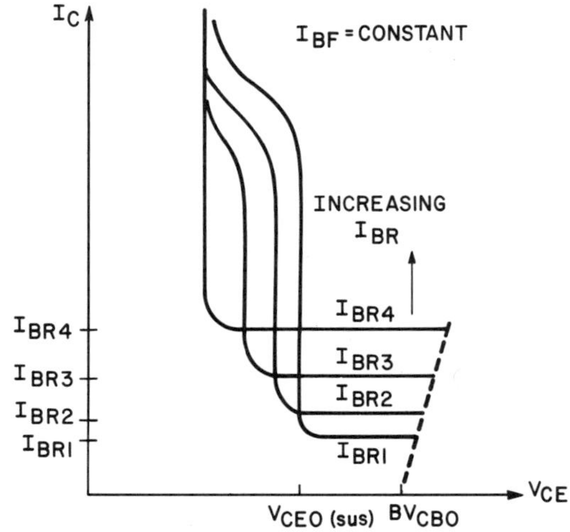

Fig. 17. Reverse bias safe operating areas of a HV transistor (RSOA) with turn-off current I_{BR} as a parameter. (After Hower [22], by permission of the publisher.)

A sudden drop in the monitored V_{CE} indicates that the second breakdown onset takes place. The vertical portion of the RSOA curves is associated with the current pinching mechanism because of the lateral voltage drop in the base induced by the applied reverse base current I_{BR}. The current is crowded at the emitter stripe center so that the critical current density for the second breakdown is exceeded. The horizontal portion of the RSOA corresponds to the situation when the applied emitter-to-collector voltage exceeds $V_{CEO(sus)}$ and when there is an additional base current component due to the collector junction avalanche breakdown.

10.7 Stable Hot Spots

Second breakdown in a transistor is initiated by the appearance of thermal instability. This occurs when the stability factor of the one-dimensional model exceeds unity, i.e.,

$$S = R_{th} V_{CE} \, \partial I_C / \partial T_j > 1, \tag{55}$$

where R_{th} is the junction-to-case thermal resistance and T_j is the junction temperature. The device case is assumed to be at a fixed temperature. Second breakdown will occur for $S > 1$ unless a stable hot spot develops which maintains the device in a quasi-stable state. The collector current constriction may stop after the current is confined to some small area [24, 25] and the second breakdown is delayed. At the moment when the hot spot is formed, the transistor current obviously becomes highly nonuniform and is concentrated mainly in the hot spot current filament. Since the area to the current flow is drastically reduced, there is a very substantial increase in the device thermal resistance, which further increases the stability factor S. On the other hand, however, the filament high current density leads to a very pronounced base widening which reduces the $\partial I_C/\partial T_j$ value and decreases the stability factor S below unity. The hot spot may extend over several emitter fingers, but has a perimeter much smaller than the actual emitter perimeter, thus leading to an increase of the transverse voltage gradient in the active base for a constant base current I_B. As a result, the current density is increased at the emitter finger edges further reducing $\partial I_C/\partial T_j$. All of these effects make the hot spot thermally stable, but it exhibits very high temperatures, sometimes exceeding 300°C. This results in transistor degradation if the device is operated for any appreciable length of time. Moreover, stable hot spots indicate that the second breakdown may be imminent.

Figure 18 [23] represents the loci corresponding to the limits of thermal instability and second breakdown. The stable hot spots occur between the two limits, and do not occur where the two loci merge below the collector current I^*. In this merged region, second breakdown takes place before the current density becomes sufficiently large for the spot restabilization. The locus of thermal instability in the I_C–V_{CE} plane can be described on the basis of the stability factor S by the relation [26]:

$$V_{CE}^* = \frac{I_C R_E + (kT_C/q)\ln(I_1/I_C)}{(k/q)[\ln(I_1/I_C) - 1]I_C R_{th}} - \frac{T_C}{I_C R_{th}}, \tag{56}$$

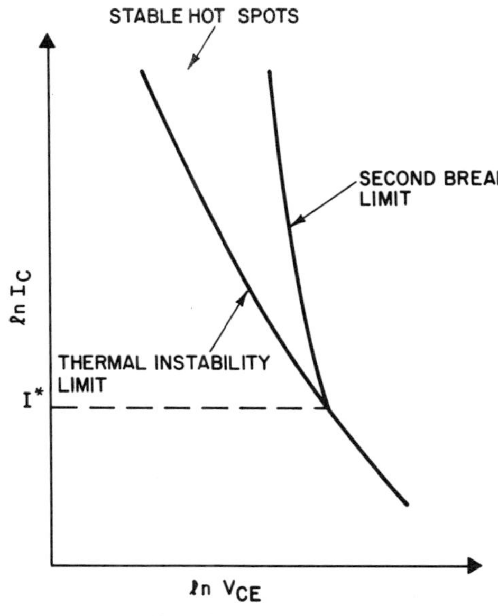

Fig. 18. Region of stable hot spots. (From Blackburn *et al.* [23].)

where V_{CE}^* is the collector–emitter voltage at which thermal instability occurs, I_C the collector current, R_E the series emitter resistance, T_C the case temperature, and R_{th} the thermal resistance. I_1 is defined by

$$I_1 = I_C \exp[-q(V_{BE} - E_g)/kT_j], \tag{57}$$

where T_j is the junction temperature. All of the above parameters can be measured. The thermal instability limit appears to be more adequate in describing the safe operating area of a transistor than the second breakdown limit.

REFERENCES

1. File Number 1165 RCA Solid State Division, Somerville, New Jersey (1979).
2. F. W. Gutzwiller and T. P. Sylvan, *AIEE Trans. Commun. and Electron. Part 1* **79,** 699 (1960–1961).
3. E. J. Diebold and W. Luft, *AIEE Trans. Commun. Electron. Part 1* **79,** 719 (1960).
4. C. F. Wheatley, *The Electronics Engineer* (January 1967).
5. B. K. Ridley, *Proc. Phys. Soc.* **82,** 954 (1963).
6. M. W. Muller, *Appl. Phys. Lett.* **12,** 218 (1968).
7. C. F. Wheatley, RCA Solid State Technical Center, Somerville, New Jersey. Private communication.
8. R. A. Sunshine and M. A. Lampert, *IEEE Trans. Electron Devices* **ED-19,** 873 (1972).
9. R. M. Scarlett, W. Shockley, and R. H. Haitz, "Physics of Failures of Electronics." p. 194. Spartan Books, Baltimore, Maryland, 1963.
10. R. M. Scarlett and W. Shockley, *IEEE Int. Conv. Record* Part 3, p. 3 (1963).
11. P. L. Hower and P. K. Govil, *IEEE Trans. Electron Devices* **ED-21,** 617 (1974).
12. B. A. Beatty, S. Krishna, and M. S. Adler, *IEEE Trans. Electron Devices* **ED-23,** 851 (1976).
13. S. Krishna and P. L. Hower, *Proc. IEEE* **61,** 393 (1973).
14. P. L. Hower and V. G. K. Reddi, *IEEE Trans. Electron Devices* **ED-17,** 320 (1970).
15. A. Blicher, "Thyristor Physics." Springer-Verlag, New York, Heidelberg, and Berlin, 1976.
16. K. Owyang and P. Shafer, *IEDM Tech. Digest*, p. 667 (1978).
17. J. H. King and J. Philips, *Proc. IEEE* **55,** 1361 (1967).
18. P. J. Kannam, *IEEE Trans. Electron Devices* **ED-23,** 879 (1976).
19. D. Navon and E. A. Miller, *Solid-State Electron.* **12,** 69 (1969).
20. F. F. Oetinger and S. Rubin, *IEEE Ann. Proc. Rel. Phys., 10th* p. 12 (April 1972).
21. C. G. Wheatley, RCA Solid State Technical Center, Somerville, New Jersey. Private communication.
22. P. L. Hower, *IEDM Tech. Notes*, p. 147 (1978).
23. D. L. Blackburn, S. Rubin, and G. J. Rogers, *IEDM Tech. Digest*, p. 151 (1976).
24. D. L. Blackburn and D. W. Berning, *IEDM Tech. Digest*, p. 671 (1978).
25. P. L. Hower, *IEDM Tech. Notes*, p. 147 (1976).
26. P. L. Hower, D. L. Blackburn, F. F. Oetinger, and S. Rubin, *Power Electronics Specialists Conf. PESC Record*, p. 234, (June 1976).

PART III

Field-Effect (Unipolar) Transistors

CHAPTER 11

JUNCTION FIELD-EFFECT TRANSISTORS (JFETs)

The field-effect transistors (FETs) are unipolar devices operating primarily by the drift of majority carriers and offering some attractive features as power devices. Because of their high input impedance, they require very low drive currents. The FET high-level drain current exhibits a negative temperature coefficient since the carrier mobility decreases with increasing temperature. The collector current in bipolar devices, on the other hand, increases rapidly with temperature with the junction bias kept constant. In order to ensure the thermal stability of bipolar transistors, it is necessary to take special precautions in their design, such as emitter and/or base ballasting. This is not necessary in the field-effect devices. Also, due to the high input impedance, there is no difficulty in connecting many FETs in parallel without special precautions.

The response times of FETs are determined primarily by the device capacitances and, to a much lesser extent, to the very short channel transit times of majority carriers. The FET transconductance, i.e., the rate of drain current change with gate voltage, becomes essentially constant at higher drain currents, whereas the bipolar transistor current gain decreases drastically at high current levels. FETs show an excellent switching time stability with varying temperature. The switching times of the bipolar transistors, on the other hand, vary appreciably with temperature and current levels because of the changes in the minority carrier lifetime.

The FET on-resistance has a positive temperature coefficient, which is advantageous for parallel connection of many devices in order to increase current and power capability. Increased on-resistance at higher temperatures leads, however, to higher power dissipation. The power FETs are not free of second breakdown common to the bipolar transistors. Even the best designed FETs have parasitic bipolar transistors present and may therefore exhibit second breakdown.

The full potential of the field-effect transistors as power devices is still to be fully realized. The prospects are good, however, thanks to the availability of very sophisticated technology already developed for large scale integration.

The field-effect transistors are divided into two classes: the JFETs with junction-type gates, and the IGFETs with insulated-type gates. IGFETs with metal–oxide gates are also known as MOS transistors.

JFETs exhibit current–voltage characteristics which are either similar to those of vacuum tube pentodes or triodes. As will be shown later, short gates and channels under some conditions produce triodelike $I-V$ curves; long gates and channels which prevent the drain voltage from electrostatically influencing the source–gate region produce pentodelike characteristics. The triode-type $I-V$ curves increase more or less indefinitely with the applied drain potential, but the pentode like characteristics exhibit drain current saturation.

Figure 1 represents schematically an n-channel JFET. The device's

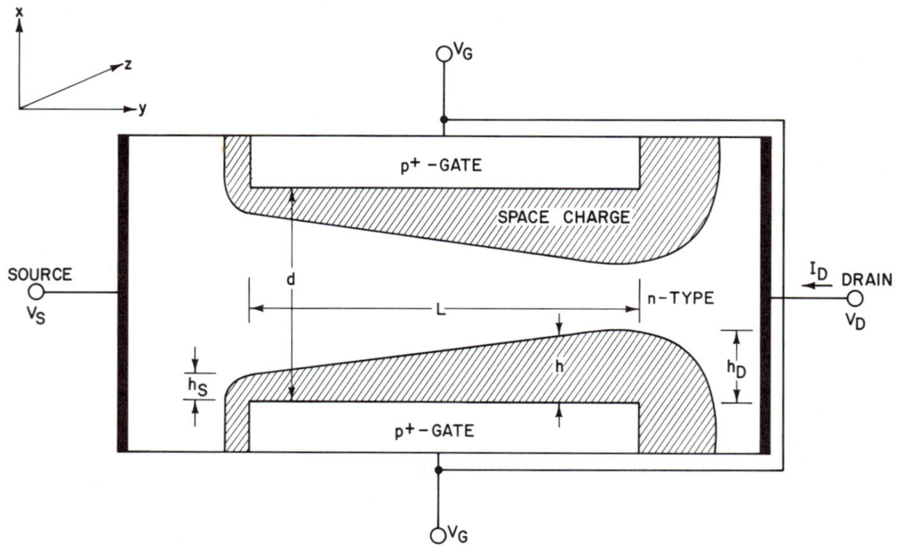

Fig. 1. Schematic representation of a junction field-effect transistor with an n-type channel.

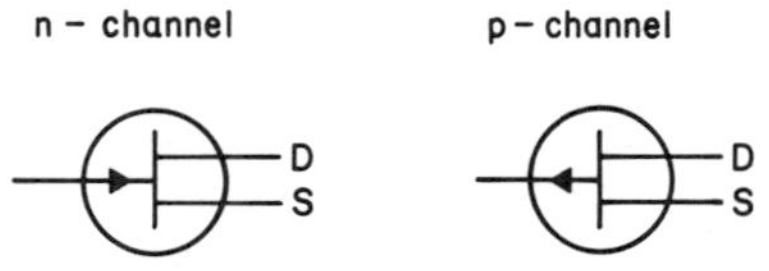

Fig. 2. *n*- and *p*-channel JFET symbols.

control electrode, the gate, consists of a p^+-n junction whose depletion region width can be changed by a variable reverse potential applied between the source and the gate. When a positive potential V_D is applied between the drain and source, electrons flow from the source to the drain. The modulation of the gate depletion region by the gate potential V_G applied between the gate and source affects the magnitude of this current by narrowing the current conducting *n*-channel. Figure 1 shows the direction of the conventional current for the *n*-channel device. In a *p*-channel device, the gate would be positive, and the conventional current flow for the negatively biased drain would be, of course, in the opposite direction to that shown for the *n*-type channel. Figure 2 shows the *n*- and *p*-channel JFETs customary symbols.

11.1 JFET Current–Voltage Characteristics

Since the conventional current flows in the *n*-channel device from the drain to the source, the lowest potential occurs at the source side of the gate due to the ohmic voltage drop, and as Fig. 1 shows, the depletion layer is the narrowest on that side. It is much wider on the drain side where the full potential appears between the p^+-region and the *n*-channel. The very narrow depletion layers in the p^+-regions are omitted in Fig. 1. For the source at ground potential, the voltage applied between the channel and the gate on the drain side is $V_D + V_G$.

The analysis of the JFET is greatly simplified by the so-called gradual channel approximation due to Shockley [1]. This approximation assumes that the gate depletion region width along the channel changes only very slowly; in addition, the effect of the electric field along the *y* direction on the depletion region is neglected (Fig. 1). Using this approximation, Bockemuehl [2] showed that *regardless of the charge distribution in the channel*, the drain current can be represented by

$$I_D = \frac{2Z\mu}{\epsilon_s L} \int_{h_S}^{h_D} [Q(d) - Q(h)]\, h\rho(h)\, dh], \tag{1}$$

where Z is the channel width in the z direction, μ the majority charge mobility,

ϵ_s the silicon permittivity, L the channel length, h the depletion layer width, $\rho(h)$ the charge density at h per unit area, and h_S and h_D are defined by Fig. 1. The electric charges $Q(d)$ and $Q(h)$ are defined by the integrals

$$Q(d) = \int_0^{d/2} \rho(x)\,dx, \tag{2}$$

$$Q(h) = \int_0^{h} \rho(x)\,dx, \tag{3}$$

where $\rho(x)$ is the charge density at x. The difference $Q(d) - Q(h)$ is the free charge per unit of the channel area so that expression (1) essentially represents Ohm's law:

$$J = \sigma E. \tag{4}$$

Equation (1) can be used, of course, to derive the drain current expression for the special case of a *uniformily* doped channel region. For this special case, however, we can use Ohm's law directly, as was suggested by Ihantola and Moll [3].

We assume that the gate is a one-sided step junction, i.e., the depletion region extends only into the lower doped n-region. For this case, the depletion layer width with the gate reverse biased is

$$h(y) = [2\epsilon_s(V_i + V_y)/qN_D]^{1/2}, \tag{5}$$

where ϵ_s is the permittivity of the semiconductor, V_i is the built-in (contact) junction potential, V_y the gate voltage, q the electron charge, and N_D the donor concentration of the uniformly doped channel. At point y along the channel on the y axis,

$$V_y = -V_G + V(y); \tag{6}$$

therefore, (5) becomes

$$h(y) = \left\{\frac{2\epsilon_s}{qN_D}[V_i - V_G + V(y)]\right\}^{1/2}. \tag{7}$$

The incremental voltage drop along the y axis (Fig. 1) is given by [3]

$$dV = I_D\,dR = \frac{I_D\,dy}{2Zq\mu_n N_D(d - 2h)}, \tag{8}$$

where dR is an increment of the channel resistance; μ_n is the electron mobility, assumed to be independent of the electric field; and Z is the channel width in the z direction perpendicular to the page. Substituting expression (7) into (8) and integrating with the voltage limits $V = 0$ at the source and $V =$

V_D at the drain gives the drain current for the reverse biased gate as

$$I_D = G_0\{V_D - \tfrac{2}{3}(8\epsilon_s/qN_Dd^2)^{1/2}[(V_i - V_G + V_D)^{3/2} - (V_i - V_G)^{3/2}]\}. \qquad (9)$$

In this expression, G_0 is the undepleted (metallurigical) channel conductance

$$G_0 = 1/R_0 = q\mu_n N_D Zd/L, \qquad (10)$$

where R_0 is undepleted channel resistance, d its width, and L its length (Fig. 1). The contact resistances and series resistances between the source and the channel and between the drain and the channel are neglected.

At low drain voltages, i.e., when $V_D \ll V_i + V_G$, Eq. (9) simplifies to

$$I_D = G_0\{1 - [8\epsilon_s(V_i - V_G)/qN_Dd^2]\}^{1/2}\, V_D. \qquad (11)$$

At low drain voltages, the current depends on the square root of the gate voltage and is linearly dependent on the drain voltage. The square root dependence is the result of the assumption that the gate is a step junction.

Figure 3 represents the I–V (output) characteristics of a typical JFET. At first, for small drain voltages, the I_D current increases rapidly and linearly; then the increase slows down in the transition region, and eventually the current reaches a maximum and remains constant (saturates) with the increasing drain voltage. The beginning of the saturation takes place when the drain voltage becomes equal to V_{Dsat}, the pinch-off voltage. At this drain potential, the width of each depletion region becomes equal to $\frac{1}{2}$ of the channel width d so that the depletion layers start touching each other at least at one small region, the channel pinch-off point. For the drain current, expression (9) is useful only up to this point; beyond it, the incremental resistance tends to infinity (Fig. 3), and the analysis loses its validity.

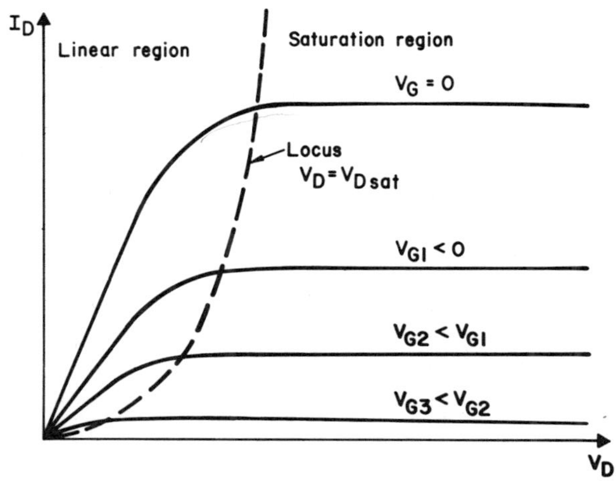

Fig. 3. Current–voltage characteristic of a JFET.

As the drain voltage is increased beyond V_{Dsat}, the current continues flowing from the undepleted part of the channel into the space-charge region in the way it flows in the collector depletion region of a bipolar transistor.

11.1.1 Saturation Drain Voltage

The voltage at which the drain current saturates V_{Dsat} can be found from expression (7) by setting the depletion layer width equal to $\frac{1}{2}$ of the undepleted channel width. The result is

$$V_{Dsat} = (qN_D d^2/8\epsilon_s) - (V_i - V_G) \tag{12}$$

for the negatively biased gate (n-channel JFET).

11.1.2 Threshold Voltage

As the gate voltage V_G is made progressively more negative (for the n-channel device), the drain current decreases (Fig. 3) and, at some gate voltage V_T, stops flowing entirely. V_T is the gate turn-off voltage, which can be determined from expression (11) by setting the drain current equal to zero. Its value is

$$V_T = V_i - qN_D d^2/8\epsilon_s. \tag{13}$$

11.1.3 Saturation Drain Current

The value of the saturation drain current can be obtained by substituting the value of V_{Dsat} from (12) into expression (9) for the drain current. The result is

$$I_{Dsat} = G_0\{\tfrac{2}{3}[(8\epsilon_s(V_i - V_G)/qN_D d^2)^{1/2} - 1](V_i - V_G) + qN_D d^2/24\epsilon_s\}. \tag{14}$$

As the drain voltage V_D increases beyond the pinch-off value V_{Dsat}, the gate-to-drain potential also increases so that the total depletion region near the drain widens; at the same time, the pinch-off point at the source side of the depletion region moves very slightly toward the source so that the undepleted channel region is only slightly shortened and its resistance somewhat lowered. The potential at the pinch-off point is fixed at V_{Dsat}, but because the channel resistance becomes a little smaller, the saturation drain current increases a little with the increased drain voltage. The slope of the I–V characteristics, which is equal to the drain conductance, will show, therefore, some finite value rather than the zero value expected from the first-order theory. The shortening of the conductive channel in the JFET with the applied drain voltage is analogous to the Early effect in bipolar transistors, where the base becomes shorter due to the growth of the depletion region when the collector voltage is increased. The effect of the channel reduction is particularily evident in JFETs with short channels on the order of a few micrometers, designed for rf applications.

In JFETs with mostly depleted channels, the magnitude of the current is limited also by the fact that as the drain voltage increases so does the mobile carrier velocity, but only up to the point at which it saturates for electrons in silicon and GaAs at $v_s \simeq 10^7$ cm sec^{-1}. Any increase of the electric field beyond about 10^4 V cm^{-1} does not change carrier velocity, and as long as the carrier total charge is not increased, the drain current remains essentially constant. The important point is that despite the existence of the saturation velocity, it would be possible to increase the current with the voltage if we had some means of increasing the charge since the current is equal to the product of the mobile charge and velocity.

A more accurate analysis of potential and charge distribution in the channel [4] shows that, particularily in the very narrow channel devices, there exists a potential barrier, due to the reverse-biased gate, which may be electrostatically lowered by the drain electric field. Lowering the potential barrier increases the injection of charges and the current flow. This phenomenon will be discussed more fully in Section 11.5 on the triodelike JFETs.

11.1.4 Channel Resistance

The channel minimum dc resistance occurs when the junction gate is unbiased so that the depletion region is very small and due only to the built-in potential V_i. The minimum channel resistance is, consequently, equal to

$$R_{\min} = L/q\mu_n N_d(d - 2h_0)Z, \qquad (15)$$

where h_0 is the depletion layer width of each gate (Fig. 1) with zero gate bias. $R_{\min}$ determines the minimum voltage drop at the rated current level and should be made small in a power device, consistently with other device parameters, in order to minimize power dissipation. In a nonideal JFET, the channel resistance is augmented by the source and drain series resistances, R_s and R_d.

11.1.5 Small Signal Conductance and Transconductance

The incremental channel (drain) conductance is defined by

$$g_d = \left.\frac{\partial I_D}{\partial V_D}\right|_{V_G = \text{const}}. \qquad (16)$$

It follows from (16) and (11) that in the linear region,

$$g_d = g_{d0} = G_0\{1 - [(8\epsilon_s/qN_D d^2)(V_i - V_G)]^{1/2}\}. \qquad (17)$$

The source and drain series resistances R_s and R_d add up to the total channel resistance. If they are taken into account, the conductance in the linear region is modified so that its effective value becomes

$$g_{d0(\mathrm{eff})} = (g_{d0}^{-1} + R_s + R_d)^{-1} \tag{18}$$

In the saturation region, when the drain voltage is increased above V_{Dsat}, more free carriers are depleted from the channel so that the length of the merged depletion region is increased but the channel length is decreased. Since V_{Dsat} remains the same, the drain current increases slightly, and the drain conductance has a finite value. In an idealized device, the conductance in the saturation region is zero.

The transconductance of a JFET is defined by

$$g_m = \left.\frac{\partial I_D}{\partial V_G}\right|_{V_D=\mathrm{const}}. \tag{19}$$

Its value can be derived by differentiation of Eq. (9), which yields

$$g_m = G_0(8\epsilon_s/qN_D d^2)^{1/2}[(V_i + V_G - V_D)^{1/2} - (V_i - V_G)^{1/2}]. \tag{20a}$$

In the saturation region, the transconductance $g_{m\mathrm{sat}}$ reaches maximum, and its value can be obtained by making $V_D = V_{\mathrm{Dsat}}$ in Eq. (20). It turns out that the transconductance in the saturation region is exactly equal to the conductance in the linear region

$$g_{m\mathrm{sat}} = g_{d0} \tag{20b}$$

and is given by (17).

In actual devices, the computation of the conductance and transconductance has to take into account device parasitic resistances. It was shown by Dacey and Ross [5] that the effective transconductance, which takes into account the presence of the source series resistance R_s, can be obtained from

$$g_{m\mathrm{eff}} = (g_m^{-1} + R_s)^{-1}. \tag{21}$$

The drain series resistance has a negligible effect on the transconductance in the saturation region.

11.1.6 g_{m0}/I_{D0} AS A FIGURE OF MERIT

The maximum value of the transconductance occurs at the gate-to-source potential equal to zero. For forward biases, high gate current would flow, and the device could not operate under the usual circumstances. The ratio of the maximum transconductance and the drain current I_{D0} for zero gate bias, i.e., g_{m0}/I_{D0}, is a useful figure of merit to compare various JFET designs for operation at low frequencies.

11.2 JFET Incremental Circuit Model

The incremental JFET circuit model in the common source configuration is represented by Fig. 4. It contains the device's intrinsic elements C_{gs}, C_{gd}, R_s, and R_d, which are respectively gate-to-source capacitance, gate-to-drain capacitance, source series resistance, and drain series resistance. In addition, there are extrinsic (parasitic) elements that come about because of the necessity of connecting the device to the outside world by mounting it on a header using leads. The extrinsic elements of the circuit model shown in Fig. 5 are the stray capacitances C'_{gs}, C'_{dg}, and C'_{ds} between the gate and source, gate and drain, and drain and source, respectively. Inductances may be neglected for moderately high frequencies. The capacitances are for two gates in parallel as in Fig. 1.

The analytical common-source small signal (incremental) drain and gate current equations are given by

$$i_d = g_m v_{gs} - C_{gs} \frac{dv_{gd}}{dt}, \tag{22a}$$

$$i_g = C_{gs} \frac{dv_{gs}}{dt} + C_{gd} \frac{dv_{gd}}{dt}, \tag{22b}$$

where v_{gs} and v_{gd} are small signal gate-to-source and gate-to-drain voltages, respectively. The current and voltage reference directions are indicated in Fig. 4. At very low frequencies, some capacitances will have high reactances and can be omitted from the equivalent circuit.

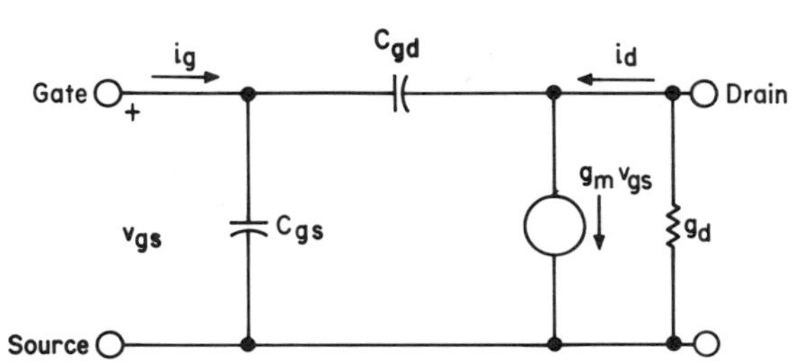

Fig. 4. Incremental equivalent circuit of a JFET. g_d is the finite drain conductance in the saturation region.

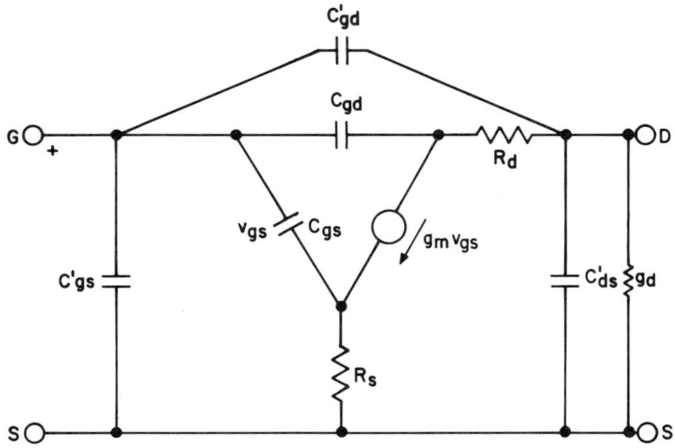

Fig. 5. Incremental equivalent circuit of a JFET with parasitic elements. g_d is the finite drain conductance in the saturation region.

11.2.1 Cut-Off Frequency

The cut-off frequency f_T of a JFET is an important figure of merit and is defined as the frequency at which the ac current through the equivalent circuit input capacitance becomes equal to the output current $g_m v_{gs}$. At f_T, the current gain is therefore equal to unity.

The input capacitance for the shorted output is given by

$$C_{in} = C_{gs} + C_{gd} \simeq C_G, \tag{23}$$

where C_G is the gate junction total capacitance. In saturation, C_{gd} is small and $C_{gs} \simeq C_G$. (More accurately, $C_{gs} = \frac{2}{3}C_G$; see Section 12.7.)

Equating the input and output currents in the absence of parasitic capacitances, one obtains

$$f_T = g_m/2\pi C_G. \tag{24}$$

The maximum possible transconductance is

$$g_m = q\mu N_D Zd/L = G_0. \tag{25}$$

The minimum gate capacitance occurs for the gate depletion region average width $h = \frac{1}{2}(d/2)$, so that

$$C_{Gmin} = 8\epsilon_s ZL/d, \tag{26}$$

where ϵ_s is the semiconductor permittivity. As a result, the cut-off frequency is given by

$$f_T \leqslant q\mu_n N_D d^2/16\pi\epsilon_s L^2. \tag{27}$$

For high frequency operation, it is necessary to make the gate capacitance C_G small, the channel short, and use *n*-doped channels since electron mobility is higher than that of holes.

11.3 Excess Gate Leakage Current and Drain Breakdown

One of the important JFET advantages is its high input impedance, which allows for the fabrication of multichannel devices consisting essentially of a large number of parallel connected single channel JFETs fabricated on the same chip. Combining in parallel many discrete JFETs should not present any problems, either, as long as the input and output impedances remain high.

The dc impedance of the gate is determined by the junction saturation current (leakage). A well-fabricated junction gate will normally have a leakage current on the order of pico- or nanoamperes, and the gate input resistance is on the order of hundreds of megohms. Surface effects may increase drastically junction leakage and special precautions should be taken to protect and passivate the surface. At the high drain biases expected in power JFETs, the dc gate current often exceeds the ideal junction saturation (leakage) current levels. This excess gate current may be traced to the electron–hole generation, due to impact ionization, in the JFET channel. Short channel devices with high electric fields are expected to be more susceptible to carrier generation.

The excess gate current in the JFET was studied by [6] using two-dimensional computer simulation. The analysis shows which areas of the device exhibit the highest electric field and/or the largest seed currents. The strongest fields exist near the gate edge (drain side) at the device surface, and generation is high even if the seed current, i.e., the reverse saturation current there, is small.

The electric field along the current path (the channel) is usually high enough for carrier-pair generation, and the seed current is the channel current itself. The generated electrons flow to the drain while the holes are collected by the negative gate and/or by the substrate if it is negatively biased with respect to the source. No actual breakdown is necessary for the abnormal increase of the reverse saturation current since the excess gate current may result from the low-level impact ionization. (Breakdown is a result of the high-level avalanche ionization.)

The analysis of Yamaguchi and Asai [6] shows that the excess gate current is equal to

$$I_G = (I^*/(1 - I^*)I_{\text{source}}, \tag{28}$$

where I^* is the so-called one-dimensional ionization integral, as defined in Section 3.1 on breakdown, and I_{source} is the current originating at the source. The holes are collected either by the gate or by the substrate; therefore, only the electron current was used for the computation of the ionization integral. The integration path for the low level of ionization was determined from the two-dimensional numerical analysis. The integration path for the breakdown condition occurring for $I^* = 1$ was taken along the surface.

The analysis and experimentation [6] showed that the excess gate current depends strongly on the impurity doping profile in the channel and in the gate and that it is possible to optimize the profile to reduce the excess gate current without deteriorating the device transconductance.

For $V_G = 0$, the gate current increases exponentially with increasing drain voltage while the device operates in the current saturation region because the avalanche multiplication factor increases when the field in the conducting channel increases. In one of the devices studied by [6], the gate current for $V_D = 10$ was on the order of 10^{-10} A; but for $V_D = 50$ V, it has increased by six orders of magnitude to about 10^{-4} A.

The FET breakdown voltage usually occurs at the gate–drain depletion region where it meets the surface and will happen prematurely unless some of the techniques described in Chapter 4 on breakdown improvement of bipolar devices are utilized. It should be noted that the drain–source breakdown is equal to

$$BV_D = BV - V_G, \tag{29}$$

where BV is the junction breakdown voltage.

11.4 Output Power from a JFET

The maximum power output P_{out} from a JFET is proportional to the maximum allowed drain current I_{Dmax} and the peak voltage swing between the drain and source

$$P_{out} \propto I_{Dmax}(BV_D - V_{Dsat}). \tag{30}$$

A good power JFET should therefore have high current capability, large drain-to-source breakdown voltage, and small V_{Dsat} at the maximum current. Devices with an n-type channel have higher maximum current capability due to higher electron mobility as compared to that of holes. This also holds for the saturated velocities v_s in the case of silicon and gallium arsenide. Shorter channels with low on-resistances are desirable if compatible with other device parameters such as punch-through or breakdown. When the best power–frequency relation is desired, then the device should be designed so that it can operate at the highest possible current and a minimum acceptable voltage [7].

11.5 JFET with Triodelike (Nonsaturating) Characteristics

JFETs with triodelike, nonsaturating characteristics are advantageous, for example, in low distortion audio power amplifiers or high frequency power amplifiers. Their structure is similar to that of the pentodelike JFET, but the current flow mechanism in the channel is strongly affected by the existence of the gate-induced potential barrier.

JFETs with nonsaturating characteristics (Fig. 6a [8]) were extensively studied by Yamaguchi *et al.* [4, and 8] and Nishizawa *et al.* [9, 10]. The two-dimensional numerical analysis of Yamaguchi *et al.* [8] obtained the electric charge and field distributions in the short and long channel JFETs. The depletion region gives rise to a potential barrier in the channel preventing current flow at low drain voltages. For high enough drain voltages, however, the potential barrier is lowered by the drain's electrostatic field so that current injection is highly enhanced in the short gate devices; in the long gate (long channel) devices the electrostatic field is not high enough to affect the barrier height. The analysis of [8] was carried out for only one channel of the multiple

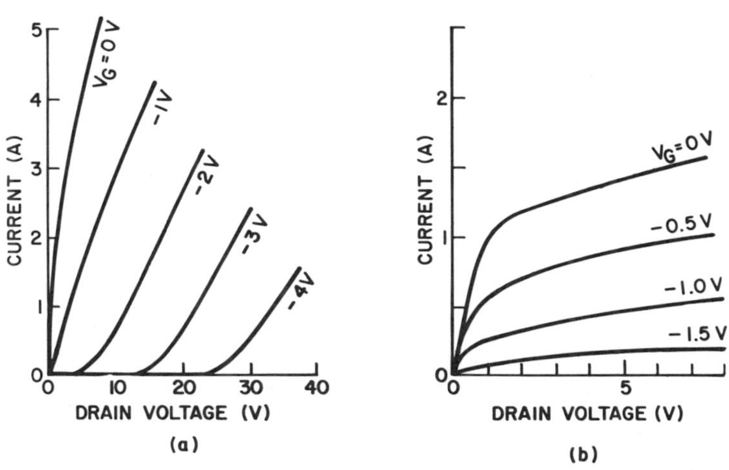

Fig. 6. (a) Triodelike current–voltage characteristics of a JFET ($a = 0.25\ \mu$m); (b) pentode-like current–voltage characteristics of a JFET ($a = 0.5\ \mu$m). (From Yamaguchi and Kodera [8].)

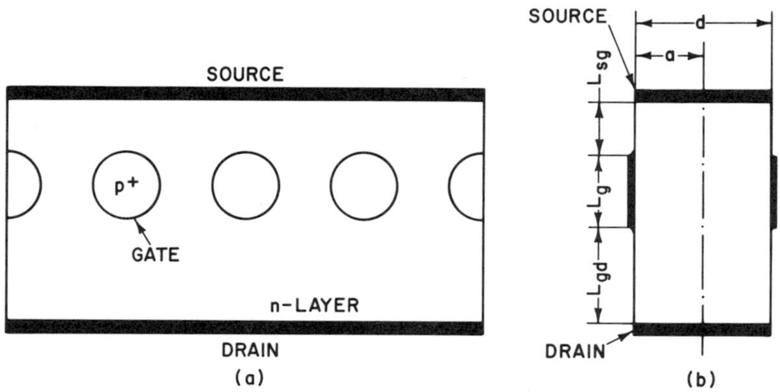

Fig. 7. (a) Cross section of a JFET with multiple embedded p^+-gates; (b) side view of the multigate JFET. (From Yamaguchi and Kodera [8].)

gate device with a vertical channel (Fig. 7a). Vertical structures for power FETs have important advantages over the lateral structures. Better silicon area utilization is one of them (see Section 11.6 and Chapter 13).

11.5.1 Effect of Channel Thickness

The channel thickness plays a decisive role in the achievement of the triodelike I–V characteristics because only very narrow channels permit complete carrier depletion at no externally applied gate potential. For a given gate-to-drain spacing L_{gd} (Fig. 7b [8]), the voltage amplification increases and actual current pinch-off voltage decreases with decreasing channel thickness d [8]. Unfortunately, small d substantially increases the resistance between the source and drain in the on-condition. An optimum value of d has to be determined, therefore, to satisfy other design requirements.

11.5.2 Gate Length Effect

Because of the electrostatic action of the drain potential, the gate length is crucial for the triodelike operation. As the gate length increases, the character of the I–V curves changes from the triodelike to pentodelike. Figure 8 [8] represents the electrostatic potential along the device axis for three different gate lengths for a device with $N_D = 10^{15}\ cm^{-3}$, $d = 2.4\ \mu m$, $L_{gd} = 2.6\ \mu m$, and $L_{sg} = 1.2\ \mu m$. In Fig. 8a, $L_g = 1\ \mu m$ and $V_G = -2V$; in Fig. 8b, $L_g = 2\ \mu m$ and $V_G = 2$ V; and finally in Fig. 8c, $L_g = 4\ \mu m$ and $V_G = -1$ V. The potential barrier in Fig. 8a is high for $V_D = 0$, but as V_D increases, it goes

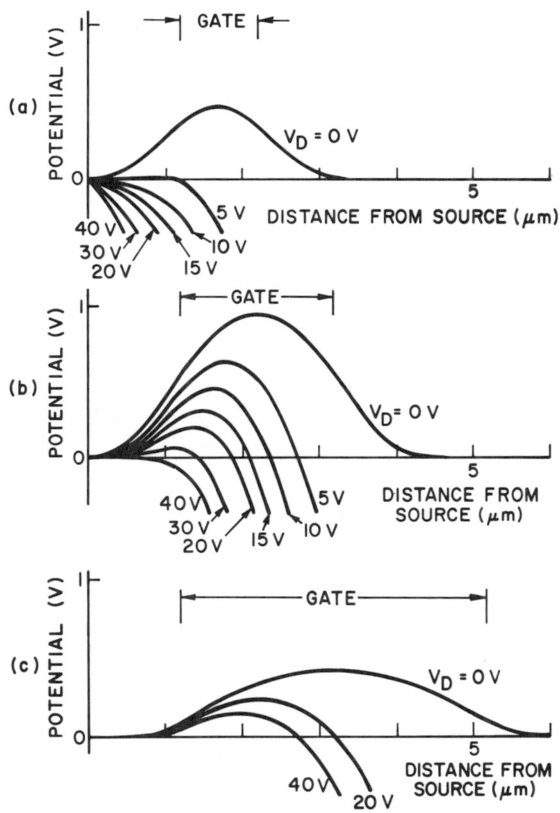

Fig. 8. The electrostatic potential along the device's symmetric line. (From Yamaguchi and Kodera [8].)

down to zero at $V_D = 5$ V. The FETs with $L_g \leqslant d/2$ behave as triodes. When the gate length is larger than $d/2$, the potential barrier vanishes at the drain voltage equal to about 35 V (Fig. 8b). For longer gate ($L_g = 4\ \mu$m) (Fig. 8c), the potential barrier is lowered by the drain's field, but still exists even at voltages higher than 40 V, and the device behaves as a pentode. It has to be pointed out, however, that FETs with large widths d behave like pentodes even if the ratio $L_g/(d/2)$ *is small.*

11.5.3 Gate-to-Drain Distance Effect

When the gate-to-drain distance L_{gd} is short, the JFET operates as a triode. Long L_{gd} results in higher output resistance characteristics because at moderate drain voltages, the electrostatic field intensity is not sufficient to significantly lower the electrostatic barrier in the channel. A high gate-to-drain breakdown voltage BV_{gd} requires sufficiently large gate-to-drain spacing and this puts the lower limit on L_{gd}.

11.5.4 Channel Resistance Effects

The extrinsic channel resistance R_S on the source side introduces a negative feedback (Fig. 9 [8]). When R_s is sufficiently increased by the addition of an external resistance, then the triodelike characteristics (dashed line) are modified and approach the pentodelike curves. According to Nishizawa *et al.* [9], the saturation of JFET's characteristics can be explained by the negative feedback action of the drain potential on the extrinsic channel resistance. This resistance eventually increases so much with drain current and voltage that the transconductance g_m [Eq. (21)] becomes

$$g'_m \simeq R_S^{-1} \tag{31}$$

and the drain current saturates. The reduction of the series resistance alone does not result in a triodelike operation if the intrinsic FET has pentodelike characteristics to start with [8].

11.5.5 Current–Voltage Relationships

The potential barrier introduced into the channel by the gate-induced pinch-off plays the role of an intrinsic gate with a height ψ_h which is a function

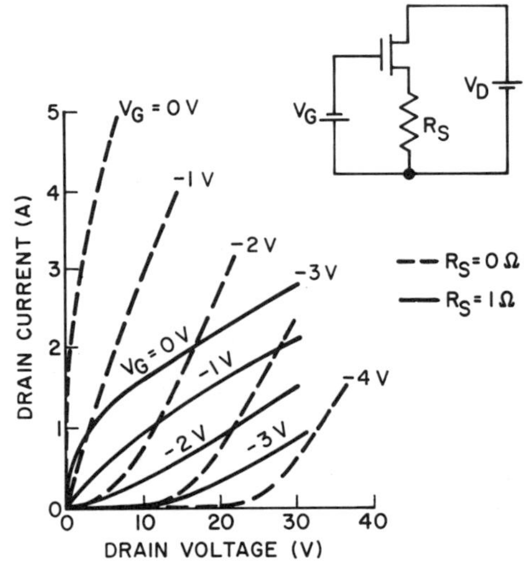

Fig. 9. Transition from the triodelike to pentodelike characteristics effected by insertion of the source series resistance R_S. (From Yamaguchi and Kodera [8].)

of the gate voltage V_G and drain voltage V_D [11]:

$$\psi_h = \eta(V_G - V_D/\mu^*). \tag{32}$$

μ^* is the transistor amplification factor defined by

$$\mu^* = \left.\frac{\partial V_D}{\partial V_G}\right|_{I_D=\text{const}} \tag{33}$$

In (32), η is a factor dependent on the device structure. It increases as the channel width is made narrower and as the channel carrier concentration is made lower. The knowledge of the barrier height permits the determination of the drain current as [11]

$$I_D = I_0 \exp[(-q\eta/kT)(V_G - V_D/\mu^*)], \tag{34}$$

where I_0 is a constant dependent on the device structure and impurity concentration, T is the absolute temperature, k is Boltzmann's constant, and q is the electronic charge. Equation (34) is valid when the series channel resistance is negligible. If this is not the case, the I–V curve will deviate from the exponential form at large drain currents. For the nonnegligible source and drain resistances R_S and R_D, the high-level drain current is expressed by [11]

$$I_D = (V_D - \mu^* V_G)/[(1 + \mu^*)R_S + R_D]. \tag{35}$$

The I–V characteristics exhibit an exponential behavior for low currents, but in a pentodelike JFET this part of the I–V characteristic is linear. At large current levels, the relationship between the voltage and current becomes nearly linear due to the effect of the channel series resistances.

The temperature coefficient of the drain current depends on the current level: it is positive in the lower current region because of the potential barrier existence in the channel; it becomes negative for the region where the characteristics become more linear when the series channel resistance comes into play.

11.6 JFET Structures

In the past, many structures of JFETs were proposed with the theoretical ability of achieving high power, high voltage, and high current capabilities. But these ideas could be realized only recently, thanks to the advances in the semiconductor technology and also due to the better understanding of the JFET's behavior.

For power levels from a few watts to hundreds of watts and for frequencies below 1 GHz, devices with vertical channels and drain regions appear to be the best suited for the achievement of the high drain breakdown voltages, low on-resistances, and good thermal dissipations. The high current capability of the JFETs may be obtained by multichannel structures with large total source, gate, and drain perimeters.

11.6.1 V-Groove JFET

A V-groove JFET (VFET) (Fig. 10 [12]) belongs to the category of non-planar devices. Only one channel is shown in the figure, but the transistor described by Salama and Oakes [12] and Mok and Salama [13] is a highly interdigitated multiple *n*-channel device. Anisotropic etching was used to etch a V-groove on (100) surface oriented silicon chips. One of the commonly used etchants is a solution consisting of KOH–H_2O and isopropyl alcohol [14]. The groove etched this way has its sides inclined at 54.7° from the horizontal. This technique has been known for several years and has been applied in large scale integration of MOS transistors. The device is fabricated on an *n*-type epitaxial layer grown on a p^+-substrate with a p^+–n junction acting as the gate. Shallow n^+-layers are diffused for subsequent metallization of the source and drain contacts. The device has two types of grooves: one type defines the channel region, and the other isolates one element of the structure from the others. As the reverse bias on the p^+-junction is increased, the depletion region moves up, cutting into the electron current path from the source to drain along the V-groove edge.

The JFET described by [13] had a chip area of 0.2 mm and a total gate width in the z direction of 8.2 mm achieved by an interdigitation of 30 channels 3 μm long each. The source and gate n^+-regions had a diffusion depth of 2 μm. The metallurigical channel distance from the apex of the groove to the p^+-region was equal to 1.0 μm. Its BV_D = 37 V and I_{Dsat} =

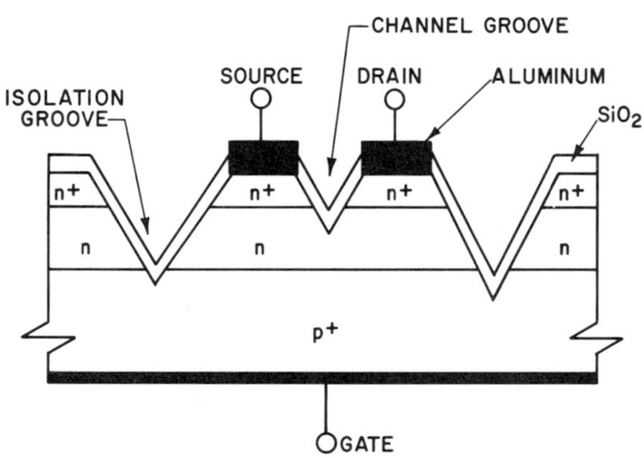

Fig. 10. V-groove power JFET. (From Mok and Salama [13].)

270 mA. This particular device was intended for vhf applications and had a cut-off frequency of 1.2 GHz with 4 W dissipation capability. This structure is capable of high packing density and can easily be scaled up to achieve higher power outputs. The device's I–V characteristics represent a transition between the triode- and pentodelike curves. Below the saturation voltages, the device behaves in a triodelike manner.

11.6.2 Static Induction Transistor (SIT)

Another vertical multichannel structure is shown in Fig. 11. This device is called a static induction transistor (SIT) by Nishizawa *et al.* [9] and has an embedded meshlike grid acting as a p^+-gate. It is very similar to a device proposed in 1964 by Teszner and Gicquel [15]. The source series resistance R_S is here minimized by putting the gate mesh-structure very close to the source. The channel length is very short, and the resulting I–V characteristics are triodelike.

One of the Japanese companies is engaged in the production of very high power SITs. At the time of this writing, they can produce JFETs with drain breakdowns above 800 V, maximum drain currents of 60 A, dc drain resistance of 2 Ω, and turn-on time of 200 ns.

11.6.3 Surface SIT

Figure 12 [10] represents another version of SIT with a surface-type gate. This is a vertical channel device with the source and gate electrodes located

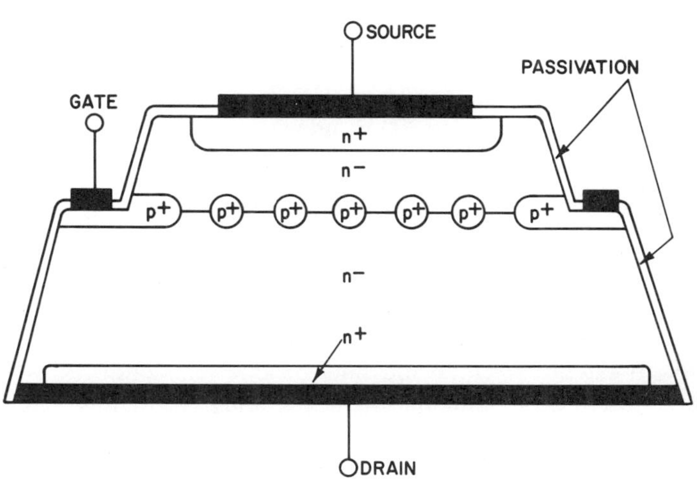

Fig. 11. Static induction transistor (SIT) with multiple gates. (From Nishizawa *et al.* [9].)

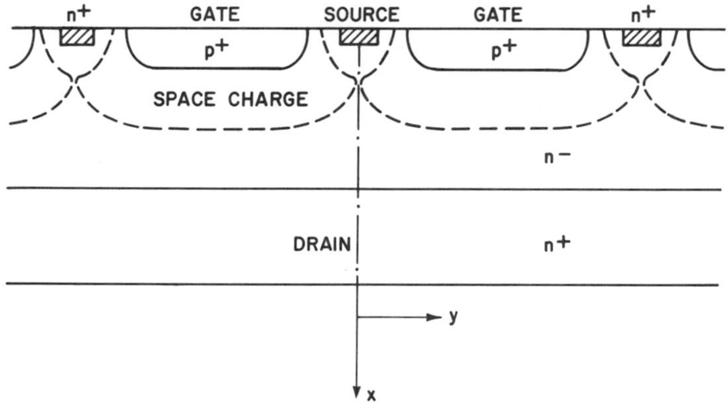

Fig. 12. Static induction transistor (SIT) with surface-type gate. (From Nishizawa and Yamaguchi [10].)

on the n^+- and p^+-layers diffused into the upper surface of the chip. The electron current flows down toward the drain electrode.

Nishizawa and Yamamoto [10] described a uhf surface-type SIT with an output of 40 W at 200 MHz and $f_{max} = 1$ GHz. They also reported a surface SIT operating with a $f_{max} = 2.5$ GHz and delivering 10 W at 1 GHz.

So far as the high frequency operation is concerned, the surface SIT is superior to the buried-gate SIT because of the very short distance between the source and gate, smaller R_S, and shorter channel length.

A structure which is similar to the surface gate SIT is shown in Fig. 13 [16]. The device is fabricated with steep, vertically walled gate regions by using orientation dependent etching with subsequent epitaxial refill. This way, a planar rather than a grooved surface is obtained, which facilitates metallization by eliminating surface discontinuities. Good contact with the gate fingers lowers the gate series resistance. Source-to-drain breakdown voltages for this device exceeded 200 V and voltage gains were greater than 30.

11.6.4 Metal Schottky-Barrier Field-Effect Transistor (MESFET)

Until about ten years ago bipolar transistors were used almost exclusively for microwave power applications. Since that time, considerable progress has been made in gallium arsenide and silicon metal semiconductor field-effect transistors (MESFETs), making them particularly useful for operating frequencies of above 4 GHz, where they are comparatively easy to match to the outside networks. At the present time, MESFETs are being used for power generation and signal amplification at frequencies up to 15 GHz.

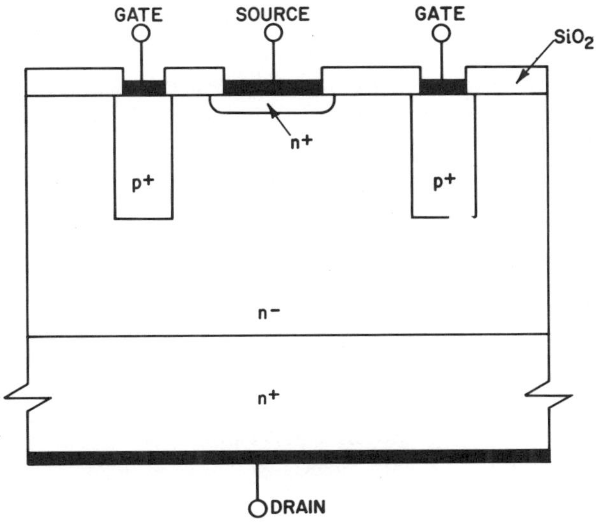

Fig. 13. Cross section of a JFET with vertical gate regions. (From Baliga and Wessels [16].)

The most commonly used material for MESFETs is GaAs, although other compound semiconductors, such as InP, are actively studied as well. The main advantage of gallium arsenide over silicon is its very high drift mobility and peak velocity. Its mobility at room temperature is 8800 cm^2 V^{-1} s^{-1} as compared to 1350 cm^2 V^{-1} s^{-1} for lightly doped single crystal silicon. The peak electron velocity $v_p = 2 \times 10^7$ cm s^{-1} is twice as high as the silicon maximum saturation velocity v_s. The GaAs peak velocity for electrons is reached for an electric field E_p of about 3×10^3 V cm^{-1}, as shown in Fig. 14 [17]. As the field is increased, the velocity decreases and reaches the saturation value of about 10^7 cm s^{-1} at the critical field $E_{crit(v)} \simeq 1.8 \times 10^4$ V cm^{-1}, equal to that of silicon.

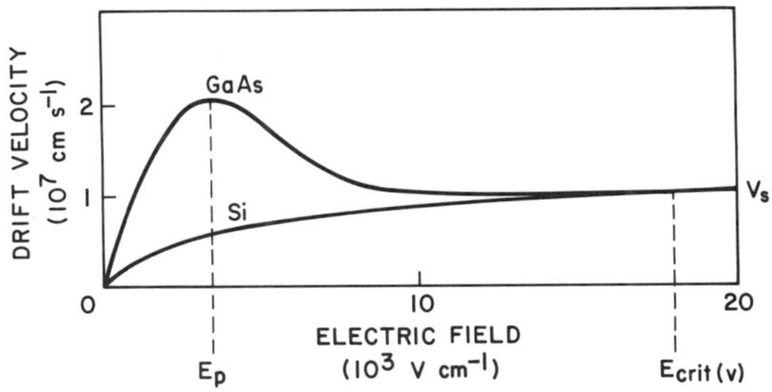

Fig. 14. Drift velocity as a function of electric field for GaAs and silicon. (From Ruch [17].)

The MESFET's structure consists of an *n*-type GaAs film deposited on top of a semi-insulating GaAs substrate (Fig. 15 [18]) doped, for instance, with chromium. The substrate provides a high conductivity thermal path to the heat sink with simultaneous electrical insulation. A part of the *n*-region is metallized by evaporation to form the Schottky barrier which plays the same role as the *p–n* junction gate in a silicon-type FET. In the case of GaAs, it is much easier to form a metal–semiconductor junction than a good *p–n junction.*

A MESFET has an advantage over the MOS transistor (Chapter 12) in that the conducting channel lies below the surface so that the high bulk mobility can be utilized rather than the lower surface mobility. In addition, surface states do not affect the gate bias as they do in the MOS transistor.

The MESFET theory of operation is considerably more complicated than that of the silicon JFET because the GaAs carrier velocity variation with the field, which exhibits a peak value, leads to Ridley–Gunn stationary charge domain formation at the drain side of the gate. Domain formation was mentioned earlier (Section 10.4) in connection with the principle of minimum entropy generation. The stationary domain formation can be explained with the help of Fig. 15 [18] on the basis of the equilibrium velocity–field characteristic. The MESFET with a 3 μm long gate is assumed to operate far into the saturated current region. The drift velocity reaches its peak value v_p close to the channel center. It falls down to v_s under the drain side of the gate edge [19], where the conducting channel is the narrowest. As a result, an electron accumulation is formed for the preservation of the current continuity. After the electrons leave this region, the channel thickness increases and so does the electron velocity because of the field intensity decrease. Since less electrons are needed to carry the same current, a positive space-charge region is formed. Thus a stationary dipole (domain) is formed, which is responsible for most of the drain voltage drop and drain voltage breakdown. The stationary domain is also responsible for the conducting channel

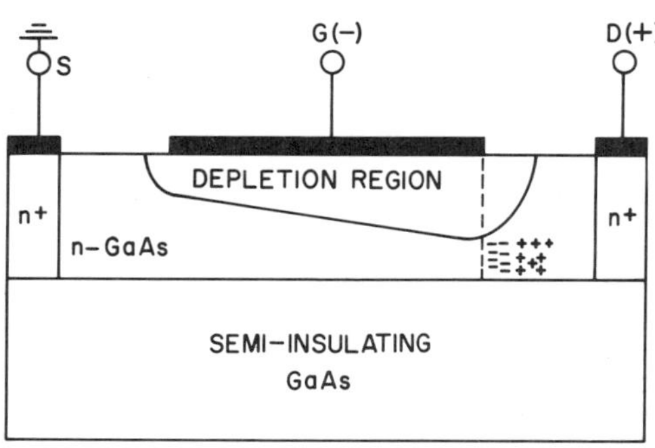

Fig. 15. Cross section of GaAs MESFET. (From Liechti [18].)

pinch-off under the transistor gate for the very short channels [19]. Practically all characteristics and device parameters are affected as compared to those obtained from the standard FET theory.

In MESFETs with very short gate lengths (below 3 μm), the electrons are in equilibrium transport condition only below the threshold field E_p. Above E_p, they are accelerated to a velocity exceeding the peak value v_p before relaxing to the equilibrium velocity. The velocity overshoot, which may be as high as $2v_p$, shortens the electron transit time in the channel. The velocity eventually drops to the saturation value v_s but only after a time which is longer than the time necessary to traverse the very short channel.

A MESFET structure operating at 4 GHz and with comparatively high drain-to-source breakdown voltage in excess of 26 V has been described by Fukuta *et al.* [20]. The device consisted of an *n*-type channel, doped with sulfur, and a chromium-doped semi-insulating GaAs substrate. The doping density and thickness of the *n*-type layer were about 6×10^{16} cm^{-3} and about 0.3 μm, respectively. The device gate length $L = 1.5$ μm, and its width $Z = 13$ mm. The output power obtained was $P_{out} = 9.6$ W, and the power added efficiency $\eta = 33.6\%$ at 18 V. The power gain was 5 dB.

GaAs MESFETs that can deliver several watts of power at frequencies as high as 10–15 GHz are now available.

A. Higashisaka *et al.* [21] reported an experimental internally matched four-chip device with 60 mm total gate width that delivered 25 W at 6 GHz and 17 W at 8 GHz.

REFERENCES

1. W. Shockley, *Proc. IRE* **40,** 1289, 1365 (1952).
2. R. R. Bockemuehl, *IEEE Trans. Electron Devices* **ED-10,** 31 (1963).
3. H. K. J. Ihantola, ASD Technical Report No. 1661-1, Stanford Electronics Laboratory (1961); H. K. J. Ihantola and J. L. Moll, *Solid-State Electron.* **7,** 423 (1964).
4. K. Yamaguchi, T. Toyabe, and H. Kodera, *IEEE Trans. Electron Devices* **ED-22,** 1047 (1975).
5. G. C. Dacey and I. M. Ross, *Proc. IRE* **41,** 970 (1953); *Bell System Tech. J.* **34,** 1149 (1955).
6. K. Yamaguchi and S. Asai, *IEEE Trans. Electron Devices* **ED-25,** 362 (1978).
7. E. O. Johnson, *RCA Rev.* **26,** 163 (1965).
8. K. Yamaguchi and H. Kodera, *IEEE Trans. Electron Devices* **ED-24,** 1061 (1977).
9. J. I. Nishizawa, T. Terasaki, and J. Shibata, *IEEE Trans. Electron Devices* **ED-22,** 185 (1975).
10. J. I. Nishizawa and K. Yamamoto, *IEEE Trans. Electron Devices* **ED-25,** 314 (1978).
11. Y. Mochida, J. I. Nishizawa, T. Ohmi, and R. A. Gupta, *IEEE Trans. Electron Devices* **ED-25,** 761 (1978).
12. C. A. T. Salama and J. G. Oakes, *IEEE Trans. Electron Devices* **ED-25,** 1222 (1978); T. D. Mok and C. A. T. Salama, *IEDM Conf. Proc. Suppl., Washington, D.C.* p. 6, (December 1976).

13. T. D. Mok and C. A. T. Salama, *IEEE Trans. Electron Devices* **ED-25,** 384 (1978).
14. J. B. Price, *in* "Semiconductor Silicon," p. 339. The Electrochemical Society, Pennington, New Jersey, 1973.
15. S. Teszner and R. Gicquel, *Proc. IEEE* **52,** 502 (1964).
16. B. J. Baliga and W. Wessels, *IEDM Tech. Digest*, p. 661 (1978); B. J. Baliga, *IEEE Trans. Electron Devices* **ED-27,** 368 (1980).
17. J. Ruch, *IEEE Trans. Electron Devices* **ED-19,** 652 (1972).
18. C. A. Liechti, *IEEE Trans. Microwave Theory Tech.* **MTT-24,** 279 (1976).
19. M. S. Shur, *IEEE Trans. Electron Devices* **ED-25,** 612 (1978).
20. M. Fukuta, T. Mimura, H. Suzuki, and K. Suyama, *IEEE Trans. Electron Devices* **ED-25,** 559 (1978).
21. A. Highashisaka, Y. Takayama, and F. Hasegawa, *IEEE Trans. Electron Devices* **ED-27,** 1025 (1980).

CHAPTER 12

INSULATED GATE FIELD-EFFECT TRANSISTORS

The most important among the insulated gate field-effect transistors (IGFETs) is the MOS device using silicon dioxide as the gate insulator. Several types of MOS transistors are in existence, with their symbols shown in Fig. 1. The most useful as power devices are the *n*- and *p*-channel enhancement-type transistors (Fig. 2), which are normally "off" and start conducting when the channel region (known also as substrate or bulk) becomes inverted at the surface. This takes place when a positive potential is applied to the gate with respect to the body of an *n*-channel transistor, and a negative potential to the gate of the *p*-channel transistor. The depletion-type devices are normally in the "on" condition, i.e., conducting until a negative (*n*-channel) or a positive potential (*p*-channel) is applied to the gate.

The basic structure of the conventional *enhancement-type n-channel* MOS transistor is shown by Fig. 2. The n^+-region on the left-hand side is the source, whereas the n^+-region on the right is the device drain. The region in between, the channel region, is known as substrate or bulk. Because of the possible confusion with epitaxial substrate, the term bulk is more in use at this time. The positive gate potential modulates the bulk surface conductivity. At the gate voltage exceeding the threshold voltage V_T, an inversion

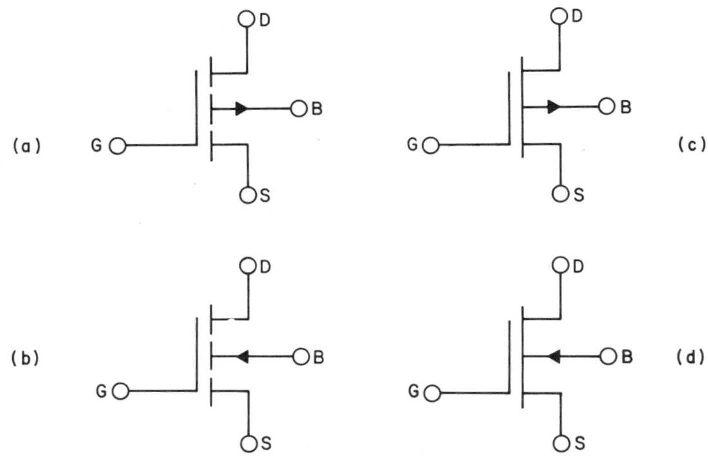

Fig. 1. MOS transistor symbols. (a) *p*-channel enhancement, (b) *n*-channel enhancement, (c) *p*-channel depletion, and (d) *n*-channel depletion.

layer (channel) is formed connecting the two electrodes. In Fig. 2 the source and the bulk are both at the ground potential. However, in some cases, there may exist a potential difference between the source and bulk.

The drain is positively biased with respect to the bulk and the source and so is the gate. The electron current flows from the negative source to the positive drain; the conventional current flows in the opposite direction. The conventional MOS structure may only be used for power devices having low drain voltages; higher power devices require high breakdown voltages and high gains. These requirements dictate the use of structures utilizing an additional drain–drift region and a vertical rather than the horizontal drain of conventional structure. The vertical structure (Fig. 3) lends itself to the better silicon surface utilization, improved thermal properties, and lower cost. The purpose of the n^--drift region is to permit the drain depletion region to spread mostly into it, thus increasing the breakdown voltage. The source and bulk region of this device are diffused from the same opening in the silicon mask (hence the name D-MOS) so that the very narrow channels necessary for high transconductance and speed can be achieved without difficulty. The bulk and source are usually shorted.

Since the *n*-channel enhancement-type device is the most desirable from the speed, gain, and power handling capability point of view, we are limiting our theoretical review to this type of device, indicating, however, how it is applicable to the *p*-type enhancement devices.

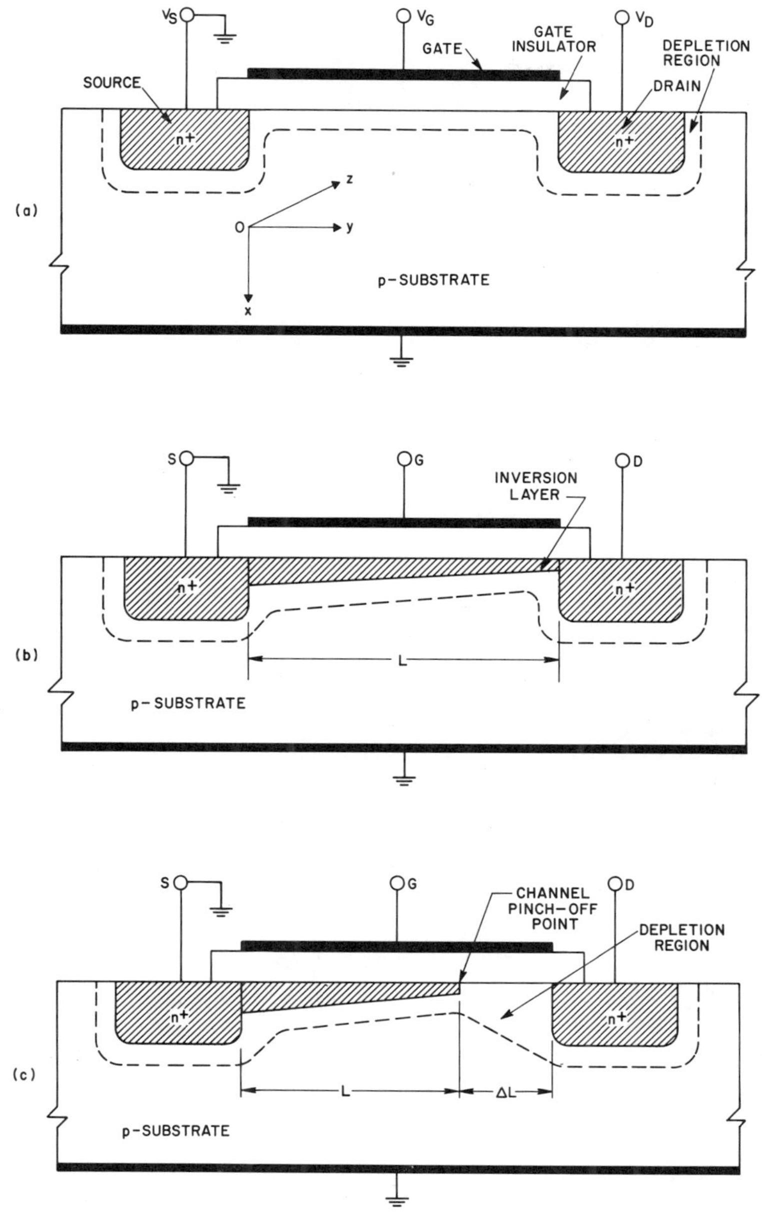

Fig. 2. Schematic cross section of an enhancement *n*-channel transistor. (a) $V_G = 0$; (b) $V_G > 0$; (c) in saturation with $V_G > 0$ and $V_D > V_{Dsat}$.

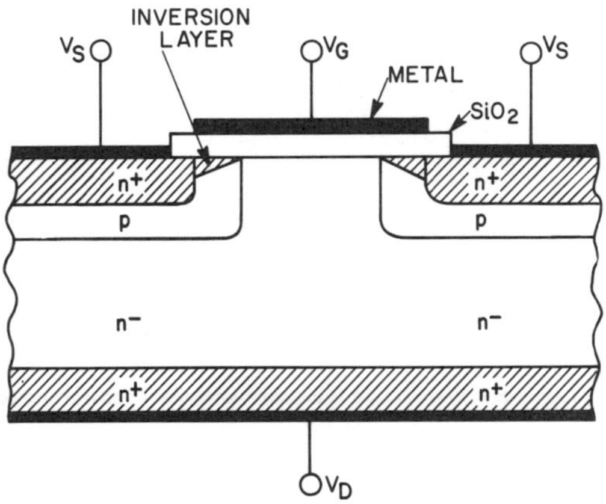

Fig. 3. *n*-channel MOST with a vertical drain.

12.1 MOS Transistor Static Characteristics

The current–voltage characteristic curves of the MOS transistor (MOST) are very similar in nature to those of the JFET (Fig. 4). They are essentially linear below the drain saturation voltage V_{Dsat}, but as the drain voltage is increased and approaches V_{Dsat}, the current saturates, i.e., stays constant to the first order of approximation.

From the analytical point of view, it is expedient to study separately the regions below and above the saturation point.

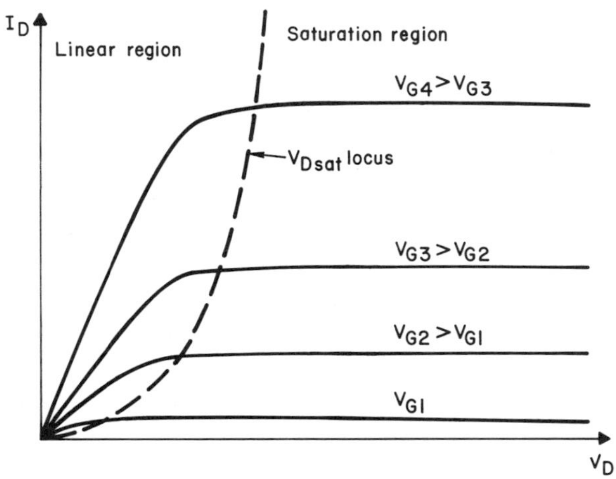

Fig. 4. Static current–voltage characteristic of an *n*-channel MOST.

12.1.1 MOS Transistor Current–Voltage Characteristics below Saturation

As for the JFET, the distributed analysis of the MOS transistor is greatly simplified by Shockley's gradual approximation [1], which assumes that the electric field along the channel in the y direction (Fig. 2) varies very slowly and is much smaller than the field normal to the silicon surface. The channel conductance is also assumed to be a slowly varying function of position. As a result of this approximation, the MOS transistor can be analyzed one dimensionally as long as the conducting channel extends from the source to the drain. Corrections to the analysis are needed as soon as a depletion region appears between the end of the conducting channel and the drain.

The flow of the current in the n-type inversion layer of an enhancement transistor (Fig. 2) is made possible by the application of the positive gate potential, which induces a negative charge Q_n in the inversion layer (channel) at the semiconductor surface. The channel charge per unit area is given by (Chapter 1)

$$Q_n = -C_i[V_G - V_{FB} - V(y) - 2\phi_F] + [2\epsilon_s q N_A(2\phi_F + V(y))]^{1/2} \quad (1)$$

or

$$Q_n = -C_i(V_G - V_T'), \quad (2)$$

where C_i is the capacitance per unit area of the insulated gate, V_G the gate potential in respect to the grounded source, V_{FB} the flat-band potential, $V(y)$ potential in the channel, a function of position y, ϕ_F the Fermi potential of the p-type substrate, ϵ_s the permittivity of the semiconductor, N_A the acceptor concentration of the uniformly doped p-type substrate (bulk), and V_T' the nonequilibrium threshold voltage.

The source and the substrate are at ground potential, and the drain voltage V_D is positive with respect to the source. If it is assumed that the flow of the drain current is only by drift with a constant electron mobility and that diffusion effects can be neglected, then the drain current in the channel [2, 3] is given by

$$I_D = -Z\mu_n Q_n(y)\, dV/dy, \quad (3)$$

where μ_n is the electron mobility, assumed to be independent of the depth x; Z is the channel width in the z direction; Q_n is the negative charge given by (1); and $-dV/dy$ is the electric field at y. Separating the variables and integrating (3) with the limits for V between 0 and V_D and for y between 0 and L (Fig. 2) yields the expression for the drain current:

$$I_D = (Z/L)\,\mu_n C_i\{(V_G - V_{FB} - 2\phi_F - V_D/2)V_D - \tfrac{2}{3}[(2\epsilon_s q N_A)/C_i^2]^{1/2}[(V_D + 2\phi_F)^{3/2} - (2\phi_F)^{3/2}]\}. \quad (4)$$

At the very low values of the drain potential, i.e., for

$$V_D \ll 2\phi_F \tag{5}$$

and

$$V_D \ll V_G - V_{FB} - 2\phi_F, \tag{6}$$

Eq. (4) becomes

$$I_D \simeq (Z/L)\,\mu_n C_i [V_G - V_{FB} - 2\phi_F - [(2\epsilon_s q N_A)/C_i^2]^{1/2}(2\phi_F)^{1/2}]\, V_D. \tag{7}$$

For the low drain voltages, we can utilize the equilibrium threshold voltage V_T [see Eq. (11) of Chapter 1], and by substituting it into (7), we obtain

$$I_D = G_0(V_G - V_T)V_D, \tag{8}$$

with

$$G_0 = (Z/L)\mu_n C_i. \tag{9}$$

G_0 is also known as the MOS transistor gain factor. Equation (8) demonstrates the linearity of the I–V characteristics at low drain voltages.

Analogous equations can be derived for the p-channel devices.

In some cases, a reverse voltage V_{BG} may be introduced between the source and bulk. For this situation, the current expression (7) should be modified to

$$I_D = (Z/L)\mu_n C_i \{(V_G - V_{FB} - 2\phi_F - V_D/2)V_D - \tfrac{2}{3}[(2\epsilon_s q N_A)/C_i^2]^{1/2}[(V_D + 2\phi_F + V_{BG})^{3/2} - (2\phi_F + V_{BG})^{3/2}]\}. \tag{10}$$

12.1.2 MOS Transistor Current–Voltage Characteristics in Saturation

For a sufficiently large value of the drain voltage V_{Dsat}, the drain current I_D reaches its maximum value I_{Dsat} (Fig. 4) and remains practically unchanged with the increasing V_D until the voltage is so high that the electric breakdown occurs. In the saturation region, the differential conductance

$$\partial I_D/\partial V_D|_{V_G = \text{const}}$$

becomes equal to zero.

By differentiating expression (4) with respect to V_D and solving for $V_D = V_{Dsat}$, we obtain

$$V_{Dsat} = V_G - V_{FB} - 2\phi_F + \frac{\epsilon_s q N_A}{C_i^2}\left[1 - \left(1 + \frac{2C_i^2(V_G - V_{FB})}{\epsilon_s q N_A}\right)^{1/2}\right]. \tag{11}$$

An analogous expression is valid for p-channel transistors.

For moderately high or high resistivity substrates, the charge density in the surface depletion region is small compared to the charge on the C_i-capacitance with thin dielectric so that

$$\epsilon_s q N_A / C_i^2 \ll 1, \tag{12}$$

and expression (11) reduces to a simple form

$$V_{Dsat} = \simeq V_G - V_T. \tag{13}$$

Low resistivity substrates with large N_A will result in lower saturation voltages than that given by (13).

The saturated drain current I_{Dsat} may be obtained by the substitution of expression (11) into expression (4) for I_D. The result is a lengthy equation, which becomes very manageable, however, for high resistivity substrates as long as the condition

$$(V_G - V_{FB} - 2\phi_F) \gg 1 \tag{14}$$

is satisfied together with (12). The saturation current for an n-channel device becomes

$$I_{Dsat} \simeq \tfrac{1}{2} G_0 (V_G - V_T)^2. \tag{15}$$

For a p-channel device, only the sign of the I_D current changes.

12.1.3 Subthreshold Current

The current arising from the majority carriers' presence in the bulk region at gate voltages below the threshold voltage is known as the subthreshold current. Its presence increases the drain-to-source leakage current when the device is in the off state. The subthreshold current consists of the drift and diffusion components and depends on the gate, drain, and substrate voltages. As the gate voltage approaches its threshold value, the subthreshold current increases very rapidly. The drain current is very substantially affected by the subthreshold current of devices with short channels. One way to decrease the excessive current flow in the off condition is to apply to the gate a bias of about 0.5 V smaller than the threshold voltage.

12.1.4 Simplified Drain Current Analysis

An approximate drain current expression can be obtained [4] by substituting into (3)

$$Q_n(y) = -C_i[V_G - V_T - V(y)], \tag{16}$$

where the threshold voltage V_T is assumed to be constant, equal to its equilibrium value, and independent of the drain voltage and position y. In contrast, the threshold voltage V_T' was used in the more accurate analysis. The drain current from (3) is therefore

$$I_D = Z\mu_n C_i[V_G - V_T - V(y)]\, dv/dy. \tag{17}$$

Separating the variables and integrating for V between 0 and V_D and for y between 0 and the channel length L, we obtain

$$I_D = (\mu_n Z C_i/L)(V_G - V_T - \tfrac{1}{2}V_D)\, V_D \tag{18}$$

valid for $V_D \leqslant (V_G - V_T)$.

The $I-V$ relationship for the saturation region is obtained by substituting $V_{Dsat} = (V_G - V_T)$ for V_D in (18). The result is

$$I_{Dsat} = \tfrac{1}{2}(\mu_n Z C_i/L)(V_G - V_T)^2 \tag{19}$$

as obtained in Section (12.1.2) for high resistivity substrates. Expression (19) is often used in the simplified approximate analysis.

12.1.5 Physical Basis for Saturation

Assuming that the gate voltage V_G is dropped entirely across the gate dielectric of thickness x_i, the gate electric field normal to the semiconductor surface is given by

$$E_i = (V_G - V(y))/x_i. \tag{20}$$

The charge density $Q_n(y)$ per unit surface area in the channel is directly proportional to $E_i(y)$ and is therefore highest at the source, where $V(y) = 0$, and the lowest at the channel's drain end, where $V(y) = V_D$. At the drain voltage

$$V_D = V_G - V_T, \tag{21}$$

the field intensity in the gate insulator in the vicinity of the drain can be expressed as

$$E_i(d) \simeq V_T/x_i. \tag{22}$$

V_T is the turn-on voltage, therefore, at the drain end at $Y = L$ (Fig. 2c); the field just slightly lower than $E_i(d)$ will only be high enough to deplete the surface beneath the gate but too low to invert it. The formation of a small depletion region therefore prevents the channel from reaching the drain and pinches-off the drain current. The drain voltage affects the charge in the channel at the drain end, but for long channels has no effect on the channel carrier density near the source. The drain voltage at the pinch-off is known as the saturation voltage V_{Dsat}, which is independent of the magnitude of the applied drain voltage. The pinch-off point position depends, however, on the magnitude of the drain voltage. As the drain-to-source voltage is increased, it moves slightly toward the source because of the increased drain depletion region width. In the case of long channels, this slight shift can be neglected so that the channel length and, consequently, its resistance, for V_G = constant, may be assumed to be unchanged. Since both the channel resistance and potential at the pinch-off point remain the same, the current stays constant and the device is in saturation.

For shorter channels, the change in the effective length is not negligible and, therefore, devices with shorter channels may not saturate. For moderately long channels, the change in the inverted region length because of the drain depletion region width variation affects the drain current in the saturation region. This effect is discussed further in Section 12.6.2 on the drain incremental conductance.

12.2 Two-Dimensional Current Flow in the Channel

The one-dimensional theory of the MOS transistor is based on Shockley's gradual approximation for which the field along the current flow axis y is very small compared to the field normal to the surface. This approximation obviously cannot apply at the pinch-off point because if the current flowing in the channel is given by

$$I_D = -Z\mu_n Q_n(y)\, dV/dy, \tag{23}$$

then at the pinch-off point, the longitudinal field $-dV/dy$ must be infinitely large since the carrier concentration at the pinch-off point drops to zero. Therefore, equations obtained with the help of gradual approximation are only valid up to some value of the surface potential beyond which the field normal to the surface becomes smaller than the longitudinal field and the current flows away from the surface as shown in Fig. 5 [5].

It was shown by two-dimensional numerical analysis [5–7] that the drain field has a dominating effect over the gate field as soon as V_D becomes greater than V_{Dsat}. At this point, the gate partially loses its control over the drain current, the drain electrode takes over, and the field normal to the channel reverses its direction. The channel current spreading has, of course, an effect on the device output impedance, which decreases inversely proportionally to the drain–junction depth [8].

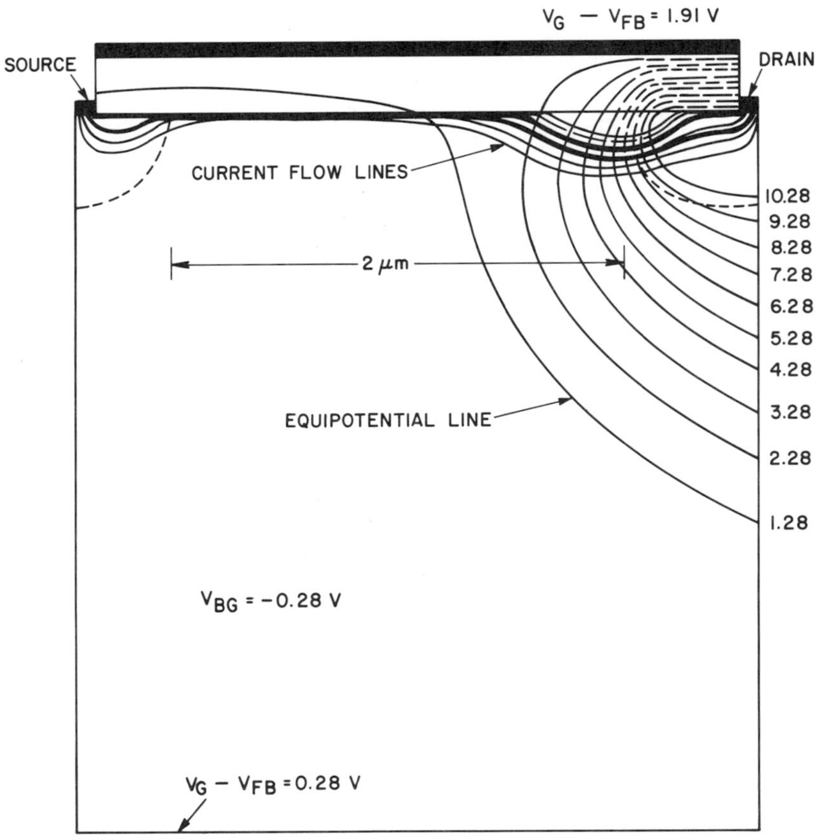

Fig. 5. Channel current flow lines and drain equipotential lines. (From Toyabe *et al.* [5].)

12.3 Threshold Voltage of Short Channel MOS Transistor

The threshold voltage V_T for channels whose length is much greater than the depth of the source and drain junctions is given by (Chapter 1)

$$V_T = V_{FB} + 2\phi_F - Q_{dmax}/C_i. \qquad (24)$$

For short channels, on the order of a few microns, V_T becomes smaller because the channel depletion region charge Q_{dmax}, coupled to the gate potential, is made smaller by the presence of the source and drain depletion regions (Fig. 2c).

A simple geometrical model for short channel MOS devices was proposed by [9] for the calculation of the threshold voltage. It is assumed that the surface depletion charge Q_{dmax} is contained in a trapezoidally shaped region (Fig. 6 [9]) and the determination of the threshold voltage requires only the computation of this charge since all other parts of expression (24) are not affected. The average charge per unit area contained in the trapezoidal region for the case of an n-channel device (p-type bulk) can be determined from Fig. 5 as being

$$Q'_{dmax} = -qN_A[(L + L')/2]x_{dmax}, \tag{25}$$

where x_{dmax} is the maximum depletion region width. L and L', the channel lengths, are defined by Fig. 6, and their sum can be determined from the geometrical considerations as

$$(L + L')/2 = 1 - [(1 + 2x_{dmax}/r_j)^{1/2} - 1]\, r_j/L, \tag{26}$$

where r_j is the junction depth. If it is rather arbitrarily assumed that the effect of Q'_{dmax} on the surface potential is uniform, the equation determining the threshold voltage becomes

$$V_T = V_{FB} + 2\phi_F - (Q_{dmax}/C_i)\{1 - [(1 + 2x_{dmax}/r_j)^{1/2} - 1]\}\, r_j/L. \tag{27}$$

According to [9], this equation accurately predicts V_T for devices with channel lengths from 1.4 to 7.4 μm.

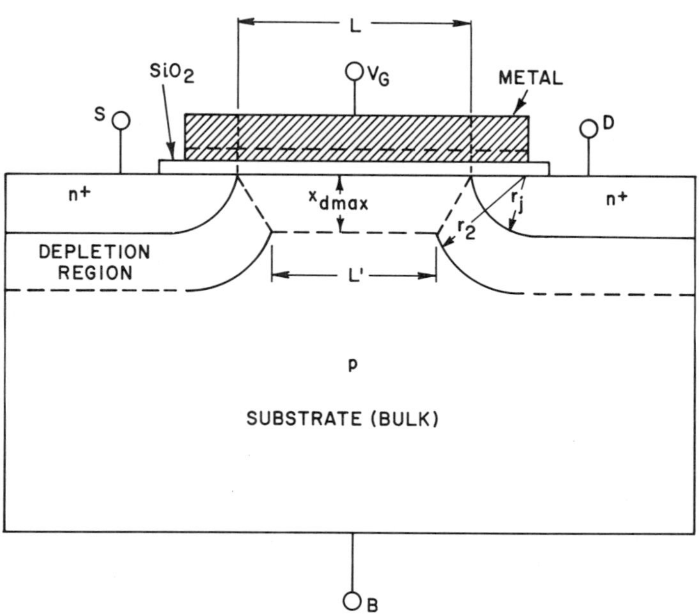

Fig. 6. Geometrical model for the computation of the threshold voltage of a short channel MOST. (From Yau [9].)

12.4 Back-Bias Effect on the Threshold Voltage

In power MOS transistors, the source and body are both usually connected to the ground. This helps in minimizing the effects of the parasitic bipolar transistor parallel to the channel and reduces the possibility of the second breakdown. In some cases, it may become necessary to introduce a reverse bias between the source and the body while keeping the source at ground potential. The reverse source–body potential appears between the channel and the body increasing the channel–body depletion region width. This, in turn, increases the total depleted charge so that the threshold voltage has to be increased to compensate for its presence by an increment

$$\Delta V_T = (2q\epsilon_s N_A/C_i)^{1/2}[(V_{BG} + V_i)^{1/2} + V_i^{1/2}], \qquad (28)$$

where V_{BG} is the source-to-body bias voltage and V_i the built-in channel–body junction potential.

12.5 MOS Transistor Threshold Control

In most applications, *n*-channel enhancement devices are more desirable than the *p*-type devices because of their higher transconductance and better frequency response. In order to obtain reasonably high threshold voltages V_T, the *n*-channel device requires, however, better processing control than the *p*-type device. The *n*-channel device threshold voltage would be negative due to the polarities of the Q_{ss}, ϕ_{ms}, and ϕ_F if not for the presence of the ionized acceptors in the surface depletion region. To obtain positive thresholds with good control, the bulk impurity concentration N_A should be greater than 10^{15} cm^{-3}. If the *p*-surface is depleted, for example, by the boron segregation into the SiO_2, the acceptor density may not be large enough to make V_T positive. It is possible to control the threshold voltage by reverse biasing the body with respect to the channel; however, this involves more expensive circuitry. Ion implantation, described in Chapter 5, provides a good technique to adjust the threshold voltage to the desired value [10], which for good control purposes should be greater than 1 V and, in most power devices, lies between 1 and 3 V. For the threshold voltage adjustment, ion implantation is usually performed by exposing the gate region; by masking other regions by, for example, photoresist; and by implanting, for instance, a boron ion through the SiO_2 layer, thus increasing the acceptor surface concentration to values above 10^{15} cm$^-$3. Ion implantation can be used either for the increase or for the decrease of the threshold voltage. The same techniques are, of course, applicable to the *p*-channel devices.

12.6 Incremental MOS Transistor Parameters

Although power devices only rarely operate at small signal levels, incremental MOS parameters are very useful for device characterization.

12.6.1 Incremental Conductance in the Linear Region

The channel incremental conductance in the linear $I-V$ characteristic region below saturation, is defined by

$$g_{\mathrm{d}} = \left.\frac{\partial I_{\mathrm{D}}}{\partial V_{\mathrm{D}}}\right|_{V_{\mathrm{G}} = \mathrm{const}} . \tag{29}$$

Differentiating expression (8) yields

$$g_{\mathrm{d}} = G_0(V_{\mathrm{G}} - V_{\mathrm{T}}), \tag{30}$$

showing the linear dependence of the incremental conductance on the gate voltage. For large drain currents, this linear dependence is not maintained since the electron mobility decreases with the gate voltage.

12.6.2 Incremental Conductance in Saturated Region (g_{dsat})

There are second-order mechanisms that affect the MOS transistor $I-V$ characteristics in the current saturation region. The ideal device would have an infinite incremental resistance in this region; a real life device deviates from the ideal in that the incremental resistance is finite. One of the mechanisms consists of the channel shortening when the drain potential increases beyond the pinch-off value V_{Dsat}. This is analogous to the base narrowing in a bipolar transistor with increasing reverse collector voltage (Early effect). In the ideal case, the movement of the depletion region with increasing V_{D} may be assumed to be negligibly small. For high resistivity substrates, high voltage, and short channels, the shift of the pinch-off point toward the source is not negligible anymore; and as the drain voltage beyond V_{Dsat} is increased, the channel becomes shorter and has a lower resistance; and the saturated current increases although the voltage at the pinch-off point is fixed at V_{Dsat}.

The shortening of the channel length L by $-\Delta L$ will cause the current in the saturation region to change from I_{Dsat} to

$$I_{\mathrm{Dsat}}^{*} = I_{\mathrm{Dsat}}[1 - (\Delta L/L)]^{-1}. \tag{31}$$

The drain depletion layer width ΔL can be determined from the relationship

$$\Delta L = (V_{\mathrm{D}} - V_{\mathrm{Dsat}})/E_{\mathrm{T}}, \tag{32}$$

where E_T is the average transverse electric field near the silicon–SiO_2 interface and consists of three components (Fig. 7 [11])

$$E_T = E_{1y} + E_{2y} + E_{3y}. \tag{33}$$

The electric field component E_{1y} is due to the fixed charge in the reverse-biased drain junction, E_{2y} is the fringing electric field due to the drain-to-gate potential difference $V_D - V_G'$, and the fringing electric field E_{3y} is due to the potential difference $V_G' - V_{sat}$ between the gate and the inversion layer. $V_G' = V_G + Q_{ss}/C_i$ is the effective gate voltage which takes into account the fixed insulator charge. E_{1y} can be determined from the p–n junction theory; the other two contributions can be approximated [11] by assuming their proportionality to the respective normal electric fields across the insulator:

$$E_{2y} = \alpha \frac{\epsilon_i}{\epsilon_s} \frac{V_D - V_G'}{x_i}, \tag{34}$$

$$E_{3y} = \beta \frac{\epsilon_i}{\varepsilon_s} \frac{V_D - V_G'}{x_i}, \tag{35}$$

where ϵ_i and ϵ_s are gate dielectric and semiconductor permittivities, respectively; α and β are the proportionality constants, approximately independent of the applied voltages and structures in use, $\alpha \simeq 0.2$ and $\beta \simeq 0.6$; and x_i is the gate dielectric thickness. The field due to the drain is determined from the abrupt junction theory and is equal to

$$E_{1y} = (qN_A/2\epsilon_s)^{1/2}(V_D - V_{Dsat})^{1/2}. \tag{36}$$

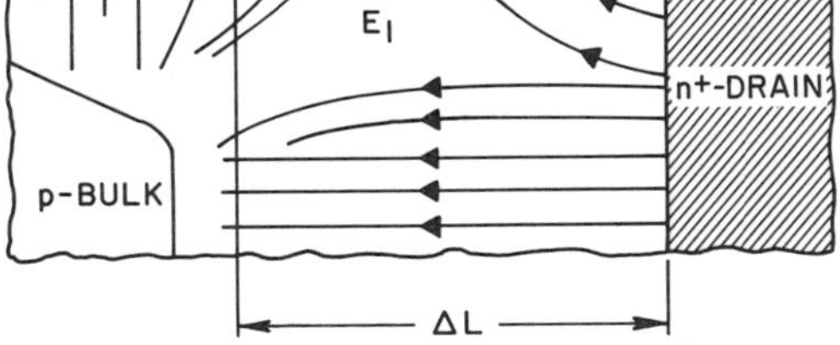

Fig. 7. Drain and gate electric fields components.

Combining (32) (36) gives [11]

$$\frac{1}{\Delta L} = (qN_A/2\epsilon_s)^{1/2}(V_D - V_{Dsat})^{-1/2} + \frac{\epsilon_i}{\epsilon_s x_i}\frac{\alpha(V_D - V'_G) + \beta(V'_G - V_{Dsat})}{V_D - V_{Dsat}} \tag{37}$$

or

$$\frac{1}{\Delta L} = \frac{1}{\Delta L(N_A)} + \frac{1}{\Delta L(x_i)}. \tag{38}$$

$\Delta L(N_A)$ and $\Delta L(x_i)$ are the channel length reductions due to the field component associated with the bulk concentration and the gate field, respectively.

The saturation conductance is defined as

$$g^*_{dsat} = \frac{dI^*_{Dsat}}{dV_D} = \frac{dI^*_{Dsat}}{d(\Delta L)}\frac{d(\Delta L)}{dV_D}. \tag{39}$$

Differentiating (31) gives

$$dI^*_{Dsat}/d(\Delta L) = I_{Dsat}/L(1 - \Delta L/L)^2, \tag{40}$$

which leads to

$$g^*_{dsat} = \frac{I_{Dsat}}{L(1 - \Delta L/L)^2}\frac{d(\Delta L)}{dV_D}. \tag{41}$$

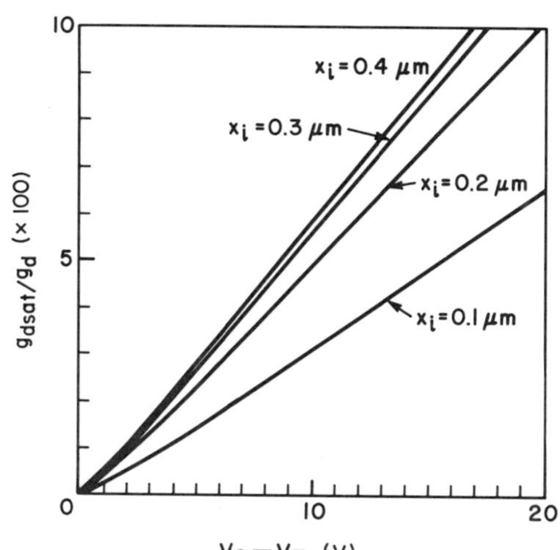

Fig. 8. Ratio of the saturation conductance g_{dsat}/g_d as a function of the effective gate voltage $V_G - V_T$ with SiO_2 thickness as a parameter. (From Frohman-Bentchkovsky and Grove [11].)

The corresponding channel conductance for small drain voltages ($V_D \ll 2\phi_F$) is given (Section 12.6.1) as

$$g_d = (Z/L)\mu_{eff}C_i(V_G - V_T). \tag{42}$$

The ratio of the saturation conductance to the low voltage channel conductance (g_{dsat}/g_d) is shown in Fig. 8 [11] as a function of the effective gate voltage $V_G - V_T$ with silicon dioxide thickness as a parameter and $N_A = 2 \times 10^{15}\ \text{cm}^{-3}$, $L = 5\ \mu\text{m}$, and $V_D = V_{Dsat} + 10$ V. Figure 9 [11] represents the same ratio as a function of the $V_G - V_T$ with N_A as a parameter. In this case, $x_i = 0.3\ \mu\text{m}$ and $L = 0.5\ \mu\text{m}$.

For the case when $x_i \gg (2\epsilon_s/qN_A)^{1/2}(V_D - V_{Dsat})^{1/2}$, i.e., when N_A is large and/or the gate insulator is thick [11],

$$\Delta L = [2\epsilon_s(V_D - V_{Dsat})/qN_A]^{1/2} \tag{43}$$

and

$$g_{dsat}^* = \left(\frac{2\epsilon_s}{qN_A}\right)^{1/2} \frac{I_{Dsat}}{2L(1 - \Delta L/L)^{1/2}(V_D - V_{Dsat})^{1/2}}. \tag{44}$$

However, for

$$x_i \ll (2\epsilon_s/qN_A)^{1/2}(V_D - V_{Dsat})^{1/2}, \tag{45}$$

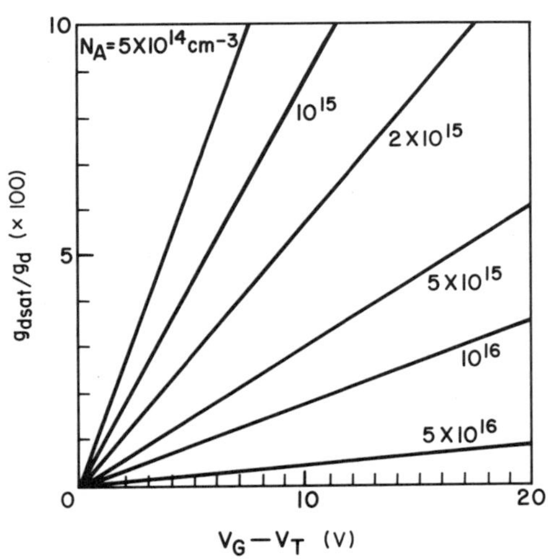

Fig. 9. Ratio of the saturation conductance g_{dsat}/g_d as a function of the effective gate voltage $V_G - V_T$ with bulk concentration N_A as a parameter. (From Frohman-Bentchkovsky and Grove [11].)

i.e., for low N_A and/or thin oxide, Eq. (37) reduces to

$$\Delta L \simeq \frac{\epsilon_s x_i}{\epsilon_i} \frac{(V_D - V_{Dsat})}{\alpha(V_D - V'_G) + \beta(V'_G - V_{Dsat})}, \tag{46}$$

and the saturation conductance becomes

$$g^*_{dsat} = \frac{\epsilon_s}{\epsilon_i} \frac{x_i}{L} \frac{(\beta - \alpha)(V'_G - V_{Dsat})}{[\alpha(V_D - V'_G) + \beta(V'_G - V_{Dsat})]^2} \frac{I_{Dsat}}{(1 - \Delta L/L)^2}. \tag{47}$$

In this last case, the saturation conductance depends on both the gate and the drain voltage. It should be pointed out that the above g^*_{dsat} expressions lose their validity for drain–source punch-through conditions.

In order to increase the drain breakdown voltage, power MOS transistors usually have an additional low-doped collector drift region (Fig. 3). In this case, only a part of the depletion region extends into the channel region (bulk), and only this part should be taken into account in the channel conductance computation.

Assuming that the channel region and the n^--drift region are both uniformly doped with the respective impurity concentrations N_A and N_D, expressions (43) and (44) should be modified to

$$\Delta L = K(2\epsilon_s/qN_A)^{1/2}(V_D - V_{sat})^{1/2} \tag{48}$$

and

$$g^*_{sat} = K\left(\frac{2\epsilon_s}{qN_A}\right)^{1/2} \frac{I_{Dsat}}{2L(1 - \Delta L/L)^2(V_D - V_{Dsat})^{1/2}}, \tag{49}$$

where

$$K = [N_D/(N_A + N_D)]^{1/2}. \tag{50}$$

Equations (48) and (49) are valid for

$$x_i \gg K(2\epsilon_s/N_A)^{1/2}(V_D - V_{sat})^{1/2}. \tag{51}$$

12.6.3 Gate Transconductance

The transconductance of a MOS transistor (MOST) is defined by

$$g_m = \left.\frac{\partial I_D}{\partial V_G}\right|_{V_D = \text{const}}. \tag{52}$$

In the linear region of the I–V characteristics, i.e., for $V_D < V_{Dsat}$,

$$g_m = G_0 V_D. \tag{53}$$

(53) is obtained by the differentiation of expression (8) for the drain current with respect to V_G. The transconductance in the saturation region of a MOST can be obtained either by differentiating expression (15) for I_{Dsat} with respect to V_G or by just replacing V_D in (53) with V_{Dsat} and obtaining

$$g_{msat} = G_0 V_{Dsat}. \tag{54}$$

V_{Dsat} is given by expression (13). For transistors with moderately or lightly doped bulk, the transconductance

$$g_{msat} \simeq G_0(V_G - V_T) \tag{55}$$

is independent of V_D and linearly dependent on V_G.

In some cases, a reverse potential V_{SB} may be applied between the source and the substrate. This back bias affects current flowing in the channel so that it is possible to define the substrate transconductance as

$$g_{mb} = \left.\frac{\partial I_D}{\partial V_{SB}}\right|_{V_D, V_G = \text{const}}. \tag{56}$$

The derivation of the g_{mb}, of course, requires the use of the drain current expression corrected for the presence of the V_{SB}.

12.6.4 Effect of Series Resistances on g_d and g_m

The conductance and transconductance of the MOS transistor are affected by the series resistances in the same way as they are in the JFET. In the linear region, the effective conductance becomes

$$g_{d(eff)} = \frac{1}{g^{-1} + R_s + R_d}. \tag{57}$$

In the saturation region, the effective transconductance changes to

$$g_{msat(eff)} = \frac{1}{g_{msat}^{-1} + R_s}. \tag{58}$$

R_s and R_d are the source and drain series resistances, respectively. In an MOS transistor, R_s can be minimized rather easily. The effect of the drain series resistance is analogous to that in a JFET (Chapter 11).

12.6.5 Saturated Carrier Velocity Effect on g_m

The effect of carrier saturation velocity on the MOS current can be determined by solving the Poisson and current continuity equations in two dimensions. This can be done only numerically and was performed, e.g., for silicon by [7] and [12]. In order to get an analytical solution, the problem was simplified by Hofstein [13] by assuming that the carrier velocity reaches its maximum value v_s only close to the drain where the charge accelerating field is the highest. The channel is divided, therefore, into two regions: one with the nonsaturated and the other with the saturated velocity. The analysis shows that the transconductance in the saturated region of $I-V$ characteristics can be expressed for large values of V_G by

$$g_{m\text{sat}} = (Z\epsilon_i \mu_n / x_i)\, E_1, \tag{59}$$

where ϵ_i is the gate dielectric permittivity, x_i is the dielectric thickness, and E_i is the field at that point of the channel at which the electron velocity reaches its limiting surface value $v_s = 6.5 \times 10^6$ cm s^{-1}. Since $\mu_n E_1 = v_s$

$$g_{m\text{sat(max)}} = ZC_i v_s. \tag{60}$$

In the limit, therefore, the transconductance in the MOS saturated current region becomes independent of the applied gate potential and channel length. This behavior was confirmed [14] in hf devices with very small channel lengths. The maximum possible transconductance per unit channel width for a MOS structure with 100 nm oxide thickness is $g_{m\text{sat(max)}}/Z \simeq 2.3$ mho cm^{-1}.

12.6.6 g_m Ultimate Limit

The conventional IGFET theory predicts the transconductance/current ratio that approaches infinity as the gate-to-channel capacitance C_i is increased by making the insulator thickness approach zero dimension. Such behavior would be completely different from any other known charge control device [15] where the theoretical maximum limit for g_m/I is q/kT. This consideration, therefore, leads to a conclusion that the potential in the channel region does not change monotonically. It was pointed out by Johnson [15] that the source-to-drain potential distribution includes a potential barrier because of the source–channel n^+-n high–low junction and can be represented as in Fig. 10 [15]. All mobile carriers must first climb the potential barrier V_{HL} before they start flowing to the drain. The high–low barrier height can be obtained from

$$V_{HL} = (kT/q)\ln(n_c/n_s), \tag{61}$$

where n_s and n_c are electron densities on the source side and the channel side interface, respectively. Carrier densities in the source and channel regions generally vary with depth from the semiconductor surface so that, in all

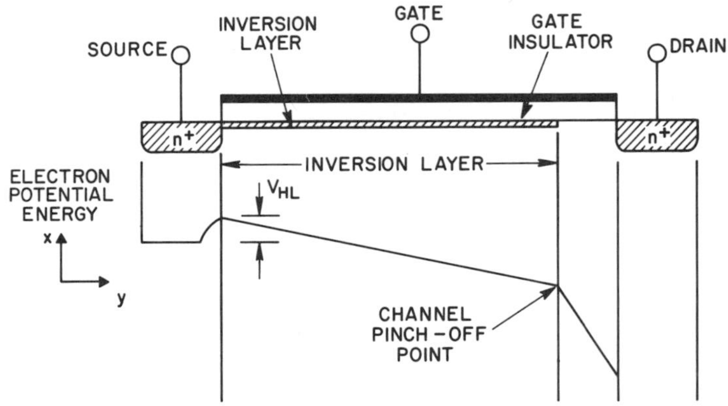

Fig. 10. Potential distribution between the source and drain, showing a potential barrier due to n^+–n-junction. (From Johnson [15].)

accuracy, V_{HL} will vary with depth as well. Ignoring this does not affect the validity of the argument, however.

A small change in V_{HL} of a few kT/q potential units can account for a large change in the current in the inversion layer because of the current's exponential dependence on V_{HL}. Now, the potential V_{HL} is in series with the voltage across the gate insulator, so when the gate insulator approaches zero thickness and the gate capacitance tends to infinity, the total input voltage approaches V_{HL} value if the ohmic voltage drop is neglected. It was shown by [15] that at this voltage

$$g_m = dI/dV_{\mathrm{G}} = dI/dV_{\mathrm{HL}} = qI/kT \tag{62}$$

and

$$g_m/I = q/kT. \tag{63}$$

The introduction of the potential barrier concept removes, therefore, the absurdity of the infinite g_m/I ratio.

12.7 MOS Transistor Incremental Circuit Model

The total charge on the transistor gate is given approximately by

$$Q'_{\mathrm{G}} = ZC_{\mathrm{i}} \int_0^L (V_{\mathrm{G}} - V_{\mathrm{T}} - V(y))\, dy \tag{64}$$

with the voltage distribution in the device channel expressed as [4]

$$V(y) = (V_{\mathrm{G}} - V_{\mathrm{T}})[1 - (1 - y/L)]^{1/2}. \tag{65}$$

Substituting (65) into (64) and integrating, we obtain

$$Q'_G = \tfrac{2}{3}ZLC_i(V_G - V_T). \tag{66}$$

The incremental gate current is obtained by differentiation of (66) as

$$\Delta I_G = i_g = dQ_G/dt = C_{gs}\,dv_{gs}/dt \tag{67}$$

with

$$C_{gs} \simeq \tfrac{2}{3}C_G = \tfrac{2}{3}ZLC_i, \tag{68}$$

where v_{gs} is the gate voltage increment, C_{gs} the gate-to-source capacitance in saturation (Fig. 11), and C_G the total gate capacitance. The dc drain current in saturation is given by

$$I_D = (\mu_n ZC_i/2L)(V_G - V_T)^2. \tag{69}$$

Therefore, the drain current increment for the gate voltage increment v_{gs} is

$$i_d = (\mu_n ZC_i/L)(V_G - V_T)v_{gs} \tag{70}$$

or

$$i_d = g_m v_{gs}. \tag{71}$$

The current flowing through the capacitance C_{gd} existing between the drain and the gate is given by

$$i_{gd} = C_{gd}\,dv_{gd}/dt, \tag{72}$$

where v_{gd} is the incremental gate to drain voltage.

The incremental current–voltage relationships for the MOS transistor are, consequently, given by

$$i_g = C_{gs}\,dv_{gs}/dt + C_{gd}\,dv_{gd}/dt, \tag{73}$$

$$i_d = g_m v_{gs} - C_{gd}\,dv_{gd}/dt. \tag{74}$$

Equations (73) and (74) are valid when all extrinsic capacitances, such as stray capacitances as well as inductances, may be neglected. The gate-to-drain capacitance is very small for the transistor in saturation since the drain and the gate are only weakly coupled electrically with the appearance of the depleted channel at the drain end (wide dielectric spacing between the drain and gate electrodes).

The incremental circuit model in the common-source configuration is represented by Figs. 4 and 5 of Chapter 11 as for a JFET. C'_{gs}, C'_{dg}, and C'_{ds} are the extrinsic (stray) capacitances. A more complete model for computer calculations would contain additional parasitic resistances due to metallization, source–drain junction depletion capacitances, and inductances for

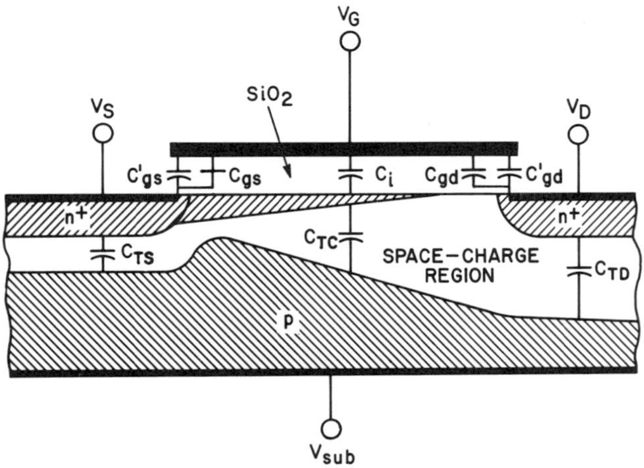

Fig. 11. Various capacitances of the MOS transistor.

very fast and microwave circuits. Many of these parasitics may be irrelevant (e.g., source–bulk junction capacitance) when the source and the substrate are tied together. At microwave frequencies, the equivalent circuit should model the channel as a distributed RC network. The lumped-element circuit is satisfactory, however, for frequencies up to about 10 GHz if the parasitic inductances are introduced. Some of the pertinent capacitances are indicated schematically in Fig. 11. C_{TS}, C_{TC}, and C_{TD} are the source, channel, and drain space-charge region capacitances, respectively. C_{gs}, C_{i}, and C_{gd} are the gate-to-source, oxide, and gate-to-drain capacitances, respectively.

12.8 Cut-Off Frequency f_{T}

A MOS transistor cut-off frequency f_{T} (also called gain–bandwidth product) is defined as that frequency at which the current through the input capacitance of the equivalent circuit is equal to the $g_{m\mathrm{sat}} v_{\mathrm{gs}}$ current of the voltage controlled current generator (Fig. 4 of Chapter 11). This takes place when the reciprocal of the input reactance becomes equal to the transconductance, i.e.,

$$\omega_{\mathrm{T}} C_{\mathrm{in}} = 2\pi f_{\mathrm{T}} C_{\mathrm{in}} = g_{m\mathrm{sat}}; \tag{75}$$

therefore,

$$\omega_{\mathrm{T}} = g_{m\mathrm{sat}}/C_{\mathrm{in}} = 2\pi f_{\mathrm{T}}. \tag{76}$$

But

$$g_{msat} = (Z/L)\mu_n C_i V_{Dsat}. \tag{77}$$

For the shorted output,

$$C_{in} \simeq C_i LZ \tag{78}$$

The substitution of (77) and (78) into (76) obtains

$$\omega_{T1} = \mu_n V_D / L^2. \tag{79}$$

The transit time of carriers through the channel of length L with velocity $\mu_n(V_{Dsat}/L)$ is

$$T_1 = L^2/\mu_n V_{Dsat}. \tag{80}$$

Consequently,

$$\omega_{T1} = T_1^{-1}. \tag{81}$$

In a more exact gain–bandwith computation, the input capacitance of the transistor operating in saturation would be taken as

$$C_{in} = C_{gs} = \tfrac{2}{3}(C_i LZ). \tag{82}$$

Expressions (79) and (80) do not apply to very short channels with large drain depletion regions where the charge transport mainly takes place in the space-charge region. If the field is high in the inversion layer, carriers may reach the saturation velocity v_s so that the minimum channel transit time becomes

$$T_1' = L/v_s. \tag{83}$$

Most power MOS transistors are designed with n^--drift regions (Fig. 6 of Chapter 13) (or p^--drift regions for the p-channel devices) necessary to support the high drain breakdown voltage. The transit time in the depleted channel and drift section causes a phase delay of the short circuit drain current with respect to the gate voltage or a delay time in the turn-on of the drain current. The delay time corresponding to this transit time was shown by [16] to be equal to $\frac{1}{2}$ of the transit time in the depletion region. Therefore, the phase delay is given by

$$\phi_d = 2\pi f(\tfrac{1}{2}\tau), \tag{84}$$

where $\frac{1}{2}\tau$ is the time delay and f the operating frequency. The total minimum time delay in the very short channel is therefore

$$T_{tot} = T_1' + \tfrac{1}{2}\tau, \tag{85}$$

and

$$2\pi f_{Tmax} = v_s/(L + \tfrac{1}{2}L') \tag{86}$$

if L' is the total depletion region length and the carrier velocity everywhere is close to v_s. When the channel is almost completely depleted,

$$T_{tot} = \tfrac{1}{2}\tau, \tag{87}$$

and

$$2\pi f_{Tmax} = 2v_s/L'. \tag{88}$$

12.9 MOS Transistors with Triodelike Characteristics

The drain depletion layer may be large enough to reach from the drain to the source depletion layer, even for $V_D = 0$, if the body resistivity is very high and the channel length short. The flow of the drain current is possible then due to the presence of the MOS channel and to the presence of the n^+–π–n^+ diode with a short and very lightly doped π-region. The π-regions with resistivities higher than a few hundred ohm centimeters are considered semi-insulating.

Electrons from the n^+-source region can be injected into the depleted π-region, either by the Schottky effect (potential barrier lowering by the drain voltage) or by tunneling.

Electrons are assumed to have an infinite lifetime in the π-region because all the recombination centers are supposed to be already filled and, consequently, cannot capture more electrons. On the other hand, the lifetime of holes is assumed to be extremely short. Under these conditions, the current consists only of injected electrons with no holes involved. This is known as single injection [17]. The injected current density is

$$J_{SCL} = -qn\mu_n D, \tag{89}$$

where SCL stands for space-charge limited, qn is the charge density, μ_n the electron mobility, and E the electric field. Poisson's equation in the depleted region between the source and drain can be written as

$$dE/dy = qn/\epsilon_s, \tag{90}$$

where ϵ_s is the semiconductor's permittivity. Combining (89) and 90) yields an expression for the space-charge limited current density

$$J_{SCL} = -\epsilon_s\mu_n E\, dE/dy. \tag{91}$$

Integration of (91) gives

$$-dV/dy = E = [(2J_{SCL}/\epsilon_s\mu n) + E_C^2]^{1/2}, \tag{92}$$

where E_C is the electric field at $y = 0$, i.e., at the source. For the drain voltages on the order of a few volts, E_C can be neglected. Integrating (92) from $y = 0$ to $y = L$ at the drain gives for the drain-to-source voltage

$$V_D = L^{3/2}(8J_{SCL}/9\epsilon_s \mu n)^{1/2} \quad (93)$$

and

$$J_{SCL} = \tfrac{9}{8}(\mu_n V_D^2/8L^3). \quad (94)$$

This is the so-called Mott–Gurney relationship [18]. The space-charge limited current takes its name from the presence of the very large electronic charge at the cathode that limits the injection of additional electrons.

The current–voltage relationship in the punch-through device deviates completely from the pentode like characteristics of the non-punch-through MOS transistor. The drain current of the punch-through device depends on the square of the drain voltage. Such a device resembles a vacuum triode [19, 20].

Current flow in the $n^+-\pi-n^+$ region is also observed with small channel lengths on *moderate* resistivity substrates if V_D is sufficiently high to obtain punch-through conditions, but the SCL current will flow only if the resistivity is above 100 Ω cm.

The presence of the inversion layer cannot be ignored, of course, and the total current flowing in the device will consist of the drift current in the inversion layer and the space-charge limited current in the punch-through region. By lowering the positive gate potential for the n-channel device or by reversing its polarity, it is possible to greatly reduce the drift component so that the drain current consists of the space-charge limited current only. The field, due to high negative gate potential, will be able to stop the flow of the SCL current flowing close to the surface, but the deeper flowing current would be unaffected. Complete current cut-off is possible [19], however, by reverse biasing the substrate with respect to the source and modulating in this way the current flow. $I-V$ characteristics of an SCL current-dominated MOS transistor would resemble the characteristics of a vacuum triode more than a pentode. It was found experimentally that triodelike characteristics may be obtained only when the ratio of the gate insulator thickness to the source–drain spacing fulfills certain conditions. As this ratio increases, the transition to the triodelike characteristics will take place at lower drain voltage values [21].

Ohmi's [22] recent one-dimensional analysis of an $n^+-p^--n^+$ or $p^+-n^--p^+$ structure in a punch-through state demonstrated the presence of a potential barrier between the n^+*-regions* (*or* p^+-regions). This barrier is similar to that of a SIT (Fig. 8 of Chapter 11).

The punch-through voltage V_{pt} can be changed by changing either the channel region length (region p^- or n^-) or the channel region impurity concentration. The barrier height is lowered, and its position is shifted toward the source with the increase in the punch-through voltage. For V_{pt} greater then three times the drain junction built-in potential ($V_{pt} > 3V_i$), the barrier height is very small, allowing high injection of majority carriers into the drain depletion region.

The picture of the potential profile between the source and drain is considerably modified by the two-dimensional analysis [22] that accounts for the effect of the gate electrode. In the computerized analysis of [22], the gate voltage was set to the flat-band condition. It turns out that the potential barrier height is different at various depths below the gate. At a depth equal to one-half the channel length L, for example, the potential barrier behavior is very similar to that demonstrated by the one-dimensional analysis, so that the barrier height is lowered by an increased drain voltage and/or decreased channel length. At a shallow depth of about 0.1 L, the potential barrier is almost entirely determined by the gate potential. The increase of the drain voltage still lowers the barrier height and shifts it toward the source, but the effect of the drain voltage on the voltage profile near the source is small. According to [22], the barrier height control exists even before complete punch-through takes place. The existence of this type of control determines the shape of the $I-V$ characteristics which show an exponential behavior. On the other hand, the analysis of [19] shows a square-law behavior of a space-charge limited current in the absence of a potential barrier.

12.10 Voltage Breakdown in MOS Transistors

The two most common mechanisms limiting the MOST performance are the gate oxide dielectric breakdown and the drain region avalanche breakdown.

12.10.1 Gate Oxide Breakdown

Thermally grown SiO_2 on a silicon substrate exhibits current conduction through the quantum-mechanical tunneling mechanism [23]. Electrons with energy close to the Fermi level of the metal electrode tunnel through the SiO_2 energy gap into its conduction band. The tunneling current is not established instataneously, however. As the gate voltage is applied, there is an initial transient current associated with charge traps and the fixed and mobile charges present in the oxide. The shape of the decay current curve can be described by at least two exponential time constants [24]. The slower decay time constant can be correlated with the release from the traps of such mobile charges as sodium ions and protons. The steady state conduction due to the tunneling mechanism can often be observed after a time on the order of several minutes to a few hours at room temperature. Experiments performed by Osborn and Weitzman [24] confirmed the tunneling mechanism as the one responsible for the oxide conduction by plotting J/E^2 versus $1/E$ (J is the current density and E the electric field in the oxide). The conduction temperature dependence was found by [24] to decrease with the increasing field and to vary with the electrode material. These results could not be explained on the basis of the usual conductance temperature dependence due to tunneling and were attributed by [24] to the temperature dependent energy barrier or effective mass. The conduction current increases with temperature; for example, it is two orders of magnitude higher for a field of ~7.0 MV cm^{-1} than the current at ~6.0 MV cm^{-1}. At the lower field (6 MV cm^{-1}), the current changes two orders of magnitude between 50 and 300°K, but at the higher field (7 MV cm^{-1}) the change is only one order of magnitude. The postmetallization anneal may reduce the conduction by three orders of magnitude, probably by reducing the density of SiO_2 defects and traps. The heat treatment may consist of annealing the oxide for about 5 min at 500°C in dry N_2.

For gate oxide fields greater than 8 MV cm^{-1}, a current instability arises that may lead to the dielectric breakdown. The breakdown of the thermally grown SiO_2 on silicon depends on the oxide thickness, substrate doping concentration, type of passivation used, and on the temperature. Below an oxide thickness of 80 nm the maximum dielectric strength varies as

$$E_{\max} \propto x_i^{-0.21}, \tag{95}$$

where x_i is the oxide thickness. Between 100 and 200 nm, SiO_2 breakdown is independent of the oxide thickness. Similar to the silicon avalanche breakdown, SiO_2 breakdown becomes somewhat higher at elevated temperatures. The doping level of the silicon substrate has a small but definite effect on the SiO_2 breakdown.

Very often, SiO_2 breaks down at fields lower than 8 MV cm^{-1} since, for example, the measurement method itself has a bearing on the final result.

The metallization and the oxide thickness and the measuring device impedance often lead to the nonshorting, self-healing breakdowns. The heat developed by the breakdown mechanism itself may be sufficient to vaporize the metal which penetrated the oxide, and the gate oxide may break more than once without any real damage. The voltage which eventually shortens the path between the gate and substrate without self-healing occurring can be considered as the one that determines the maximum dielectric strength.

Because of the great variability in the oxide quality, it is a good practice to design the oxide thickness with at least a 50% safety factor.

Polycrystalline silicon is often used in place of metals as the gate electrode material because of its refractory nature and gate self-alignment capability. The comparison of polycrystalline silicon gate electrodes with aluminum electrodes [25] shows that these two materials result in the same oxide breakdown.

12.10.1.1 Gate Oxide Protection

Static electricity can easily damage the oxide, particularly in devices with small gate capacitances; hence, there is a need to protect MOS transistors, e.g., by Zener diodes connected in parallel with the gate to avoid device destruction upon exposure either during testing or operation.

Very wide (large Z) channel power transistors are less sensitive because their gate capacitances are large.

For a constant gate charge Q, the voltage developed across the gate insulator is inversely proportional to the gate capacitance C_G:

$$V_G = Q/C_G. \tag{96}$$

For instance, a device with $C_G = 1$ pF and $x_i = 100$ nm would break down for $Q = 8 \times 10^{-11}$ C, but with the gate capacitance of 100 pF, $Q = 8 \times 10^{-9}$ C would be required to reach the breakdown voltage. As already noted, the gate can be protected by a Zener diode integrated on the same chip with the transistor. Another protecting scheme consists of an integrated n^+–p–n^+ punch-through diode for an n-channel device or p^+–n–p^+ diode for a p-MOST in parallel with the gate. The protective diode punches through before the main transistor oxide can be destroyed.

12.10.2 Drain-to-Source Voltage Breakdown and Second Breakdown

In the long n-channel MOS transistors, the drain-to-source breakdown voltage increases with increasing positive gate voltage. Similarily, in the p-channel MOS transistors, the breakdown is increased when the gate is made increasingly negative. This type of breakdown, which depends entirely on the avalanche breakdown of the drain junction, is illustrated by Fig. 12a. On the other hand, the situation is quite different in the case of the short channel n-MOS transistor. Here, the breakdown characteristics (Fig. 12b) exhibit a negative differential resistance region similar to that obtained with

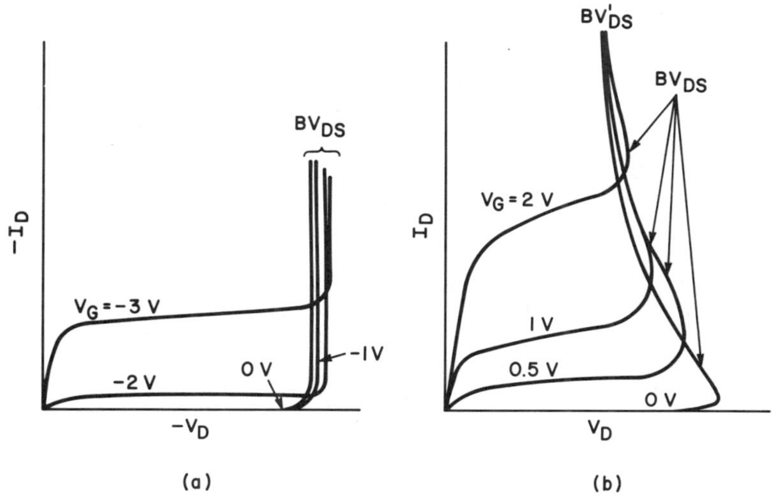

Fig. 12. (a) Drain-to-source breakdown BV_{DS} in a p-channel MOST; (b) drain-to-source negative resistance breakdown BV_{DS} in a short n-channel MOST. BV'_{DS} is the sustain breakdown voltage. (From Toyabe *et al.* [5].)

bipolar transistors in the common-emitter configuration. In addition, the increase in the positive gate potential results in the decrease of the breakdown voltage.

The appearance of the negative resistance characteristics with the sustain voltage BV'_{DS} is due to the presence of a parasitic n^+-p-n^+ bipolar transistor in parallel with the main MOS transistor [5] The source current I_s flowing into the drain depletion region undergoes avalanche multiplication, similar to the drain saturation current. The excess current over the initial I_S current (Fig. 13) flows through the substrate parasitic resistance R_{sub} and forward biases the source–bulk junction, increasing the injected current. Due to this regenerative effect, the breakdown characteristics snap back as illustrated by Fig. 12b [26]. This situation was analyzed by Toyabe *et al.* [5] and Kennedy, and Philips [26].

The drain current under avalanche multiplication conditions can be expressed as the sum of two components: one due to I_S and the other one to the drain junction saturation current I_{D0}. If M_1 and M_2 are the avalanche multiplication factors for these two currents, respectively, then we can write

$$I_D = M_1 I_S + M_2 I_{D0}. \tag{97}$$

In order to determine the values of M_1 and M_2, Toyabe *et al.* [5] solved Poisson's and current continuity equations in two dimensions by computer.

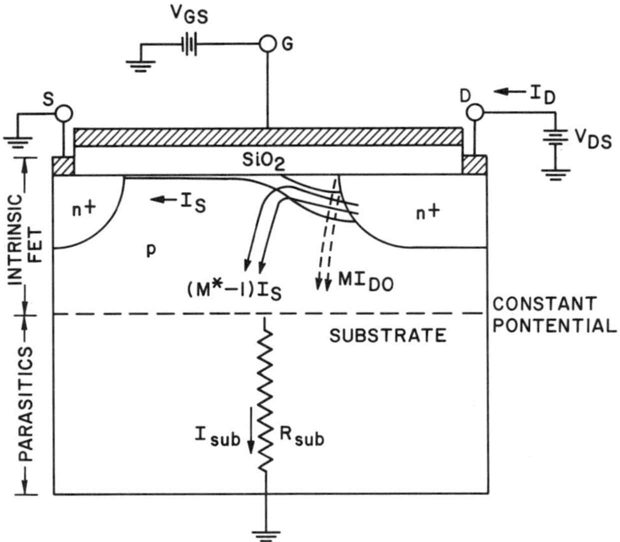

Fig. 13. Parasitic substrate resistance R_{sub}. (From Toyabe *et al.* [5].)

The field magnitudes obtained this way permitted first the calculation of the electron and hole ionization rates and then that of the multiplication factors M_1 and M_2 by carrying out the integration of ionization integrals along the curvilinear channel current path, determined by locating the maximum current density points.

In the case of the *p*-channel device (Fig. 12a [5]) the breakdown occurs when the multiplication factor M_2 goes to infinity, as expected for the normal breakdown of the drain junction. The same is true for the long *n*-channel MOS transistor.

In the case of the short *n*-channel MOS transistor, a negative resistance-type breakdown (Fig. 12b) occurs since the source current I_S also becomes the seed current for avalanche multiplication. Since I_S is much larger than the saturation current I_{D0}, the $M_2 I_{D0}$ term in Eq. (97) can be neglected, and the excess substrate current I_{sub} can be expressed as

$$I_{sub} = I_D - I_S = (M_1 - 1)I_S. \tag{98}$$

The I_{sub} current develops a forward voltage drop across the source–substrate junction (Fig. 13) [5] given by

$$V_{BS} = I_{sub} R_{sub}, \tag{99}$$

if there is no external back–gate voltage applied. When V_{BS} exceeds ~ 0.7 V, the source junction starts heavily emitting minority carriers into the substrate. As a result, the channel current is increased, which causes the excess substrate current to increase also. This way a positive feedback loop is created, and the drain-to-source breakdown voltage BV_{DS} is decreased; moreover, the differential negative resistance current-voltage characteristics may appear.

Eliminating I_{sub} from (98) by using (99) results in

$$(M_1 - 1)I_S = -V_{BS}/R_{sub}. \tag{100}$$

The onset of the positive feedback occurs for the p–n junction offset voltage of 0.65 V, consequently, for $V_{BS} = 0.65$ V. The quantities M_1 and I_S were obtained [5] by two-dimensional analysis as a function of V_D. BV_{DS} is that value of the drain voltage which satisfies Eq. (100). In the case where there is an external back-gate voltage V_{BG} applied between the bulk and the source, Eq. (100) becomes

$$(M_1 - 1)I_S = (V_{BG} - V_{BS})/R_{sub} \tag{101}$$

Using the results of the two-dimensional numerical analysis, Toyabe and Asai [27] developed an empirical expression for the drain-to-source breakdown valid for MOS transistors with conventional structures.

For the gate voltages exceeding the threshold voltage V_T by 1 V, the expression for the drain-to-source breakdown voltage is

$$BV_{DS} = 2.65 \times 10^5 (V_{BG} + 0.65)^{1/3} x_j^{\;1/3} L^{4/9} x_i^{\;2/9} N_A^{-1/3} N_A'^{1/3}, \tag{102}$$

where x_j is the drain junction diffusion depth; L the channel length; x_i the gate oxide thickness; N_A and N_A' channel region and deep substrate acceptor concentrations, respectively.

For high R_{sub} values and/or high channel currents, the solution of Eqs. (100) and (101) yields two different current levels for a given set of V_D and V_G, as expected for the negative resistance V–I characteristics.

Power transistors usually have a high resistivity drain drift region to support the requisite high voltage. Moreoever, field plates may also be used to enhance the breakdown voltage, as shown, for example, in Fig. 5 of Chapter 13. Under these conditions, expression (102) is not valid; however, the two-dimensional analysis of Toyabe *et al.* [5] may still be applicable with appropriate modifications.

Generally, BV_{DS} of any MOST will be higher for higher drain junction breakdown voltages and will be lower for shorter channel lengths.

12.10.2.1 *p-Channel versus n-Channel MOS Transistors*

As mentioned previously, the p-channel devices exhibit breakdowns without negative resistance characteristics. This is attributed by [5] to the hole ionization rate which according to [28] is one order of magnitude smaller than that of electrons. A smaller ionization rate results in a smaller multiplication factor M and smaller I_{sub} current. Since the n-type substrate usually has lower resistance than the p-type substrate thanks to the higher electron mobility, the resulting voltage drop that forward biases the source–bulk junction is not high enough to cause any appreciable injection of holes into the substrate.

Figure 14 [5] shows the breakdown voltage BV_{DS} of a p-channel MOST and the sustain voltage BV'_{DS} of the n-channel device as a function of the effective channel length L_{eff}. The p–MOST breakdown is almost independent of L_{eff}. The MOS transistors investigated had channel lengths from 0.3 to 5.0 μm, oxide thicknesses of 50 nm, and channel widths of $Z = 13$ μm. The n-MOST bulk resistance was 20 kΩ, and that of the p-MOST 10 kΩ. The Toyabe *et al.* [5] results require still further experimental confirmation, particularly in regard to the p-channel MOS transistors.

12.10.2.2 Second Breakdown

The drain-to-source avalanche breakdown with differential negative resistance may lead to filamentary current flow and second breakdown. This was studied by Nakagiri and Iida [29] and Krishna [30]. Nakagiri and Iida found that, consistent with the parasitic bipolar transistor model, high p-region resistivities increase the source–bulk forward biasing and decrease the voltage at which the second breakdown occurs. Short channel regions also lower the drain-to-source breakdown because of the increased parasitic transistor current gain h_{FE}. According to [29], the current gain shows a linear dependency on $(\log L)^{-1}$ with L being the channel region length. The

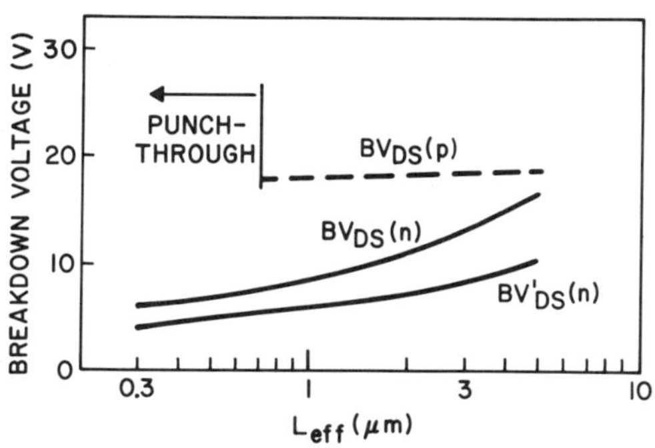

Fig. 14. Breakdown voltage BV_{DS} for p-channel MOST (broken line). Solid lines are for the n-channel breakdown voltage BV_{DS} and n-channel sustain voltage BV'_{DS} as a function of the effective channel length. (From Toyabe *et al.* [5].)

comparatively slow increase of the gain with the decreasing channel length is attributed by [29] to the inversion layer shunting effect.

In order to increase the MOS transistor capability to withstand second breakdown, it is necessary to take some precautionary steps, such as shorting the source and bulk electrodes, which lowers the parasitic transistor's gain; using the highest acceptable *p*-region doping level; and designing with the longest possible channels compatible with the transconductance and/or frequency requirements. Taking these precautionary steps may still not be enough to avoid second breakdown. Thus it is advisable to protect the MOST by appropriate drain voltage clamping devices.

It should be noted that, so far, no second breakdowns were reported for the *p*-channel devices.

12.10.3 Drain Breakdown in Cut-Off Condition

When the drain current flow in the *n*-MOST consists only of the saturation current I_{D0}, i.e., when the gate is biased much below V_T and is negative, the drain breakdown is that of a planar junction with the gate as an overlapping electrode. The electric field is then the highest in the corner region (Fig. 7) at the intersection of the n^+ drain region with the surface.

For the negative gate, the drain and gate fields add up and reduce the drain–bulk breakdown. The effect of the negative gate voltage is also to accumulate the silicon surface at the drain electrode, reducing the width of the drain depletion region at the surface.

12.11 Hot-Electron Effects

When an electron moves in the electric field above 10^4 V cm^{-1}, it gains more energy from the field than it loses to the silicon crystal lattice in the scattering process. Therefore, its temperature starts exceeding the lattice temperature: it becomes hot.

Under such conditions, the electron energy in the *n*-channel, of an MOS transistor may exceed the energy of the conduction band edge.

Hot electrons may have sufficient energy to overcome surface potential barriers, to be injected into the gate SiO_2 layer, and to be trapped there, causing excess gate current, threshold voltage instability, and transconductance degradation [31].

The hot electrons in a MOST originate in the substrate (bulk) and/or in the channel in the presence of high electric fields due to small dimensions and/or high voltages.

The substrate hot-electron effect (SHE) arises [31] from the electrons generated in the channel–bulk depletion region, particularly when there is a large reverse potential V_{SB} applied between the source and bulk. In power devices, $V_{SB} = 0$ as the source and bulk are normally shorted. For $V_{SB} = 0$, there will still be many carriers generated in the channel–bulk space-charge region with sufficient energies to overcome the surface barrier. Also, some of the electrons present in the p-bulk neutral region may move to the Si–SiO_2 interface and gain enough energy from the high electric field present in the surface depletion region.

The substrate hot-electron effect is enhanced for large incident (seed) currents and heavily doped substrates [31]. The incident current is increased, for example, by high temperature generated leakage. A forward biased source–bulk junction will have a similar effect.

Channel hot electrons gain their high velocity and energy from the high fields existing near the drain–channel region junction. Most of them reach the drain electrode, but some are injected into the gate oxide, where they may be trapped. Channel hot-electron emission (CHE) may take place when the device is operating in the subthreshold linear or saturation regions. In the linear region, the gate voltage is higher than the drain voltage; therefore, the field is oriented from the gate to the channel, and electrons are attracted toward the gate. In the saturation region, at some point along the channel, the field normal to the silicon dioxide reverses its direction due to the drain field presence; consequently, electrons are deflected from the oxide–silicon interface. The position of this point will vary depending on the device geometry and drain voltage. As a result, depending on the situation, more or less electrons will be emitted into the oxide.

Drain bias necessary to develop a field sufficiently high to heat the electrons is smaller for devices with shorter channels because of the shorter depletion region and because of the presence of the mobile space charge. A higher channel region concentration results in a higher channel field. A shallow drain region diffusion with a small radius of curvature leads to a higher field as well. Generally then, device geometry has a very pronounced effect on the excess gate CHE current.

To reduce the hot-electron effects on the threshold voltage, it is necessary to minimize the number of the positive oxide traps. The number of existing traps is due sometimes to the oxide growing techniques. Electron or ion beam litography will also have a pronounced effect. Annealing of the oxide at about 400°C is a good way to reduce the number of traps. At this time, trap density on the order of 10^{10} cm^{-2} appears to be the achievable minimum.

REFERENCES

1. W. Shockley, *Proc. IRE* **40,** 1289, 1365 (1952).
2. H. K. J. Ihantola, *ASD Technical Report* No. 1661-1 (1961).
3. H. K. J. Ihantola and J. L. Moll, *Solid-State Electron* **7,** 423 (1964).
4. J. C. T. Huang, *IEEE Trans. Electron Devices* **ED-16,** 775 (1969).
5. T. Toyabe, K. Yamaguchi, S. Asai, and M. S. Mock, *IEEE Trans. Electron Devices* **ED-25,** 825 (1978).
6. A. Popa, *IEEE Trans. Electron Devices* **ED-19,** 774 (1972).
7. J. E. Schroeder and R. S. Muller, *IEEE Trans. Electron Devices* **ED-15,** 954 (1968).
8. G. Merckel, J. Borel, and N. Z. Cupcea, *IEEE Trans. Electron Devices* **ED-19,** 681 (1972).
9. L. D. Yau, *Solid-State Electron.* **17,** 1059 (1974).
10. E. C. Douglas and A. G. F. Dingwall, *IEEE Trans. Electron Devices* **ED-21,** 324 (1974).
11. D. Frohman-Bentchkovsky and A. S. Grove, *IEEE Trans. Electron Devices* **ED-16,** 108 (1969).
12. K. Yamaguchi, *IEEE Trans. Electron Devices* **ED-26,** 1068 (1979).
13. S. Hofstein, *in* "Field Effect Transistors" (J. T. Wallmark and H. Johnson, eds.), p. 129. Prentice Hall, Englewood Cliffs, New Jersey, 1966.
14. J. H. Sigg, G. D. Vendelin, T. P. Cage, and J. Kocsis, *IEEE Trans. Electron Devices* **ED-19,** 45 (1972).
15. E. O. Johnson, *RCA Rev.* **34,** 80 (1973).
16. J. M. Early, *Bell System Tech. J.* **33,** 517 (1954).
17. M. A. Lampert, *Proc. IRE* **50,** 1781 (1962).
18. N. F. Mott and R. W. Gurney, "Electronic Processes in Ionic Crystals," p. 172. Oxford Univ. Press (Clarendon), London and New York, 1948.
19. P. Richman, *IEEE Trans. Electron Devices* **ED-16,** 759 (1969).
20. J. A. Geurst, *Solid-State Electron.* **9,** 129 (1966).
21. E. S. Rittner and G. F. Neumark, *Solid-State Electron.* **9,** 895 (1966).
22. T. Ohmi, Technical Report No. 43. The Research Institute of Technical Communication, Tohoku University, Sendai, Japan (1979).
23. M. Lohzlinger and E. H. Snow, *J. Appl. Phys.* **40,** 278 (1969).
24. C. M. Osborn and E. J. Weitzman, *J. Electrochem. Soc.* **119,** 603 (1972).
25. D. W. Ormond, *J. Electrochem. Soc.* **126,** 162 (1979).
26. D. P. Kennedy and A. Philips, Jr., *IEDM Tech. Digest*, p. 160 (1973).
27. T. Toyabe and S. Asai, *IEEE Trans. Electron Devices* **ED-26,** 453 (1979).
28. W. C. Niehaus, T. E. Seidel, and D. E. Iglesias, *IEEE Trans. Electron Devices* **ED-20,** 765 (1973).
29. M. Nakagiri and K. Iida, *Jpn. J. Appl. Phys.* **16,** 1187 (1977).
30. S. Krishna, *Solid-State Electron* **20,** 875 (1977).
31. P. E. Cottrell, P. R. Troutman, and T. H. Ning, *IEEE Trans. Electron Devices* **ED-26,** 520 (1979).

CHAPTER 13

MOS POWER TRANSISTOR STRUCTURES AND DESIGN CONSIDERATIONS

The two major MOS power transistor applications are in the power amplifier and power switching circuits. When the best performance from the frequency response or the distortion and from the power dissipation point of view are expected, the design considerations for both types of service may be different.

13.1 MOST as a Power Amplifier

A MOS transistor is essentially a square-law device: and to a good approximation, its current can be expressed, as we have seen, by

$$I_D = \tfrac{1}{2}G_0(V_G - V_T)^2. \tag{1}$$

The absence of the odd-order terms from the MOST characteristic implies that there will be no appreciable intermodulation distortion, i.e., no interaction between two high frequencies within the pass band when the device is used as a linear amplifier.

MOST transfer characteristics are approximately linear for small signals as the output current i_d is related to input voltage v_{gs} by

$$i_d = g_m v_{gs}. \tag{2}$$

This property, if conserved over a sufficiently wide current range, makes MOST superior to bipolar devices, which require much more negative feedback than the MOSTs to suppress distortion. MOS power transistors should, be therefore, very suitable for operation as dudio amplifiers. The basic principle of operation is illustrated by the simplified circuit of Fig. 1.

Another MOST application of audio amplification consists of the pulse width modulation technique feasible with MOST, but impractical with bipolar transistors because of the required high carrier frequency.

The maximum usable voltage of the MOST is somewhat below the drain-to-source breakdown BV_{DS}; the maximum voltage swing allowed for the operation in saturation is $BV_{DS} - V_{Dsat}$, where V_{Dsat} is the drain saturation voltage at the maximum drain current I_{Dmax}. The power output of the device is, consequently,

$$P_{out} \propto I_{Dmax}(BV_{DS} - V_{Dsat}). \tag{3}$$

It is therefore desirable to make BV_D much higher than V_{Dsat} to obtain high voltage swings and high power outputs.

The punch-through conditions should be avoided in the amplifier because it affects the saturated I–V characteristics and makes the device deviate from its square-law characteristics, leading to the intermodulation distortion.

In order to achieve high current capability, the device should have large channel width Z, which is usually obtained by interdigitation of the gate and source or by a cell-like structure. All series resistances should be minimized. The source series resistance appears in the input circuit of the common-source amplifier and acts as a negative feedback element reducing device amplification. Moreover, source and drain resistances dissipate power. The

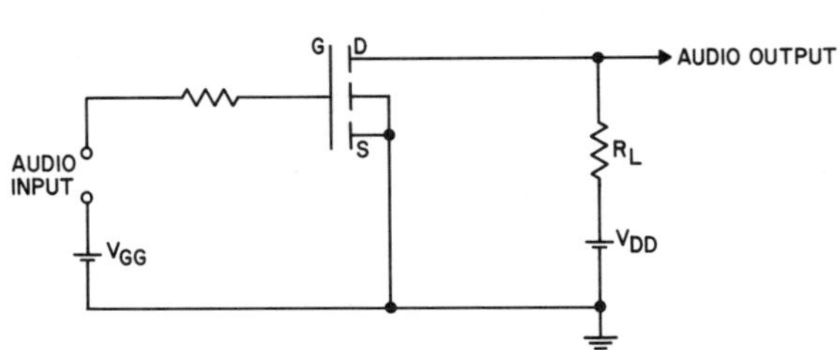

Fig. 1. Simplified schematic of the MOST audio amplifier in the common-source configuration.

dc on-resistance of the present-day MOS transistors is usually on the order of a few tenths of an ohm for devices operating at the 20–40 A range.

The higher the operating frequency of the MOST amplifier is, the more important the gain degrading effect of the parasitic gate-to-source and gain-to-drain capacitances. Due to the Miller effect, the drain-to-gate capacitance C_{gd} appears at the input of the device equivalent circuit with a multiplication factor equal to the stage gain plus one. The overlap capacitances can be minimized by using self-aligned polycrystalline gates, for example. Generally, the drain-to-gate capacitances are efficiently minimized in power devices with such structures as the off-set gate devices and vertical drain transistors, where the drain electrode is well decoupled from the gate. The use of thicker oxide under the gate metal in areas other than those directly over the active channel regions is also helpful in reducing the parasitic capacitances.

Transistors with *n*-channels as amplifiers are superior in many respects to the *p*-channel devices because of the higher electron mobility, better frequency response, and higher transconductance.

Materials other than silicon have been used to improve the MOST transconductance and increase its operating temperature beyond the $\sim 175°C$ limit of silicon devices. The most prominent of these materials is the III–V compound GaAs. Its high electron mobility and wide energy gap make it particularly attractive for this purpose. Not much progress has been made in this area after the first successful results obtained by [1] with the GaAs *n*-channel depletion-type transistor. The difficulty rests here primarily with the presence of the very high density of the interface states between GaAs and the gate insulator. MESFETs, described in Chapter 11, do not have this problem since their gate depletion region extends well below the GaAs surface.

13.2 MOS Power Transistor as a Switch

MOS power transistors are capable of switching large currents with voltage drops comparable to but still somewhat higher than those of bipolar transistors with silicon dies of the same active area. The voltage drop resulting from the presence of the on-state resistance R_{on} is one of the most important parameters of a switch, as it determines the amount of dissipated energy due to Joule heating.

Figure 2 shows the idealized switching waveforms of the input gate-to-source voltage V_G and the output current I_D. As the gate voltage is increased, the current appears only after a short delay time t_d, and when the threshold

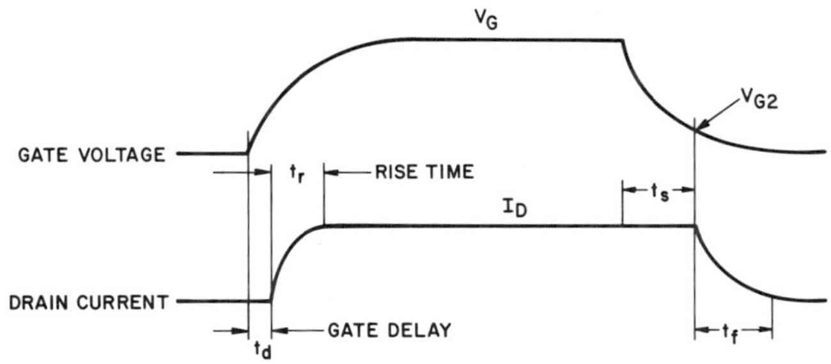

Fig. 2. Idealized gate voltage V_G and drain curent I_D switching waveforms.

voltage V_T is exceeded, it starts growing rapidly. At the end of the rise time period t_r, the drain current reaches its maximum value. When the gate biasing voltage is removed, the device capacitances start discharging. After the relative short gate capacitance storage time t_s, the gate voltage reduces to the V_{G2} value at which the current is leaving the MOST saturation region. After reaching this point, the gate voltage continues to drop and so does the drain current until, at $t = t_f$, the gate voltage is reduced to the threshold value V_T at which the turn-off practically ends.

The power MOST is a majority carrier device and does not exhibit charge storage like a bipolar transistor in the base and collector regions. It does store, though, an appreciable charge in the input capacitance. This charge is mainly determined by the total gate area and is much smaller than the charge stored in a bipolar device. Consequently, a MOST is a potentially much faster device than a bipolar transistor. Typical turn-on and turn-off times for high current (10 A) high voltage (100–200 V) well-designed MOSTs are on the order of 100 ns and depend, of course, on the circuit used. The delay time $t_d \simeq 50$ ns, and storage time $t_s \simeq 100$ ns. The switching times show negligible variation with temperature; they change, though, appreciably in bipolar devices because of the minority carrier lifetime dependence.

The maximum obtainable switching frequencies are on the order of a few hundred kilohertz as compared to about 50 kHz achievable with bipolar transistors.

The delay time t_d of the MOS transistor can be found from the input capacitance charging equation:

$$V_G(t) = V_{GG}[1 - \exp(-t/R_{gen}C_{in1})], \tag{4}$$

where C_{in} is the input capacitance and

$$C_{in1} = C_{gs} + C_{gd}. \tag{5}$$

R_{gen} is the internal resistance of the current pulse generator and V_{GG} is the

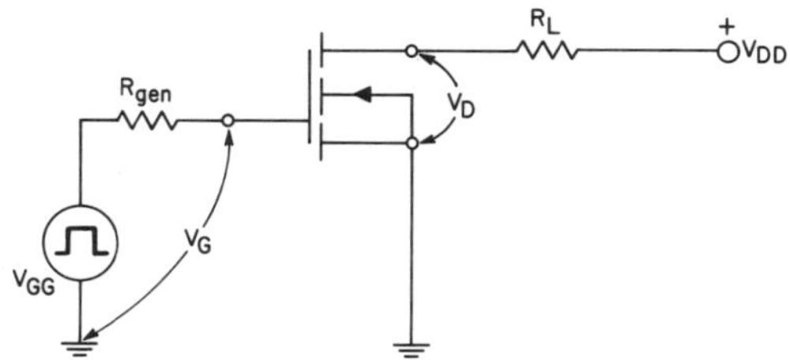

Fig. 3. MOST switching circuit. V_{DD} is the voltage of the drain power supply. (From Evans and Hoffman [2].)

peak voltage (Fig. 3 [2]). The turn-on delay time ends when the gate potential reaches its threshold value V_T and from (4), is therefore equal to

$$t_d = C_{in1} R_{gen} \ln(1 - V_T/V_{GG}). \tag{6}$$

During the rise time, the input capacitance C_{in2} is different from C_{in1} because of the Miller effect. Its value changes with the device varying voltage gain, but can be assumed to be constant without too much error. The rise time is given, therefore, as [2]

$$t_r = C_{in2} R_{gen} \ln[1 - (V_{G2} - V_T)/(V_{GG} - V_T)], \tag{7}$$

where V_{G2} is the gate voltage at the end of the rise time interval. C_{in1} and C_{in2} can be determined by plotting the gate-to-source voltage V_G as a function of the gate charge Q_G (Fig. 4) and using the relationships

$$C_{in1} \simeq Q_{G1}/V_{G1}, \tag{8}$$

$$C_{in2} \simeq Q_{G2}/V_{G2}, \tag{9}$$

which assume approximate constancy of C_{in1} and C_{in2} during the delay

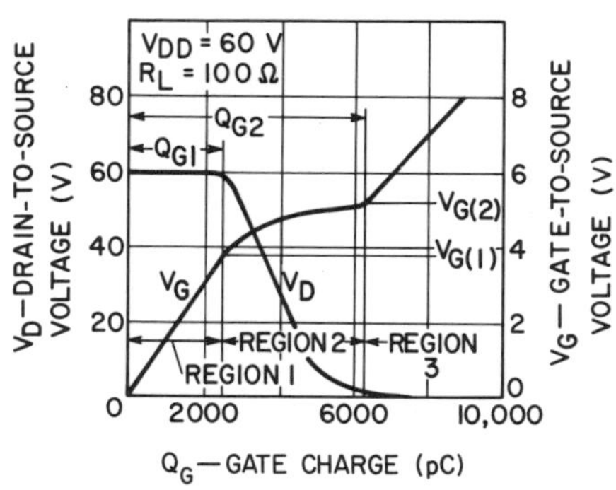

Fig. 4. Gate V_G and drain V_D voltage as functions of gate charge Q_G of a VMOS power transistor. (From Evans and Hoffman [2].)

and rise times, respectively. The plot (Fig. 4) of the gate and drain voltages V_G and V_D versus the gate charge Q_G is obtained by using a pulsed constant current drive for the gate [2]. It is divided into three regions. The first one ends at $Q_G = Q_{G1}$ and the second one at $Q_G = Q_{G2}$. The third region is characterized by the high values of the gate voltage and its charge and the low values of the drain voltage V_D. In this region, the MOS transistor operates in saturation, and there is no significant change in the drain current I_D with the drain voltage. Consequently, the Miller effect is negligible, and

$$C_{in3} \simeq C_{gs} + C_{gd} \tag{10}$$

However, $C_{in3} > C_{in1}$ because of the very different voltages appearing at C_{gd}. In region 1 the device bulk is still in the depletion state, whereas in region 3 the bulk surface is highly inverted.

The computation of the turn-off times may be performed in a way similar to that of the turn-on times [2]. If the device operates in the overdrive region (region 3), where the gate voltage is much greater than the threshold voltage V_T, the turn-off delay becomes appreciable because of the excess charge on the gate.

The gate circuit energy required to keep the device in the on- or off-condition is usually negligible, but changing the device from one state to another requires an appreciable amount of energy on the part of the driving device. This energy is necessary to charge the MOST input capacitance C_{in} and is equal to

$$W = \tfrac{1}{2}C_{in}\,\Delta V_G^2 \tag{11}$$

where ΔV_G is the total gate voltage swing.

It should be noted that the turn-off transient can be seriously affected by the presence of a parasitic n^+–p–n^-–n^+ transistor in parallel with the intrinsic MOST. Under some switching conditions, when the drain junction breaks down electrically, the parasitic transistor's emitter (the MOST source) may become forward biased, and the turn-off time would be that of a bipolar device rather than that of a MOST.

13.3 Power MOST Structures

The MOS power transistors may be divided into two basic classes: two- or three-dimensional structures. The two-dimensional lateral devices are similar to the conventional MOS transistors with an extended high resistivity drain region. This is necessary for HV performance.

In the three-dimensional devices, the extended drain region, also known as the drift region, is vertical; and the drain electrode is located at the chip bottom. The three-dimensional structure allows for better silicon chip area utilization since only the source and drain electrodes are left there.

13.3.1. Two-Dimensional (Lateral) Structures

13.3.1.1 The Offset Gate MOS Transistor

The offset gate MOS transistor depicted by Fig. 5a [3] represents a device which, conceptually, was proposed years ago to reduce the parasitic drain-to-gate feedback capacitance. The device of Fig. 5a is a *p*-channel MOST provided with a polysilicon gate and ion-implanted extended drain region. The gate is offset, i.e., it does not cover fully the region between the source and the highly doped p^+-drain. The presence of the extended drain *p*-region permits the drain depletion region to spread into it at high voltages without the risk of punching through to the source. The *p*-region, of course, does not become inverted when the gate is negatively biased, and it introduces an additional series drain resistance that should be minimized to avoid appreciable power losses. This can be accomplished by making the channel width Z and, consequently, the silicon chip area very large. It is obviously not an economical solution since only a small number of large area devices can be processed on a single silicon wafer. This situation can be considerably improved upon by using a more conductive ion-implanted offset region with a field plate over the gate edge as shown by Fig. 5b. The purpose of the field plate is to reduce the peak field at the gate edge and thus to improve the drain breakdown.

The ion-implanted layer acts as a pinched resistor and drops all the drain voltage above the pinch-off voltage. The analysis of Okabe *et al.* [3] shows that the intensity of the peak field near the gate edge is smaller for smaller pinch-off voltages that can be achieved by the use of a field plate, for example. With the field plate present, the electric field now has two peaks: one at the gate edge, as before, and the other at the field-plate edge.

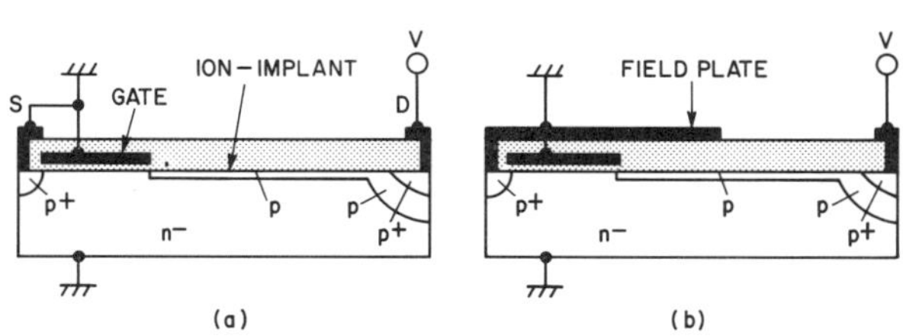

Fig. 5. Offset gate *p*-channel MOST with ion-implanted extended drain *p*-region. (a) Without a field plate, and (b) with a field plate. (From Okabe *et al.* [3].)

Their magnitudes can be made, however, much lower than the peak field without the plate.

Using this approach, Okabe *et al.* [3] fabricated complementary *p*- and *n*-channel devices with drain breakdowns higher than 250 V and with 12 A current capability. The channel lengths for the *p*-MOST and *n*-MOST were 7 and 8 μm, respectively. The channel widths were 61 and 30 cm, respectively. Both types were fabricated on a 6 × 6 mm silicon chip. A pair of these devices in a push–pull arrangement was capable of delivering 300 W at 100 KHz.

13.3.1.2 Lateral DMOST (LDMOST)

The lateral DMOS transistor is the predecessor of the most present-day two- and three-dimensional power MOST structures [4, 5]. The DMOS takes its name from the fabrication technique consisting of double diffusion through the same oxide opening. The short and well-controlled channel length results from the subsequent diffusions of the *p*-channel region (bulk) and the *n*-source impurities under the same oxide layer (Fig. 6 [5]). The very short channel length L is necessary not only for high frequency response but also for achieving high device transconductance and gain. The device current capability increases with the increase of the channel aspect ratio Z/L. The high voltage capability is realized by the presence of a lightly doped n^--extended drain region with a field plate. The drain depletion layer at high voltages extends a little into the channel, but mostly into the n^--drift region, which can be made as long as necessary to achieve the requisite breakdown voltage. In this structure, as in the offset gate, the channel length becomes independent of the drain breakdown voltage requirement, a degree of freedom not available in the conventional MOS transistors. As in the offset gate device, the lateral deposition of electrodes has a disadvantage of poor silicon area utilization, particularly for drain voltages above 200 V.

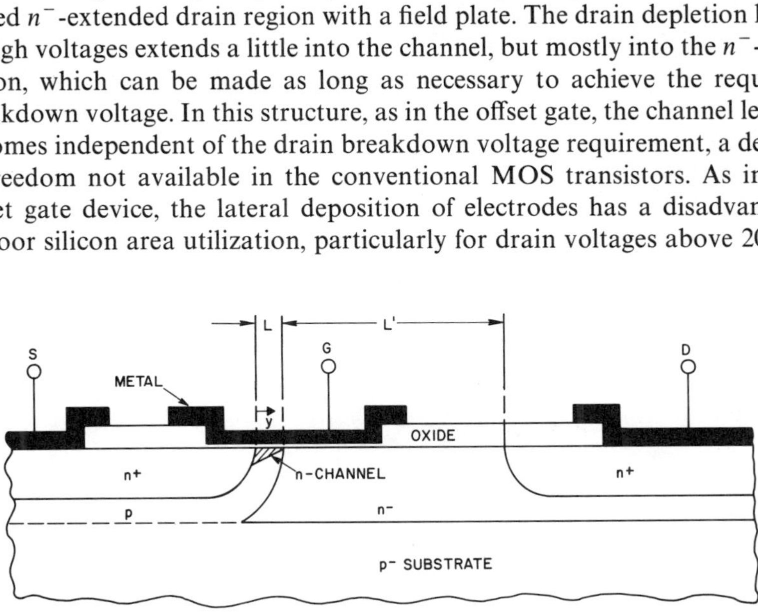

Fig. 6. Lateral *n*-channel DMOST with n^--drain drift region. (From Sigg *et al.* [5].)

The packing density for multichannel or multicell structures, necessary for high current capability, is smaller here than that of designs with vertical drains.

The LDMOS transistors described by Sigg *et al.* [5] and Fong *et al.* [6] are *n*-channel devices fabricated on ⟨100⟩ oriented silicon crystal. This orientation shows about 25% higher mobility than the ⟨111⟩ orientation and has a much lower density of surface states.

The basic processing steps for a LDMOST are shown by Fig. 7 [5]. The

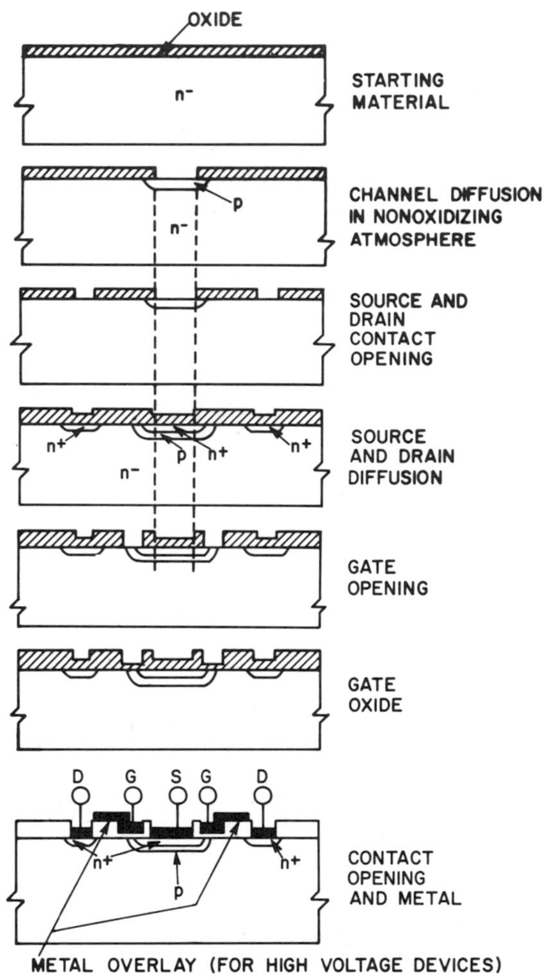

Fig. 7. Basic processing steps of a lateral DMOST. (From Sigg *et al.* [5].)

starting n^--silicon wafer (or n^--epilayer if epitaxial material is used) is first oxidized; then an opening is etched in the oxide to permit the diffusion of the p-channel region (p-bulk). A nonoxidizing atmosphere is used for the boron diffusion to preserve the sharp oxide edges. The same oxide opening serves to diffuse the source. The drain region diffusion takes place through the previously etched openings simultaneously with the n^+-source diffusion. After that, the gate contact region is opened and oxidized to the desired thickness. Next, the oxide from the source and drain openings is etched away, and metallization is applied to all electrodes.

Using a similar fabrication technique, Fong *et al.* [6] were able to fabricate devices with the gain–bandwidth product $f_T = 1.1$ GHz, drain breakdown voltage of ~120 V, and pulse current capability of 3.5 A (80 μs pulses with 1% duty cycle). The on-resistance of their devices was 2.5 Ω. The devices had an excellent gate aspect ratio Z/L, with $L = 1$ μm and $Z = 2.5$ cm, which resulted in a transconductance of ~600 mS. Let us recall that the maximum obtainable transconductance is given by

$$g_{m\max} = ZC_i v_s \tag{12}$$

and is equal to about 850 mS.

13.3.2 Three-Dimensional Structures

As noted before, the three-dimensional MOST transistor structures with vertical drains permit the achievement of very high packing densities. Offset gate and DMOS transistors can be easily provided with vertical drains.

13.3.2.1 Offset Gate MOST with Vertical Drain

An offset gate transistor was described by Yoshida *et al.* [7] and is shown schematically by Fig. 8. This p-channel device is fabricated on the n-type epitaxial silicon with heavily doped polysilicon gates. A second-level metallization, consisting of aluminum, covers the entire top surface and is used as the contact to the source regions. An SiO_2 layer separates the two types of metallization. The source and drain (Fig. 8) are arranged in a checkerboard (meshlike) fashion. In the device described by Yoshida *et al.*, the mesh size was 22 μm, channel length $L = 8$ μm, and channel width $Z = 94$ cm. The device was fabricated on a 5×5 mm^2 chip and had a breakdown voltage of 100 V and a drain current capability I_D of 20 A at $V_G = -10$ V. Its transconductance was $g_m = 3500$ mS, and the on-state resistance was 0.5 Ω at $I_D = 10$ A.

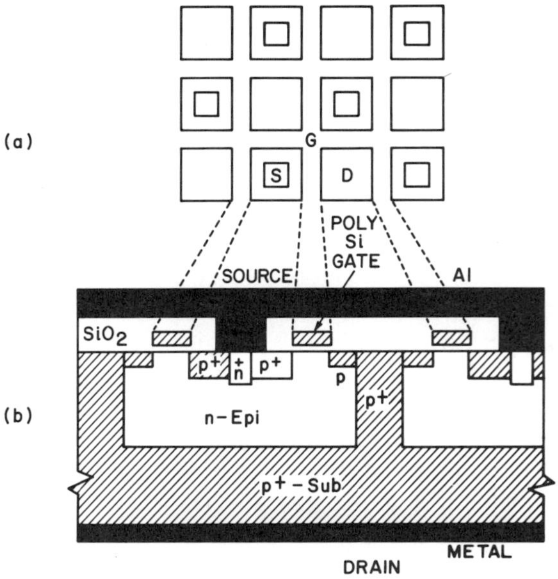

Fig. 8. (a) Top view and (b) a schematic cross section of the offset gate *p*-channel MOST with vertical drain. (From Yoshida *et al.* [7].)

13.3.2.2 DMOST with Vertical Drain (VDMOST)

Figure 9 shows a DMOST with vertical drain (VDMOST) with lateral disposition of the source and gate electrodes and a vertically oriented drain, consisting of an n^--epilayer on the n^+-substrate. The current flow lines represent the case when the drain voltage is low, but the current is high. Self-aligned polycrystalline gates are used here as the first level and aluminum as the second level of metallization. Aluminum film (not shown in the figure) serves as a contacting metal for the source regions. The typical doping profile of such a device is shown in Fig. 10. n^+-, p-, and n^--regions represent the source, channel region (bulk), and drain drift n^--region, respectively. In this particular design [8], the p^+-diffused areas play a double role: they short the source–bulk n^+–p junction and minimize the breakdown effects of the parasitic n^+–p–n^-–n^+ transistors. Moreover, the p^+–n^+ junctions play the role of Zener diodes and are expected to break down electrically before the drain n^-–p junctions. The p^+-diffusion region improves, therefore, device second breakdown characteristics.

For good silicon area utilization, the device consists of multiple cells of rectangular shape. Some other shapes, such as hexagons, may be somewhat more advantageous from the "real estate" utilization point of view. Devices

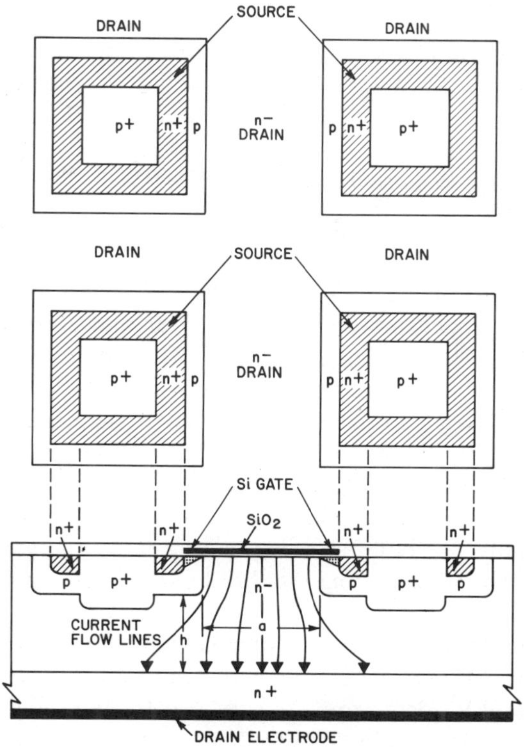

Fig. 9. Top view and a schematic cross section of a VDMOST with multiple cell structure.

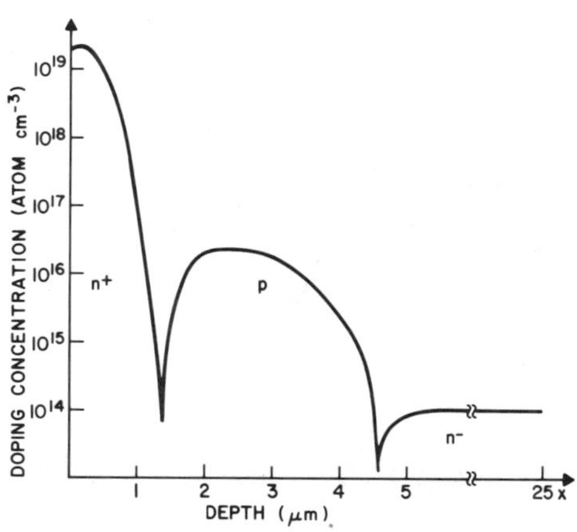

Fig. 10. Typical doping profile of a VDMOST.

of this geometry are commercially available at this time. They exhibit, for example, 400 V drain breakdown, are capable of operation with 4 A drain current, and dissipate 75 W. This particular device has the on-state resistance of $\sim 1\ \Omega$ at $V_G = 10$ V and a transconductance of 3 S.

13.3.2.3 Vertical V-Groove Transistor (VVMOS)

The vertical V-groove transistor (VVMOS) transistor is a nonplanar DMOS device whose cross section is shown in Fig. 11 [9]. Here again, the drain electrode is at the device's bottom, but the source and gate electrodes are located on the upper surface. To etch the V-shaped groove, an anisotropic etching is used in ⟨100⟩ oriented silicon [10]. The grooves are accurately aligned with the ⟨110⟩ direction. The grooves' walls are 54.74° from the horizontal. The device shown in Fig. 11 is an *n*-channel enhancement-type MOST fabricated by double diffusion for the definition of the very short channel length on the order of a few micrometers or less. The gate region is defined by V-groove etching which exposes the gate–channel region. The device is normally off, but starts conducting upon the application of the positive gate potential exceeding the threshold voltage, similar to any other enhancement-type device.

The packing density for the multiple channel or multicell geometries is good for the VVMOST since only two electrodes are placed on the top surface. A vertical n^--drift region is provided for the required high breakdown voltage and low drain–gate feedback capacitance. Both sides of the groove conduct the current, which is advantageous from the on-resistance point of view.

A VVMOS transistor with drain-to-source breakdown voltage greater

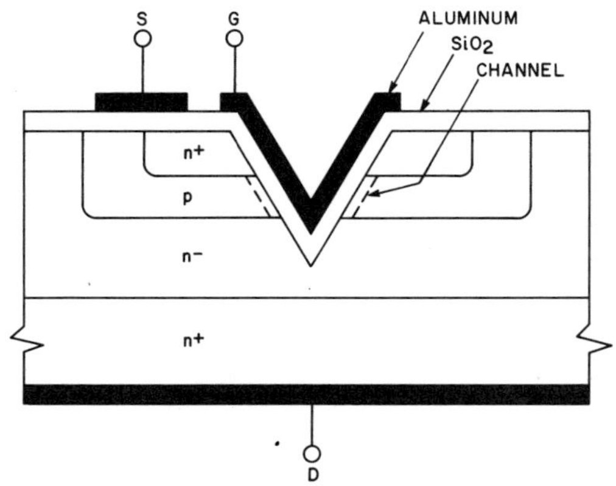

Fig. 11. Schematic cross section of a VDMOST. (From Salama and Oakes [9].)

than 45 V was fabricated and described by Lisiak and Berger [11]. With an active device area of 2.0 mm^2, the on-resistance of this device was about 5.5 Ω. Lower resistances, on the order of 0.05 Ω were achieved by [11] for devices with 30 V breakdowns. The schematic cross section of this device is shown by Fig. 12 [11]. The device has two layers of metallization. The top metallization fully covering the upper surface consists of aluminum contacting the n^+-surface and the p-type bulk regions. The second level of metallization for the gate electrodes consists of highly doped polycrystalline silicon separated from the aluminum and the bulk by SiO_2 (black area).

A fully ion-implanted VVMOS with a cut-off frequency of 1.5 GHz ($I_D = 200$ mA, $V_{DS} = 20$ V) was reported by Fuoss and Verma [12]. The channel region consisted of a boron implant, and the source implant was arsenic. The doping concentration versus depth for this device is shown by Fig. 13 [12].

13.3.2.4 Truncated V-Groove MOST (VUMOST)

In the V-groove device, the anizotropic etching stops when the two inclined walls meet. Stopping the etching process before this happens results in a truncated V-groove shape, i.e., in a trapezoidal U-shaped notch [9, 13, 14]. The device is represented by Fig. 14 [9]. Similar to the V-groove device, the channel length is defined by double diffusion through the same oxide opening. The main advantage of this structure over the V-groove device is its lower on-resistance owing to its better current spreading in the n^--drift region. The definite disadvantage is a more critical etching process, which, in contrast to the V-groove etching, is not self-limiting.

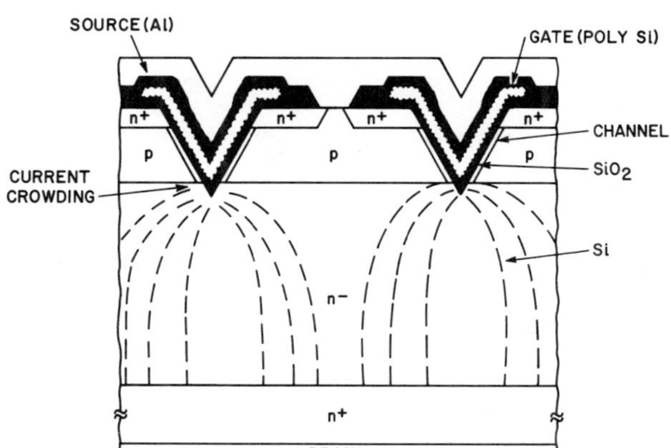

Fig. 12. VVMOST with polysilicon gate and aluminum as source metallization. SiO_2 is indicated by black area. (From Lisiak and Berger [11].)

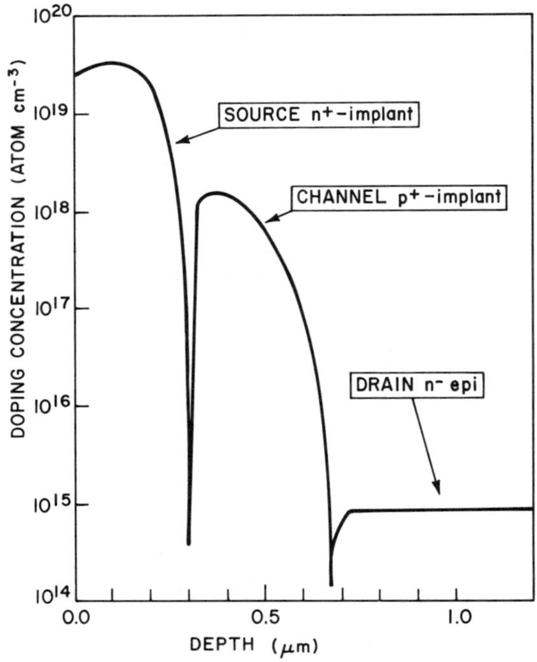

Fig. 13. Doping profile of the fully ion-implanted microwave transistor. (From Fuoss and Verma [12].)

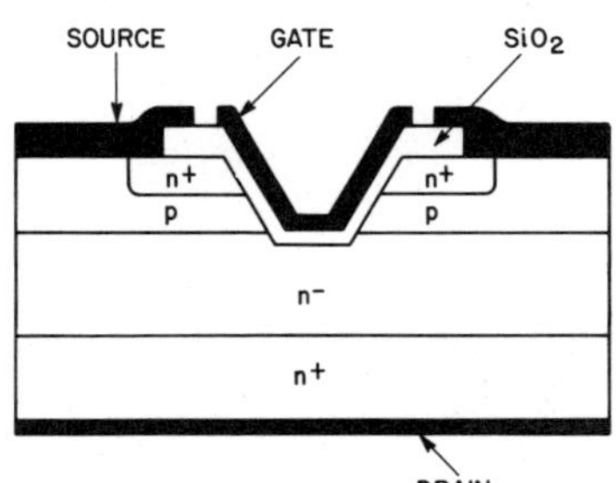

Fig. 14. Schematic cross section of a VUMOST. (From Salama and Oakes [9].)

13.4 Design Considerations for Power DMOST

The design of power DMOST must take into account a number of various parameters. The most important ones are the threshold voltage, the transconductance, the on-resistance, the various capacitances, electrical breakdowns, and temperature effects. These aspects of the DMOST designs are considered in what follows.

13.4.1 DMOST Threshold Voltage

In the DMOST, the bulk region net doping concentration is not uniform, as was assumed in the simplified theory of Chapters 1 and 12, but is a function of depth x from the device surface. The threshold voltage in this case is determined by the maximum impurity concentration in the bulk. For the n-channel device, this concentration is designated by $N_{A(max)}$. A computerized model for the DMOS threshold voltage was developed by Pocha *et al.* [15], who used a one-dimensional solution to the two-step diffusion process consisting of the predeposition and driving steps.

The diffusion profiles along the device surface in the channel region (bulk) (y direction) and along the vertical direction x differ from each other [15, 16]. This is illustrated by Fig. 15 [15], which first shows the source and bulk impurity diffusion fronts (Fig. 15a) and then the impurity profiles in the x direction (Fig. 15b). The impurity profiles in the y direction along the surface are shown by Fig. 15c. Due to the relatively deep boron diffusion used for the p-region, its profile is Gaussian. The relatively shallow source phosphorus diffusion (interrupted line) has a profile intermediate between

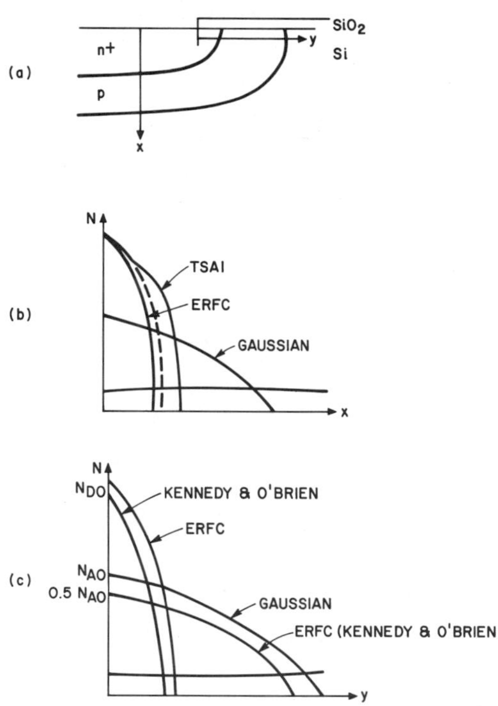

Fig. 15. (a) Source and bulk (channel region) diffusion fronts. (b) impurity profiles in x direction, and (c) impurity profiles in y direction. (From Pocha *et al.* [15.)

the complementary error function and the anomalous profile of the shallow phosphorus diffusion of Tsai [17].

The impurity profiles are quite different in the *y* direction mainly due to two effects: (a) because of the impurity redistribution at the SiO_2–Si interface and (b) because along the surface the impurity diffusion can be considered as originating from an infinite source since the oxide opening is very long as compared to the actual diffused distance. Lateral effect leads to a prediction [16] that the boron drive-in diffusion would result in a complementary error function profile with 50% of the original surface concentration N_{A0} (Fig. 15c). The results of the Pocha *et al.* [15] computations are represented by Fig. 16, where the threshold voltage V_T is plotted versus $N_{A(max)}$ with the gate oxide thickness as a parameter. Figure 16 illustrates the spread in the threshold voltage of each oxide thickness when there is a $\pm 10\%$ deviation in the impurity diffusion length. For this particular set of plots, Q_{ss} was assumed to be constant along the channel at the surface and equal to 10^{11} cm^{-2}. The major effect is due to variations in the boron predeposition step. The sensitivity of the threshold voltage to the boron predeposition was about three times greater than to all other steps combined.

The control of the impurity concentration is much improved when ion implantation is used in place of thermal diffusion so that $\pm 2\%$ rather than $\pm 10\%$ scatter can be expected in the impurity diffusion length.

Other results of the study [15] indicate that processing variation effects can be minimized for sufficiently long channels (longer than 2 μm). In another study concerned with the threshold voltage of a VVMOS transistor, Lane and Salama [18] found that for an oxide thickness of 100 nm, channel length 2 μm, and $Q_{ss}/q = 4 \times 10^{11}$ cm^{-2}, a threshold voltage of about 1 V requires

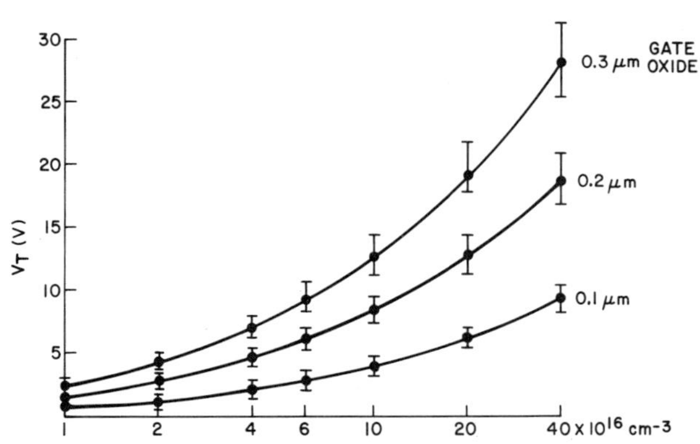

Fig. 16. Threshold voltage V_T as a function of $N_{A(max)}$ of the channel region, with the gate oxide thickness as a parameter. (From Pocha *et al.* [15].)

a peak channel doping $N_{A(max)} = 5 \times 10^{16}$ cm^{-3}. The peak doping in the channel is reduced by impurity redistribution during the gate oxidation. The original peak doping, though, is restored if the inert atmosphere annealing is performed after the oxidation. It was also determined that a variation of $\pm 10\%$ in the n^--epitaxial layer resistivity resulted in $\pm 16\%$ change in the threshold voltage at the n^--doping levels of about 10^{15} cm^{-3}.

13.4.2 DMOS TRANSISTOR CONDUCTANCE AND TRANSCONDUCTANCE

The DMOST conductance at low drain voltages close to the origin of the V–I characteristics is determined by the active channel conductance. At higher currents, the conductance is greatly affected by the drain drift region resistance and can be calculated from expression (57) of Chapter 12. For very small V_D values, the conductance g_d is a linear function of the gate voltage V_G with a slope of

$$G_0 = Z\mu C_i/L. \tag{13}$$

On the other hand, the initial slope of the transconductance g_m versus $(V_G - V_T)$ curve will not be equal to G_0 as expected from the simplified theory. This comes about because of the body effects [18].

In the saturation region, the short channel power MOSTs operate with carrier velocities equal to the scattering velocity v_s. Under these conditions, the device transconductance is given by

$$g_{m(max)} = ZC_i v_s. \tag{14}$$

For electrons, the scattering velocity v_s at the surface is equal to $\sim 6.5 \times 10^6$ cm s^{-1} for n-channels on $\langle 100 \rangle$ oriented silicon crystal. Initially, before the electric field reaches its critical value, g_m increaseses with $V_G - V_T$ with a slope proportional to μ_n [13]. Further voltage and field increase causes the increase of the carrier velocity until the limiting value v_s is reached at which the transconductance attains its maximum.

Figure 17 [13] shows the results of g_m measurements versus $(V_G - V_T)$ made by Sun and Plummer [13] for VDMOS transistors with 0.5, 3, and 7 Ω cm epitaxial n^--layer resistivities. V_{TE} is the threshold voltage of the enhancement-type transistor. These devices show $g_{m(max)}$ magnitudes of about 50 to 60 mS, i.e., lower than expected from expression (14). One of the reasons for this behavior is the presence of the drain drift resistance. Its effect is the clearest for the 7 Ω cm epilayer device. Because of the series resistance presence, increased value of the drain voltage are needed to maintain the device in the saturation region as the current increases. This in-

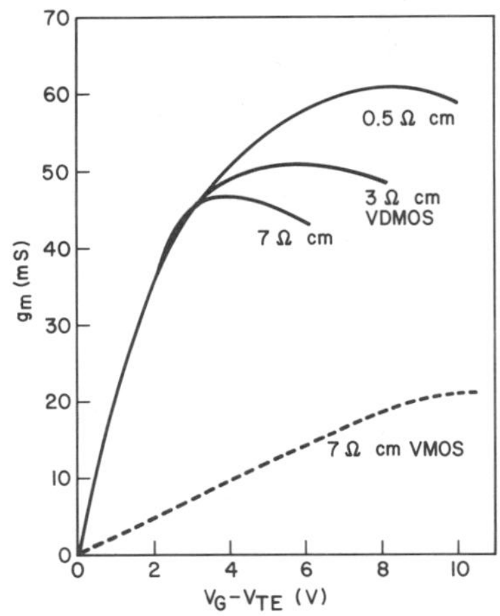

Fig. 17. Transconductance g_m as a function of the effective gate voltage $V_G - V_{ET}$. (From Sun and Plummer [13].)

creases power dissipation and the device temperature and therefore reduces the low- and high-field carrier mobilities and the transconductance [20].

For comparison, Fig. 17 also shows g_m versus $(V_G - V_T)$ for a VVMOS structure with a 1.5 longer effective channel length than that of the VDMOSTs. Since this device was fabricated on ⟨111⟩ oriented surface, the electron mobilities were lower than for the VDMOSTs. This resulted in a much lower device transconductance and required higher gate voltages for electrons to reach the limiting velocity. The thermal effects do not let the VVMOST reach its maximum transconductance value. Very short current pulses were required to partially avoid the thermal effects and somewhat increase the transconductance. In general, the transconductance is higher in LDMOS and VDMOS than in the VVMOS devices [13] with equal channel lengths because of higher electron mobilities in all fields for the ⟨100⟩ oriented silicon crystal.

The transconductance exhibits a negative temperature coefficient as compared to the bipolar transistor current gain, which increases with temperature (Section 13.6.2). This negative temperature coefficient is what makes the MOS power device superior to a bipolar device from the thermal runaway point of view.

13.4.3 DMOS Transistor On-Resistance

The MOST resistance in the on-state is of the utmost importance for the power devices. This resistance determines the device dissipation and in good transistors should be comparable to the bipolar transistor's on-resistance. The on-resistance consists of several components, such as the channel resistance and drift region resistance.

The effective channel resistance of an enhancement MOST is given by

$$R_E = \frac{L}{\mu(V_G)ZC_i} \frac{1}{(V_G - V_{TE})}, \tag{15}$$

where $\mu(V_G)$ is the V_G-dependent electron mobility and V_{TE} is the threshold voltage of the enhancement-type device. For HV devices with high resistivity n^--regions, R_E is usually negligible.

13.4.3.1 LDMOS Transistor On-Resistance

The LDMOST resistance of the n^--region should be considered as the spreading resistance (Fig. 18 [19]). The current from the inverted channel flows into the surface accumulation layer formed under the gate at the n^--region surface. From here, it spreads into the n^--drift region, finally reaching the drain. The spreading resistance computation can be carried out by considering the source and the drain as two current carrying lines with length Z in an infinite homogeneous medium (Fig. 19 [19]). The source and drain lines' effective radii are r_1 and r_2, respectively. The effective center-to-center spacing L' is equal to the effective n^--epilayer length. The two currents flow in opposite directions. The spreading resistance of the n^--layer is given [19] by

$$R_1 = \frac{\rho}{\pi Z}\left[\ln\left(\frac{L' - r_1}{r_1}\right) + \ln\left(\frac{L' - r_2}{r_2}\right)\right]. \tag{16}$$

In many cases, the accumulation layer resistance is negligible. The effective length L' can be found from the device mask dimensions and the lateral extents of the *p*- and *n*-type diffusions. Sun and Plummer [13] considered, as an example, a device with $L' = 24\ \mu$m. They found that $r_1 = 0.5\ \mu$m

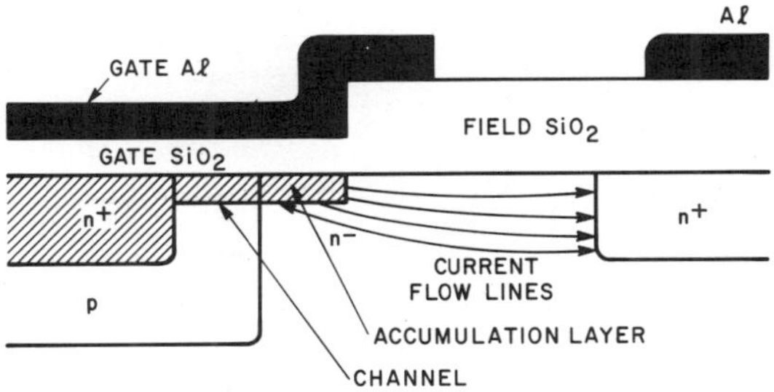

Fig. 18. Approximate current spreading in a LDMOST. (From Pocha and Dutton [19].)

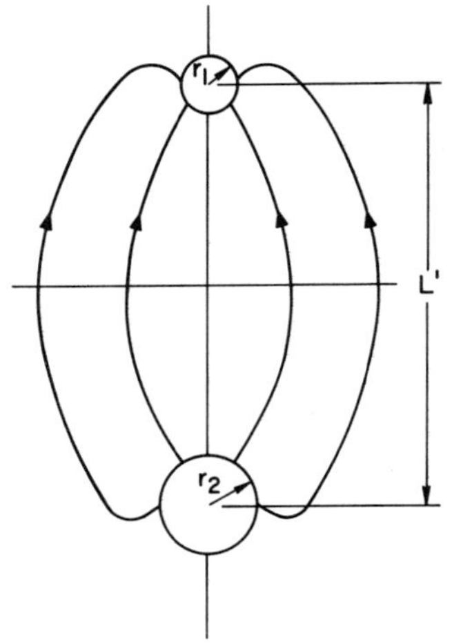

Fig. 19. Source and drain considered as two current carrying lines for the LDMOST on-resistance computation. (From Pocha and Dutton [19].)

and $r_2 = 2\ \mu\text{m}$ gave a good agreement between the calculated and measured resistance values. The results for three transistors with the n^--region resistivities of 0.5, 3, and 7 Ω cm are shown by Fig. 20 [13] as a plot of $R_{on} = R_1 + R_E$ versus $(V_G - V_{TE})$ with the drain voltage $V_D < 100$ mV. The agreement between the theoretical and observed resistance values of these short channel devices (L $\simeq 2\ \mu$m) is very good with the exception of the region close to the threshold voltage, where the channel resistance rather than the spreading resistance dominates and the accuracy of expression (16) is inadequate. At the low gate biases, the accumulation layer is not sufficiently conducting to avoid current crowding at the channel end.

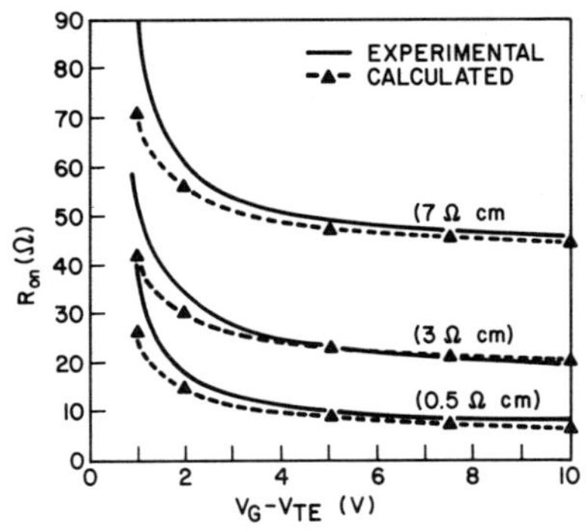

Fig. 20. LDMOST on-resistance as a function of effective gate voltage $V_G - V_{TE}$ for various n-region resistivities. (From Sun and Plummer [13].)

13.4.3.2 VDMOST On-Resistance

When a positive potential is applied to the drain electrode with respect to the shorted source and bulk (Fig. 21 [21]), the p–n^- bulk–drain junction becomes reversed biased. The two symmetrically located depletion regions are pinching off the n^--region and the current flow. This amounts to an addition of a parasitic JFET between the two adjacent p-regions as shown in Fig. 22 [21]. This JFET increases the drain resistance. To minimize this effect, it is necessary to keep the two adjacent p-regions sufficiently far apart. At low drain voltages, i.e., when the device operates at the high current levels, the drain voltage is low, and the p–n^- junction depletion regions are very narrow and can be disregarded. This situation is represented by Fig. 23a (from Sun and Plummer [13]). Since the gate metal extends beyond the channel over the n^--drift region, its surface becomes accumulated and becomes highly n^+ for highly positive gates. The electron current flowing from channel 1 follows the path of the least resistance, i.e., through accumulation layer 2, which distributes it over the n^--region. After leaving the

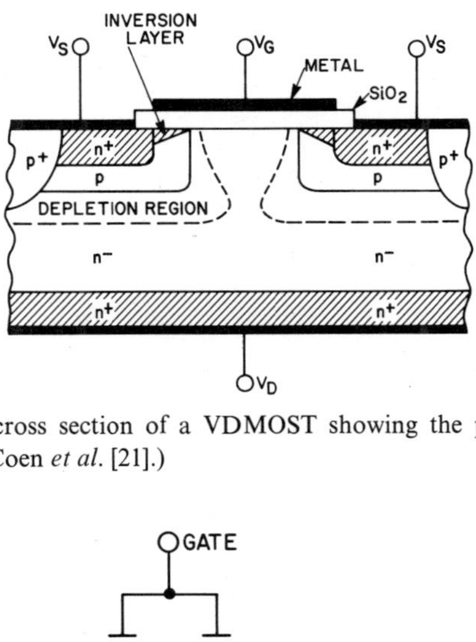

Fig. 21. Schematic cross section of a VDMOST showing the pinching-off effect of the parasitic JFET. (From Coen *et al.* [21].)

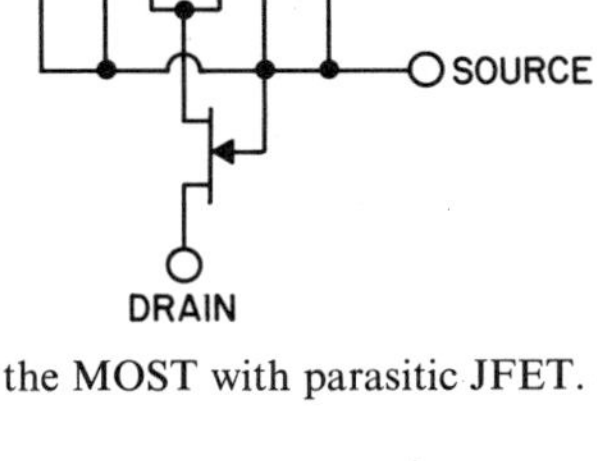

Fig. 22. Schematic of the MOST with parasitic JFET. (From Coen *et al.* [21].)

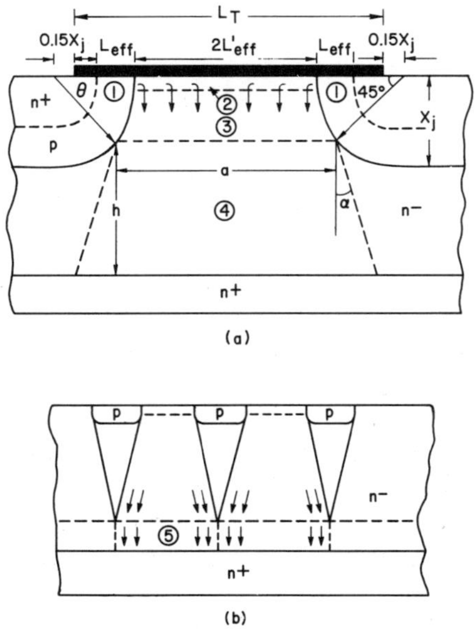

Fig. 23. (a) VDMOST schematic cross section showing various regions for R_{on} calculation (b) Region 5 for the case of overlapping trapezoids. (From Sun and Plummer [13].)

accumulation layer, the current flows downward and sideways toward the drain electrode, first passing through the pinch-off region 3 of the parasitic JFET. The channel resistance is given, as for the lateral DMOS, by expression (15). The surface accumulation resistance R_D is actually the channel resistance of a depletion model MOS transistor consisting of a p-source, n^--bulk, and p-drain and can be obtained from

$$R_D = \tfrac{1}{3}[(Z/L'_{eff})C_i\mu_D(V_G)(V_G - V_{TD})]^{-1}. \tag{17}$$

In this equation L'_{eff} is the effective channel length of the depletion mode transistor (Fig. 23a), $\mu_D(V_G)$ is the electron mobility in the accumulation layer and is dependent on V_G, and V_{TD} is the depletion parasitic transistor threshold voltage. The factor of $\frac{1}{3}$ arises from the lumped effect of the change of the current flow from horizontal to the vertical direction. The electron mobility in the accumulation layer is different from that in the inversion layer, but it changes the same way with the applied gate bias. The value used by [13] was $\mu_D = 1050\ \text{cm}^2\ \text{V}^{-1}\ \text{s}^{-1}$ for the $\langle 100\rangle$ oriented surface at $V_G = 0$.

The parasitic JFET region is represented by region 2 with the p–n^- junction modeled as a portion of a circle with the origin offset from the original mask by $0.15x_j$. The offset takes into account the difference between the lateral and vertical diffusions. The resistance of the JFET region with $\theta = 45°$ was shown by Sun and Plummer to be [13]

$$R_{\mathrm{JFET}} = \frac{2\rho}{Z}\left\{\left[1 - \left(\frac{2x_j}{L}\right)^2\right]^{-1/2} \tan^{-1}\left[(0.414)\left(\frac{L + 2x_j}{L - 2x_j}\right)^{1/2}\right] - \frac{\pi}{8}\right\}, \quad (18)$$

where ρ is the n^--region resistivity and $L = L_T + 0.3x_j$ (Fig. 23a).

The series resistance of the n^--epilayer is represented in Fig. 23a by the trapezoidal region 4, which starts at the point where angle $\theta = 45°$. The resistance of region 4 can be obtained [13] from

$$R_4 = (\rho/Z)(\tan\alpha)^{-1} \ln[1 + 2(h/a)\tan\alpha]. \quad (19)$$

α and h are defined in Fig. 23a. For $h \geqslant a$, $\alpha = 28° - h/a$, but for $h < a$, $\alpha = 28° - a/h$. Region 5 (Fig. 23b) arises when the geometry chosen is such that the adjacent trapezoids overlap. In the merged region, the current is assumed to be uniform over the entire area; therefore, the R_5 resistance can be computed from a simple geometry.

The total on-resistance is represented by the sum of all the above components

$$R_{\mathrm{on}} = R_{\mathrm{E}} + R_{\mathrm{D}} + R_{\mathrm{JFET}} + R_4 + R_5. \quad (20)$$

For low voltage devices with $\rho < 1\ \Omega$ cm, the term R_{E}, due to the channel resistance, dominates; for high voltage devices, other terms are more important, however. For instance, for $\rho = 10\ \Omega$ cm, the device on-resistance becomes

$$R_{\mathrm{on}} \simeq R_{\mathrm{JFET}} + R_4. \quad (21)$$

Figure 24 [13] shows R_{on} versus $V_G - V_{TE}$ with ρ as a parameter for devices comparable to the LDMOS transistors discussed in Section 13.6.1 and $N_{\mathrm{A(max)}} = 5 \times 10^{16}$ cm^{-3}. Again, the discrepancy between the theoretical and experimental data at low gate voltages is attributed to the inability of the accumulation layer on the top of the n^--region to prevent current crowding at the channel's end.

The spacing L_T between the adjacent p-type channel regions seriously affects the R_{on}. Sun and Plummer [13] made R_{on} measurements varying R_T for devices with comparatively small channel width $Z = 200\ \mu$m (Z for Figs. 20 and 24 was 3440 μm). The small Z accounts for the very high values of R_{on} of Fig. 25, which shows the very high R_{on} sensitivity to L_T for devices with the high resistivities necessary for high breakdown voltages. At low resistivities, the effect of the spacing L_T is negligible.

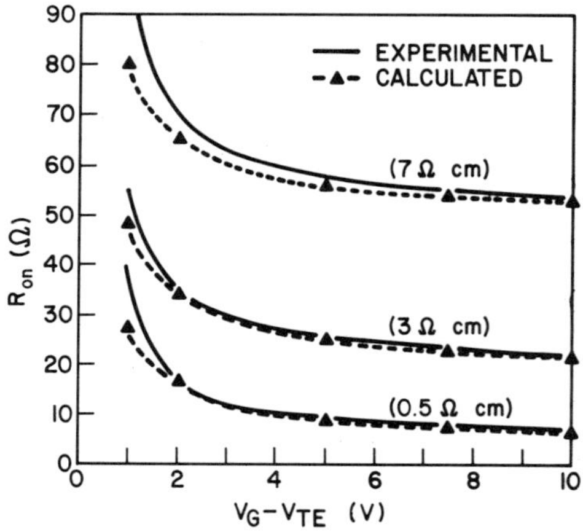

Fig. 24. VDMOST on-resistance plotted versus the effective gate voltage for various n^- region resistivities. (From Sun and Plummer [13].)

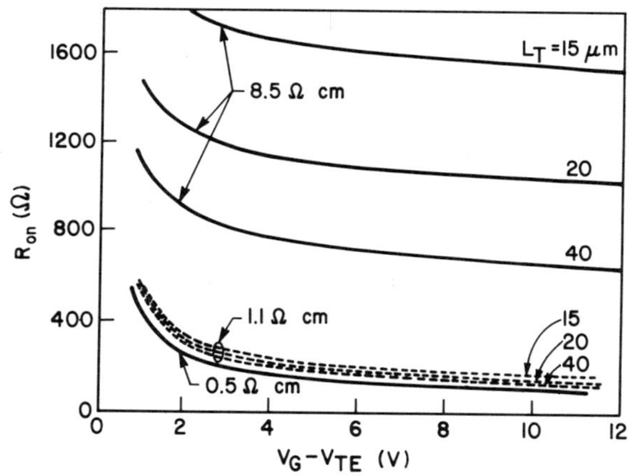

Fig. 25. VDMOST on-resistance plotted versus $V_G - V_{TE}$ with the spacing L_T between the adjacent *p*-channel regions as a parameter. (From Sun and Plummer [13].)

13.4.3.3 VVMOS Transistor On-Resistance

In the VVMOST, the thin gate oxide extends beyond channel region 1 (Fig. 26 [13]) into regions 2 and 3 of the n^--epi drain region. Upon the application of the positive gate potential, the surface of the channel region (bulk) inverts, and the surface of the n^--region under the gate accumulates similarly to the situation in the DMOST. The current flow lines of Fig. 26 clearly indicate the resulting improvement in the current spreading leading to a lower series resistance. The total resistance here, as in the VDMOS, is given by expression (20). The channel resistance R_E and accumulation

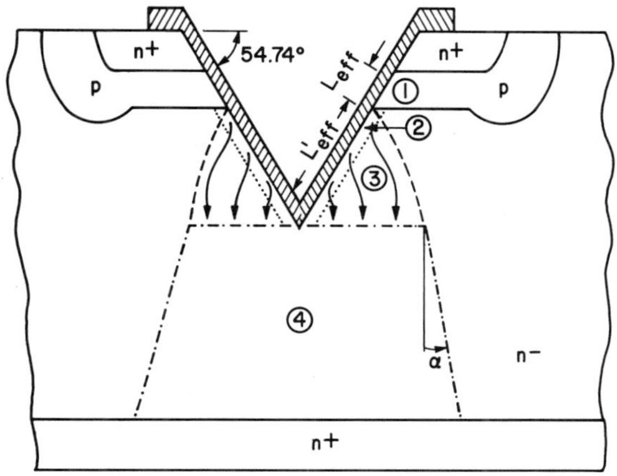

Fig. 26. Cross section of a VVMOST showing various regions for the on-resistance calculation. (From Sun and Plummer [13].)

layer resistance R_D are given by the same expressions as in the case of VDMOS. In the VVMOS, however, the mobilities in the channel and accumulation layers are only equal to about 80% of the values used for VDMOST fabricated on ⟨100⟩ oriented silicon crystal, but are reduced by the same factor with increasing gate field. The resistance R_3 for the VVMOS is obtained from the geometrical considerations as

$$R_3 \simeq 0.48(\rho/Z). \tag{22}$$

R_4 and R_5 are determined in the same way as for the VDMOST. Figure 27

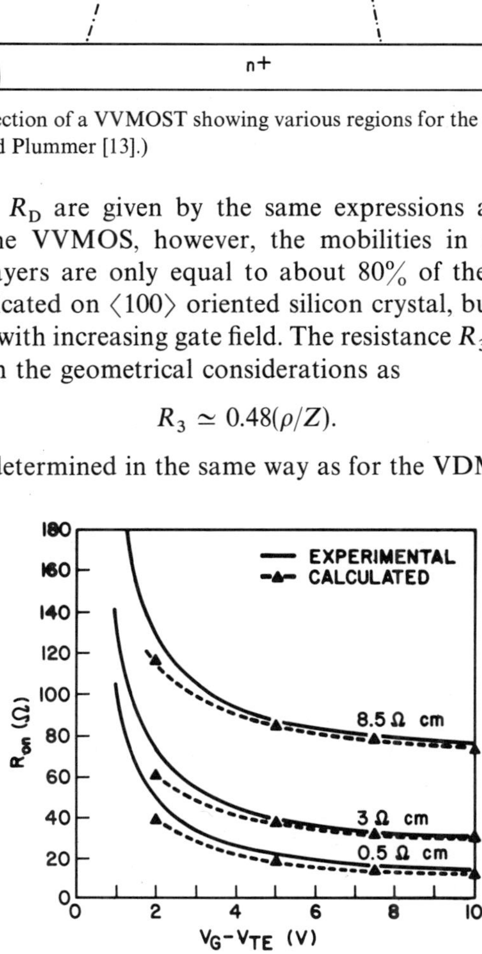

Fig. 27. VVMOST on-resistance plotted versus $V_G - V_{TE}$ with n^--region resistivity as a parameter. (From Sun and Plummer [13].)

[13] compares the theoretically obtained on-resistance values with the experimental results. The resistivity ρ of the n^--region serves as a parameter. The deviation of the experimental results from the theoretically obtained R_{on} values at low gate voltages is attributed, as for other DMOS devices, to the ineffective accumulation layer. Figure 28 [13] represents the effect of the V-groove depth on the on-state resistance with p-region depth kept constant and equal to 4.5 μm. R_{on} is normalized to R_{on} value obtained for the *V-groove spread* L of 10 μm. The on-resistance increases with decreasing L because the accumulation layer becomes less effective in current spreading. For $L = 7$ *μm*, the resistance goes to infinity since the V-groove does not penetrate into the n^--region anymore.

13.4.3.4 VUMOS Transistor On-Resistance

For low gate voltages, the U-shaped device has no advantage over the VMOST since the channel resistance dominates. For high gate voltages, the groove with the flat bottom has a significantly lower R_{on} than the analogous V-groove device. This is due to the current spreading effect of the accumulation layer at the groove bottom.

13.4.3.5 Comparison of the On-Resistance of Various Structures

For devices operating at low and moderate voltages, i.e., for the case when the channel resistance and the parasitic JFET resistance dominate over other resistive components, the LDMOS and VDMOS transistors are

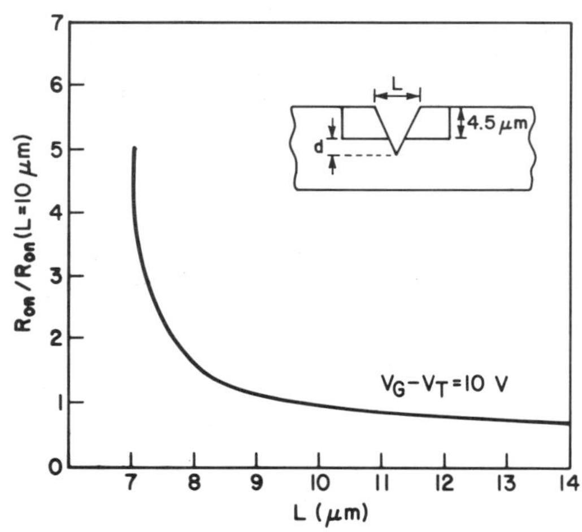

Fig. 28. VVMOST normalized on-resistance plotted versus V-groove spread L. (From Sun and Plummer [13].)

superior to the VVMOS and VUMOS transistors because of higher electron mobility. For the high voltage case, the n^--region resistance dominates over all other resistive components of Eq. (20), and the VUMOST is slightly better than the other designs because for this device, the distance for the current flow from the accumulation layer to the n^+-drain region is shorter than in the case of the VDMOST.

As pointed out by Sun and Plummer [13], the purpose of a judicious design is to obtain the lowest possible on-state resistance with a minimum chip area. Therefore, the layout considerations may become more important for the optimum device design than the somewhat better resistance of a given structure.

13.5 Gate Capacitance of MOST with Drift Region

The gate capacitance of MOS transistors with n^--drift regions differs from that of the conventional MOST. Referring to Figs. 6, 9, 14, 21, and 23, we can appreciate that the gate oxide and metal extend beyond the channel region over the n^--drift region. Due to this overlap, the DMOST gate-to-drain capacitance is higher than in the conventional MOST with the same channel length. This capacitance is not only larger for geometrical reasons but also because of the presence of many mobile electrons due to the current flowing in the n^--drift region [22]. The increase of the capacitance with current appearance is attributed by [22] to the fact that some part of the drift region close to the n^+-region may remain undepleted at low drain voltages. Consequently, electrons leaving the channel depleted part enter the undepleted region and are, therefore, slowed down. Low electron velocity is equivalent to electron surface accumulation which results in the considerable gate capacitance increase. In devices which permit wide current spreading, as in the VDMOST the current density is usually low enough for this effect to be negligibly small.

13.6 DMOST Voltage Breakdown

13.6.1 Avalanche Breakdown

A DMOS transistor has high voltage capability thanks to the presence of the high resistivity drain drift region allowing the drain depletion region to spread into it. An additional improvement in the breakdown takes place due to the gate metal overlapping the p–n^--drain junction and forming a field plate. The drain-to-source breakdown voltage is often determined with the gate shorted to the source, i.e., for $V_G = 0$. The purpose of the field plate is to improve the breakdown voltage as discussed in Chapter 4. It was observed by Declercq and Plummer [23] that the drain breakdown voltage is severely reduced if the field plate extends up to the highly doped n^+-drain region of the lateral DMOS. This reduction is attributed by [23] to the avalanche multiplication process occurring under the field plate close to the n^+-region. The generated carriers cannot create a stable surface inversion layer in that region because of the p–n junction proximity and are injected toward the source and drain decreasing the drain-to-source breakdown. This phenomenon is remedied by keeping the field plate edge away from the n^+-region as shown in Fig. 18.

In a practical device, it is not sufficient to assure good drain-to-source breakdown voltage. In addition, it is necessary to take care of the voltage breakdown and its stability at the device periphery close to the chip scribe line. This may be accomplished by using one of the techniques described in Chapter 4 on p–n junction breakdown improvement methods. Temple *et al.* [24], e.g., applied in their devices the junction termination extension method consisting of an ion-implanted region (Fig. 35 of Chapter 4) and were able to achieve breakdown voltages close to the ideal values. Coen *et al.* [21], on the other hand, used for their VDMOS a series of floating p-diffused guard rings placed around the device periphery to minimize the surface field and increase the breakdown voltage (Fig. 29 [21]). A semi-insulating, polycrystalline, oxygen-doped silicon film (Sipos) is applied over the floating guard rings in order to avoid carrier trapping at the surface. The Sipos layer is completely covered by a layer of phosphorus-doped oxide, deposited at low pressure for final passivation. Breakdowns of better than 80% of the ideal plane junction breakdown were achieved by this technique with excellent high temperature reverse-bias stability. The scribe line of the silicon chip is at the drain potential; consequently, the maximum voltage appears between the scribe line and the source electrode. The total surface depletion layer width of the guard rings, at breakdown, should be made at least equal

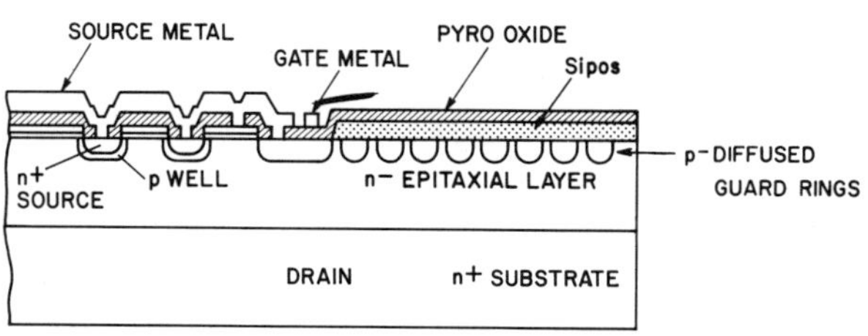

Fig. 29. VDMOST with floating p-diffused guard rings placed around the device periphery. Sipos layer is used for surface stability. (From Coen *et al.* [21].)

to the n^--layer thickness since it is designed for the same potential difference. The multiple guard-ring technique uses a reasonable amount of the silicon chip area.

13.6.2 Punch-Through Voltage

The drain punch-through voltage can be calculated from the one-dimensional Poisson equation by assuming, for simplicity, that the impurity profile in the p-channel region has a *linear* distribution while the n^--region is uniformly doped. With these assumptions, Pocha *et al.* [25] found the punch-through voltage to be equal to

$$V_{\mathrm{pt}} = \frac{qL^2}{\epsilon_s}\left[\frac{N_{\mathrm{A(max)}}}{6} + \frac{N_{\mathrm{A(max)}}^2}{8N_{\mathrm{D}}}\right], \tag{23}$$

where $N_{\mathrm{A(max)}}$ is the maximum doping concentration in the p-channel region and N_{D} is the donor concentration of the n^--region. Fong *et al.* [6] assumed a uniform distribution in the p-region with the doping concentration equal to $\frac{1}{2}N_{\mathrm{A(max)}}$. For this approximation, the factor 8 in the last term of (23) is changed to 4. An accurate computation of the punch-through voltage would require the use of the actual p-region impurity profile in Poisson's equation.

13.7 DMOST Second Breakdown

MOS power transistors are finding applications in such high voltage switching circuits as inverters, current regulators. The V–I locus of the power switch in such circuits is a square loop curve discussed in Chapter 10 in Section 10.5.2 on the reverse second breakdown. When turning off from the high conducting to the off state, the device is supporting, simultaneously, for a short period of time, a full current with full drain potential. During this short time interval, the extra dissipated power is usually of no special concern. What is important, however, is the possibility of the occurrence of the second breakdown. It was suggested by Krishna [26] that the second breakdown in a lateral DMOST (Fig. 30), similar to a bipolar device, may occur due to impact ionization followed by an avalanche injection, which starts for the critical current density (Section 10.3.2)

$$J_{\mathrm{crit}} = qv_sN_{\mathrm{D}} + \epsilon_s v_s E_{\mathrm{crit}}/L', \tag{24}$$

where N_{D} and L' are the donor density and the length of the drain space-charge region, respectively (Fig. 30 [26]). $E_{\mathrm{crit}} \simeq 1 \times 10^5\ \mathrm{V\ cm^{-1}}$ is the

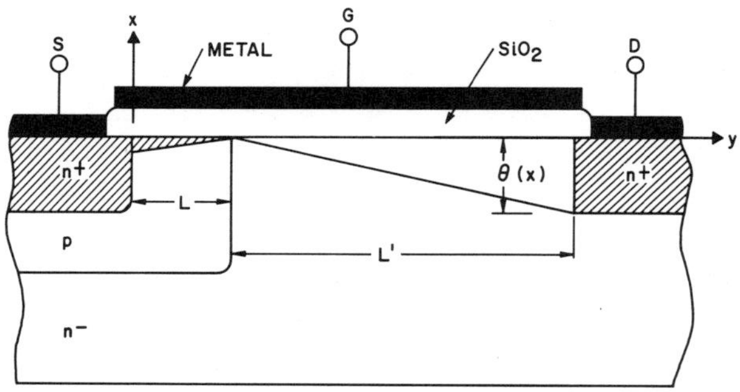

Fig. 30. Current spreading in the LDMOST drain depletion region. $\theta(x)$ is the spreading distance. (From Krishna [26].)

critical field for avalanche breakdown in silicon at 300°K. The channel current in the saturation region can be expressed by

$$J_{DS} = \frac{\mu_n C_i}{2L} V_G^2 \frac{1}{\theta(x)}, \tag{25}$$

where L is the channel length and the distance $\theta(x)$ (Fig. 30 [26]) accounts for the current spreading as it is transported across the drain depletion region. The value of $\theta(x)$ in a lateral DMOST may be on the order of 10^{-2} to 10^{-1} μm. To reduce the problem to one dimension, the minimum current density J_{DS} occurring at the drain contact is used, as that prevails over the entire depletion layer length. This assumption is in the right direction from the device conservative design point of view.

With no current flowing in the drain space-charge region, the avalanche breakdown will take place at

$$BV = E_{crit} L'. \tag{26}$$

With the current flowing, the avalanche injection and second breakdown will take place for $J_{DS} = J_{crit}$. Combining Eqs. (24) and (25) yields for the second breakdown [26]

$$V_{SBR} \simeq \frac{E_{crit}^2 \epsilon_s v_s}{J_{DS} - q v_s N_A}. \tag{27}$$

In this simplified analysis, an approximation was made that the critical field E_{crit} is independent of the mobile charge density in the drain depletion region. The second breakdown V_{SBR} is several times lower than the avalanche breakdown BV for reasonable values of the spreading distance $\theta(x)$.

A power MOST with vertical drain (VDMOST) allows for a very wide spreading of the current flow and can be expected, therefore, to be superior to the LDMOST from the second breakdown point of view. As already noted in Section 12.10.2.2, the second breakdown onset may be delayed by shorting together the source and channel regions (bulk) and using the longest possible channel that is compatible with the transconductance and/or frequency requirements.

13.8 Temperature Effects on MOS Transistor Characteristics—Safe Operating Area (SOA)

13.8.1 Mobility

The MOS transistor drain currents both in the linear and saturated regions are directly proportional to the gain factor $G_0 = (Z/L)\mu_{\text{eff}}C_{\text{i}}$ and are functions of the threshold voltage V_{T}. The effective carrier mobility μ_{eff} in the channel decreases with increasing temperature due to the enhancement of various scattering mechanisms. At temperatures higher than $+125°$C, the mobility temperature dependence can be expressed as $\mu_{\text{eff}} \propto T^{-3/2}$ [27]. In the temperature range of -55 to $+125°$C, the effective mobility varies less rapidly, and its temperature dependence follows the $\mu_{\text{eff}} \propto T^{-1}$ relationship [28] which is valid for holes and electrons.

13.8.2 Threshold Voltage

The threshold voltage of a MOS transistor is given by

$$V_{\text{T}} = 2\phi_{\text{F}} + \phi_{\text{ms}} - \frac{Q_{\text{ss}}}{C_{\text{i}}} \mp \frac{(2\epsilon_{\text{s}}qN)^{1/2}}{C_{\text{i}}}(2\phi_{\text{F}})^{1/2}, \tag{28}$$

where the upper minus sign is for the p-channel and the lower plus for the n-channel transistor. In (28) N is the bulk concentration. In expression (28) all the terms are to various degrees temperature dependent. The least affected is the Q_{ss} term. Threshold voltage variations with temperature were studied by [29]. Figure 31 [29] of V_{T} versus temperature represents the results of this theoretical and experimental work for the p- and n-channel devices. The experimental work was done with p-MOST whose silicon bulk concentration was $N_{\text{D}} = 10^{15}$ cm^{-3}, $Q_{\text{ss}}/C_{\text{i}} = 0.35$ V, and the oxide thickness $x_{\text{i}} = 100$ *nm*. The n-MOST had a bulk with concentration of $N_{\text{A}} = 3 \times 10^{16}$ cm^{-3} and $Q_{\text{ss}}/C_{\text{i}} = 0.35$ V, and $x_{\text{i}} = 100$ nm. Both devices were fabricated on (100) oriented silicon crystal. The threshold voltage change for

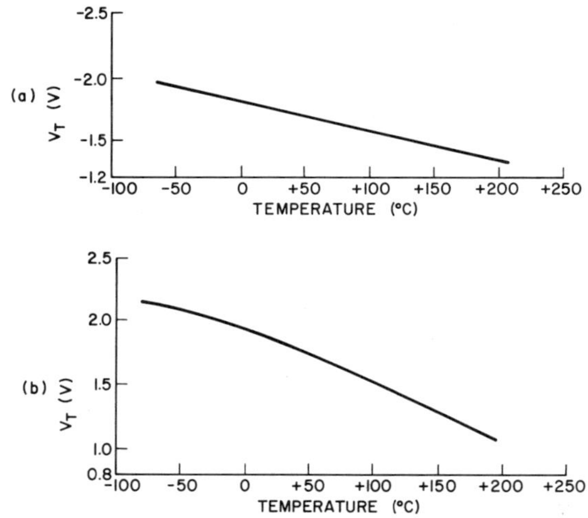

Fig. 31. Threshold voltage of *n*- and *p*-channel MOSTs. (a) With bulk concentration 10^{15} cm^{-3} and (b) with bulk concentration 3×10^{16} cm^{-3}. [From R. Wang, T. Dunkley, T. A. DeMassa, and L. F. Jelsma, Threshold voltage variations with temperature in MOS transistors, *IEEE Trans. Electron Devices* **ED-18,** p. 386 (© 1971 IEEE).]

either the *n*- or *p*-channel transistors studied was -4 mV $°C^{-1}$ for the impurity concentration of 3×10^{16} cm^{-3}, and -2 mV $°C^{-1}$ for the impurity concentration of 10^{15} cm^{-3}. The threshold voltages of devices with heavily doped bulk are more sensitive to temperature variation than the lightly doped substrates.

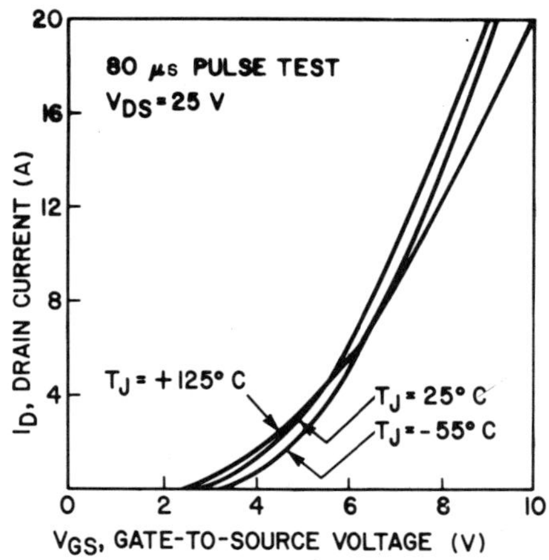

Fig. 32. Drain current I_D at $-55°C$, 25°C, and 125°C. (Courtesy of International Rectifier [30].)

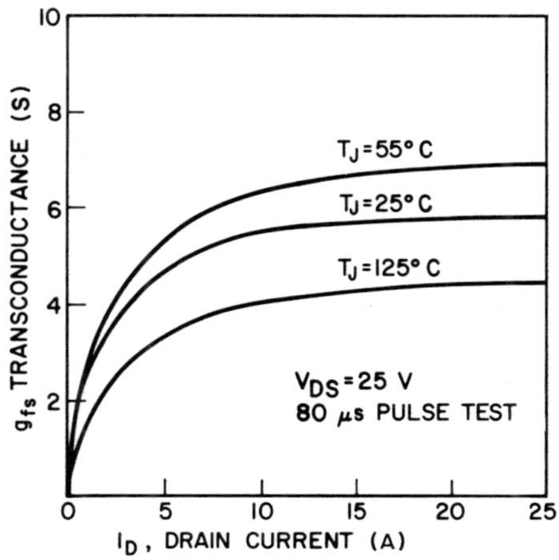

Fig. 33. Transconductance $g_{fs} = g_m$ at $-55°C$, $25°C$, and $125°C$. (Courtesy of International Rectifier [30].)

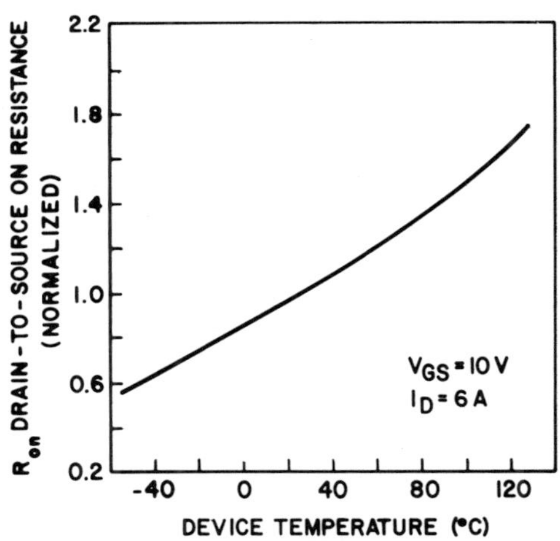

Fig. 34. MOST on-resistance versus temperature. (Courtesy of International Rectifier [30].)

13.8.3 Drain Current, Transconductance, and On-Resistance versus Temperature

The change of the carrier mobilities and threshold voltage with temperature affect the drain current I_D, the transconductance, and the on-resistance. Figures 32–34 [30] show how all these quantities change with the temperature variation for typical power MOS transistors. The drain current temperature coefficient is positive at low current levels and *negative* at high currents [31]. The positive coefficient at low current levels is determined by the threshold voltage shift. There is more current flowing for the same applied gate voltage as the threshold voltage goes down with temperature. At higher current levels, the current decreases because of the decrease of the effective mobility. The crossover points generally do not seem to be the same for various devices.

The device transconductance decreases while the on-resistance increases with temperature and has a temperature coefficient of about +0.7% per degree Celsius. The transconductance temperature coefficient is on the order of −0.2% per degree Celsius.

13.8.4 Maximum MOS Transistor Safe Operating Area (SOA)

The safe operating area of the typical power MOS transistor is shown in Fig. 35 [30], where the drain current I_D is plotted against the drain-to-source voltage V_{DS}. The dc or pulse drain currents are limited only by the power dissipation slope [I_D = (maximum allowable dissipation)/V_{DS}] and maximum allowable temperature. There are no secondary slopes as in a bipolar transistor with second breakdown. This assumes, however, that the device is protected from the second breakdown either by an external or by a built-in clamp like the Zener diode, which breaks in advance of the MOS transistor drain breakdown.

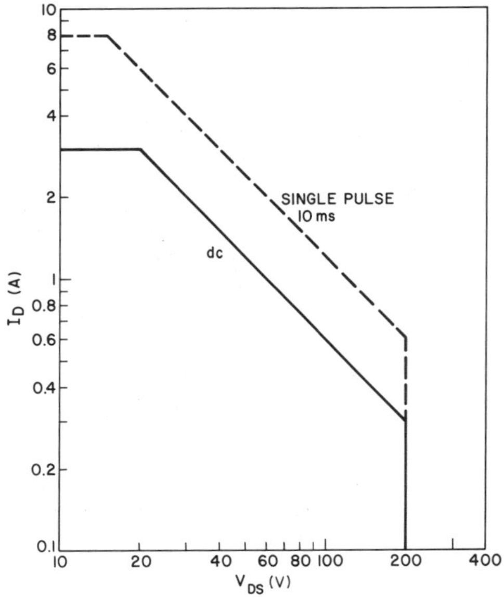

Fig. 35. Typical Safe Operating Area (SOA) of a power MOST, with device protected from second breakdown. (Courtesy of International Rectifier [30].)

REFERENCES

1. H. W. Becke, R. Hall, and J. P. White, *Solid-State Electron.* **8,** 813 (1965).
2. A. Evans and D. Hoffman, *Siliconix Appl. Note* **AN79-3,** 4–26 (1979).
3. T. Okabe, I. Yoshida, S. Ochi, S. Nishida, and M. Nagata, *IEEE Trans. Electron Devices* **ED-27,** 334 (1980).
4. Y. Tarni, Y. Hayashi, and T. Sekigawa, *Proc. Conf. Solid State Devices 1st, Tokyo* p. 105 (1969).
5. H. J. Sigg, G. D. Vendelin, R. P. Cauge, and J. Kocsis, *IEEE Trans. Electron Devices* **ED-19,** 45 (1972).
6. E. Fong, D. C. Pitzer, and R. Zeman, *IEEE Trans. Electron Devices* **ED-27,** 322 (1980).
7. I. Yoshida, M. Kubo, and S. Ochi, *IEEE J. Solid-State Circuits* **SC-11,** 472 (1976).
8. R. Denning and H. Becke, RCA Solid State Technical Center, Sumerville, New Jersey. Private communication.
9. C. A. T. Salama and J. J. Oakes, *IEEE Trans. Electron Devices* **ED-25,** 1222 (1978).
10. K. E. Bean, *IEEE Trans. Electron Devices* **ED-25,** 1185 (1978).
11. K. P. Lisiak and J. Berger, *IEEE Trans. Electron Devices* **ED-25,** 1229 (1978).
12. D. Fuoss and K. Verma, *IEDM Tech. Digest* p. 657 (1978).
13. S. C. Sun and J. D. Plummer, *IEEE Trans Electron Devices* **ED-27,** 356 (1980).
14. S. Kay *et al. IEDM Tech. Digest* p. 97 (1979).
15. M. D. Pocha, A. G. Gonzalez, and R. W. Dutton, *IEEE Trans. Electron Devices* **ED-21,** 778 (^{4}974).
16. D. P. Kennedy and R. R. O'Brien, *IBM J. Res. Develop.* **9,** 179 (1965).
17. J. C. Tsai, *Proc. IEEE* **57,** 1499 (1969).
18. W. A. Lane and C. A. T. Salama, *IEEE Trans. Electron Devices* **ED-27,** 349 (1980).
19. M. D. Pocha and R. W. Dutton, *IEEE J. Solid-State Circuits* **SC-11,** 718 (1976).
20. D. Sharma, J. Gautier, and G. Merckel, *IEEE J. Solid-State Circuits* **SC-13,** 378 (1978).
21. R. W. Coen, D. W. Tsang, and K. P. Lisiak, *IEEE Trans. Electron Devices* **ED-27,** 340 (1980).
22. E. F. Stickvoort, *IEEE Trans. Electron Devices* **ED-25,** 1388 (1978).
23. M. J. Declercq and J. D. Plummer, *IEEE Trans. Electron Devices* **ED-23,** 1 (1976).
24. V. A. K. Temple, R. P. Love, and P. V. Gray, *IEEE Trans. Electron Devices* **ED-27,** 343 (1980).
25. M. D. Pocha, J. D. Plummer, and J. D. Meindl, *IEEE Trans. Electron Devices* **ED-25,** 1325 (1978).
26. S. Krishna, *Solid-State Electron.* **20,** 875 (1977).
27. O. Leistico, A. S. Grove, and C. T. Sah, *IEEE Trans. Electron Devices* **ED-12,** 248 (1965).
28. L. Vadash and A. S. Grove, *IEEE Trans. Electron Devices* **ED-13,** 863 (1966).
29. R. Wang, T. Dunkley, T. A. DeMassa, and L. F. Jelsma, *IEEE Trans. Electron Devices* **ED-18,** 386 (1971).
30. Data Sheet PD-9.302 International Rectifier, Semiconductor Division, El Segundo, California (1979).
31. N. Nagata, *Jpn. J. Appl. Phys.* **16,** 212 (1977).

INDEX

V